INTRODUCTION TO MODERN POWER ELECTRONICS

INTRODUCTION TO MODERN POWER ELECTRONICS

SECOND EDITION

Andrzej M. Trzynadlowski

WILEY

A JOHN WILEY & SONS, INC., PUBLICATION

Published by John Wiley & Sons, Inc., Hoboken, New Jersey.
Published simultaneously in Canada.

For general information on our other products and services or for technical support, please contact our Customer Care Department within the United States at (800) 762-2974, outside the United States at (317) 572-3993 or fax (317) 572-4002.

Wiley also publishes its books in a variety of electronic formats. Some content that appears in print may not be available in electronic formats. For more information about Wiley products, visit our web site at www.wiley.com.

Library of Congress Cataloging-in-Publication Data:

Trzynadlowski, Andrzej.
  Introduction to modern power electronics / Andrzej M. Trzynadlowski. – 2nd ed.
      p. cm.
  Includes bibliographical references and index.
  ISBN 978-0-470-40103-3
1. Power electronics.   I. Title.
  TK7881.15.T79   2010
  621.31′7–dc22

                                                          2009034630

Printed in the United States of America

10   9

For my wife, Dorota, and children, Bart and Nicole

CONTENTS

PREFACE

This book is intended primarily for a one-semester introductory course in power electronics at the undergraduate level. However, as it contains a comprehensive overview of modern tools and techniques of electric power conditioning, the book can also be used in more advanced classes as a complementary text. Practicing engineers wishing to refresh their knowledge of power electronics, or interested in branching into that area, are also envisioned as potential readers. Students are assumed to have working knowledge of electric circuit analysis and basic electronics.

Since the first edition of the book was published eleven years ago, power electronics has enjoyed robust progress. Although most topologies of power converters had been known, novel applications and control techniques abound. Utilizing advanced and powerful semiconductor switches, power converters reach ratings of several kilovolts, kiloamperes, and tens of megavolt-amperes. Recent energy challenges, associated with the threat of climate change, geopolitical and environmental issues, and the growing scarcity and cost of fossil fuels, set off intensive interest in alternative sources of clean energy. As a result, power electronic systems become increasingly important and ubiquitous. Changes in this second edition reflect the dominant trends of modern power electronics, such as the growing practical significance of PWM (pulse width modulation) rectifiers, matrix converters, and multilevel inverters and the use of power converters in renewable energy systems and powertrains of electric and hybrid vehicles.

In contrast with most books, which begin with a general introduction devoid of detailed information, Chapter 1 constitutes an important part of the teaching process. Employing a hypothetical generic power converter, basic principles and methods of power electronics are explained thoroughly. Therefore, regardless of the content sequence that an instructor wishes to adopt, Chapter 1 should be covered first.

Chapters 2 and 3 provide a description of semiconductor power switches and supplementary components and systems of power electronic converters. The reader should be aware of the existence and function of those auxiliary but important parts, although the book is more focused on the power circuits, operating characteristics, control, and applications of converters.

The four fundamental types of electrical power conversion—ac to dc, ac to ac, dc to dc, and dc to ac—are covered in Chapters 4 through 7, respectively. Chapters 4 and 7, on rectifiers and inverters, are the longest, reflecting the great importance of those converters in modern power electronics. Chapter 8 is devoted to switching dc power supplies. Chapter 9, added in the second edition, covers applications of power electronics in selected clean energy systems.

Each chapter begins with an overview and includes a brief summary following the main body. Numerical examples, homework problems, and computer assignments complement most chapters. Several relevant and easily available references are provided after each of them. Three appendixes conclude the book.

The book is accompanied by a series of forty-eight PSpice circuit files that constitute a virtual power electronics laboratory, available at ftp://ftp.wiley.com/public/ sci_tech_med/power_electronics. The files contain models of most power electronic converters covered in the book. The models constitute a valuable teaching tool, giving the reader the opportunity to tinker with converters and visualize their operation.

In contrast to many contemporary engineering textbooks, the book is notably concise. Still, covering all the material in a one-semester course requires the students to perform substantial homework. The teaching approach suggested consists of presenting the basic issues in class and letting students broaden their knowledge by reading assigned materials, solving problems, and performing PSpice simulations.

I want to express my gratitude to the reviewers of the book proposal, whose valuable comments and suggestions are appreciated. My students at the University of Nevada, Reno, who used the first edition for so many years, provided very constructive critiques as well. Finally, my wife, Dorota, and children, Bart and Nicole, receive apologies for my long preoccupation, and many thanks for their unwavering support.

A. M. TRZYNADLOWSKI

1 Principles and Methods of Electric Power Conversion

In this introductory chapter we provide a background for the subject of the book. The scope, tools, and applications of power electronics are outlined. The concept of generic power converter is introduced to illustrate the principles of operation of power electronic converters and types of power conversion performed. Components of voltage or current waveforms and the related figures of merit are defined. Two basic methods of magnitude control, that is, phase control and pulse width modulation, are presented. Calculation of output current waveforms is explained. Single-phase diode rectifiers are described as simple examples of practical power converters.

1.1 WHAT IS POWER ELECTRONICS?

Modern society, with its conveniences, relies strongly on the ubiquitous availability of electric energy. Electricity performs most of the physical labor, provides heating and lighting, activates electrochemical processes, and facilitates information collecting, processing, storage, and exchange.

Contemporary *power electronics* can be defined as a *branch of electrical engineering devoted to conversion and control of electric power using electronic converters based on semiconductor power switches.* The existing power systems deliver an alternating-current (ac) voltage of fixed frequency and magnitude. Typically, homes, offices, stores, and similar small facilities are supplied from single-phase, low-voltage power lines, and three-phase supply systems with various voltage levels are available in industrial plants and other large commercial enterprises. The 60-Hz fixed-voltage electric power used in the United States (50 Hz in most other parts of the world) can be thought of as *raw power*, which for many applications must be *conditioned*. Power conditioning involves *conversion*, from ac to dc (direct current), or vice versa, and *control* of the magnitude and/or frequency of voltages and currents. Using electric lighting as a simple example, an incandescent bulb can be supplied directly with raw power. However, a fluorescent lamp requires an electronic ballast, which starts the arc and controls the current to maintain the light output desired. The ballast is thus

Introduction to Modern Power Electronics, Second Edition, by Andrzej M. Trzynadlowski
Copyright © 2010 John Wiley & Sons, Inc.

a power conditioner, necessary for proper operation of the lamp. If used in a movie theater, incandescent bulbs are supplied from an ac voltage controller that allows gradual dimming of the light just before the movie begins. Again, this controller is an example of a power conditioner, or *power converter*.

The birth of power electronics can be traced back to the dawn of the twentieth century, when the first mercury-arc rectifiers were invented. However, for conversion and control of electric power, *rotating electromachine converters* were generally used. An electromachine converter was simply an electric generator driven by an electric motor. If, for example, adjustable dc voltage was to be obtained from fixed ac voltage, an ac motor operated a dc generator with controlled output voltage. Conversely, if ac voltage was required and the supply energy came from a battery pack, a speed-controlled dc motor and an ac synchronous generator were employed. Clearly, the convenience, efficiency, and reliability of such systems were inferior to those of today's *static power electronic converters*, involving motionless and direct energy conversion.

Today's power electronics began with the development of the *silicon-controlled rectifier* (SCR), also called a *thyristor*, by the General Electric Company in 1958. The SCR is a unidirectional semiconductor power switch that can be turned on (closed) by a low-power electric pulse applied to its controlling electrode, the gate. The voltage and current ratings of SCRs are the highest of those of all types of semiconductor power switches. However, the SCR is inconvenient for use in dc-input power electronic converters because, when conducting a current, it cannot be turned off (opened) by controlling the gate. Thus, the SCR is a *semicontrolled* switch. Within the last three decades, several types of *fully controlled* semiconductor power switches turned on and off by an electric signal have been introduced to the market. These switches, as well as the SCR, are described in detail in Chapter 2.

Widespread introduction of power electronic converters to most areas of distribution and use of electric energy is common in all developed countries. The converters condition the electric power for a variety of applications, such as electric motor drives, uninterruptible power supplies, heating and lighting, electrochemical and electrothermal processes, electric arc welding, high-voltage dc transmission lines, active power filters and reactive power compensators in power systems, and high-quality supply sources for computers and other electronic equipment.

It is estimated that at least half of the electric power generated in the United States flows through power electronic converters, and an increase in this share to almost 100% is expected in the following few decades. In particular, a thorough revamping of the existing U.S. power system is envisioned within the FACTS program initiated by the Electric Power Research Institute. Introduction of power electronic converters to all stages of power generation, transmission, and distribution allows a dramatic increase in the system's capability without investment in new power plants and transmission lines. The important role of power electronics in renewable energy systems and electric and hybrid vehicles is also worth stressing. Therefore, it is safe to say that practically every electrical engineer encounters some power electronic converters in his or her professional career.

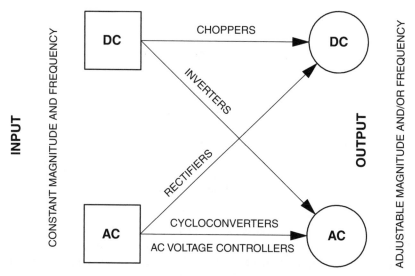

Figure 1.1   Types of electric power conversion and corresponding power electronic converters.

Types of electric power conversion and the corresponding converters employed in the contemporary power electronics are shown in Figure 1.1. For example, ac-to-dc conversion is accomplished using rectifiers, which are supplied from an ac source whose output voltage contains a significant fixed or adjustable dc component. Individual types of power electronic converters are described and analyzed in Chapters 4 through 8. Basic principles of power conversion and control are explained in the following sections of this chapter.

1.2   GENERIC POWER CONVERTER

Although not a practical apparatus, the hypothetical *generic power converter* shown in Figure 1.2 is a useful teaching tool to illustrate the principles of electric power conversion and control. It is a two-port network of five switches. Switches S1 and S2 provide *direct connection* between the input (supply) terminals, I1 and I2, and the output (load) terminals, O1 and O2, respectively, while switches S3 and S4 allow *cross-connection* between these pairs of terminals. A voltage source, either dc or ac, supplies the electric power to a load through the converter. Practical loads usually contain a significant inductive component, so a resistive–inductive load (an RL load) is assumed in subsequent considerations. To ensure a closed path for the load current under any operating conditions, a fifth switch, S5, is connected between the output terminals of the converter, and closed when switches S1 through S4 are open. It is assumed that the switches open or close instantaneously.

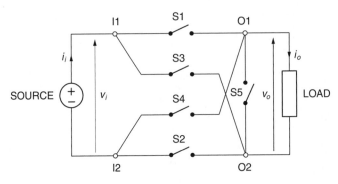

Figure 1.2   Generic power converter.

The supply source is an ideal voltage source, and as such it may not be shorted. Also, the load current may not be interrupted. It would cause rapid release of the electromagnetic energy accumulated in the load inductance, and a high and potentially damaging overvoltage would occur. Therefore, the generic converter can only assume the following three states:

State 0. Switches S1 through S4 are open and switch S5 is closed, shorting the output terminals and closing a path for the lingering load current, if any. The output voltage is zero. The input terminals are cut off from the output terminals, so the input current is also zero.

State 1. Switches S1 and S2 are closed and the remaining switches are open. The output voltage equals the input voltage, and the output current equals the input current.

State 2. Switches S3 and S4 are closed and the remaining switches are open. Now the output voltage and current are reversed with respect to their input counterparts.

To illustrate voltage and current waveforms, specific values of the input voltage and the RL load of the generic converter were employed. The amplitude of the input voltage was taken as 100 V for both the ac and dc voltage considered, and the load resistance and inductance were assumed to be 1.3 Ω and 2.4 mH, respectively. These data were needed for the preparation of subsequent figures and in example calculations in the next section. However, for generality, the waveforms are shown without the magnitude scale.

Let us assume that the generic converter is to perform the ac-to-dc conversion. The sinusoidal input voltage, v_i, whose waveform is shown in Figure 1.3, is given by

$$v_i = V_{i,p} \sin(\omega t) \tag{1.1}$$

where $V_{i,p}$ denotes the peak value of the voltage and ω is the input radian frequency. The output voltage, v_o, of the converter should contain a possibly large dc component.

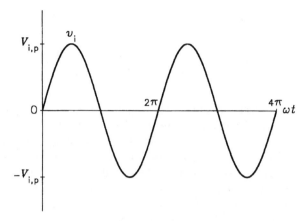

Figure 1.3 Input ac voltage waveform.

Note that the output voltage is not expected to be of ideal dc quality, since such voltage and current are not possible to obtain in the generic converter as well as in practical power electronic converters. The same applies to the ideally sinusoidal output voltage and current in ac-output converters. If within the first half-cycle of the input voltage the converter is in state 1 and within the second half-cycle in state 2, the output voltage waveform will be as depicted in Figure 1.4; that is,

$$v_o = |v_i| = V_{i,p} \, |\sin(\omega t)| \,. \qquad (1.2)$$

The dc component is the average value of the voltage. Power electronic converters performing ac-to-dc conversion are called *rectifiers*.

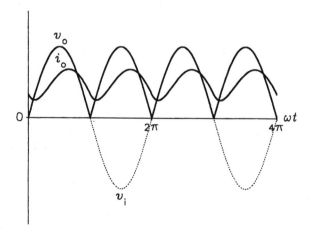

Figure 1.4 Output voltage and current waveforms in a generic rectifier.

The output current waveform, i_o, can be obtained as a numerical solution of the load equation

$$L\frac{di_o}{dt} + Ri_o = v_o. \tag{1.3}$$

Techniques for analytical and numerical computation of voltage and current waveforms in power electronic circuits are described at the end of the chapter. Here, only general features of the waveforms are outlined. The output current waveform of the generic rectifier considered is also shown in Figure 1.4, and the consecutive states of the converter are indicated there. It can be seen that this waveform is closer to an ideal dc waveform than is the output voltage waveform, because of the frequency-dependent load impedance. The kth harmonic, $v_{o,k}$, of the output voltage produces the corresponding harmonic, $i_{o,k}$, of the output current such that

$$I_{o,k} = \frac{V_{o,k}}{\sqrt{R^2 + (k\omega_o L)^2}} \tag{1.4}$$

where $I_{o,k}$ and $V_{o,k}$ denote rms (root-mean-square) values of the current and voltage harmonics in question, respectively. In the rectifier considered, the fundamental radian frequency, ω_o, of the output voltage is twice as high as the input frequency, ω. The load impedance (represented by the denominator on the right-hand side of Eq. (1.4)) for individual current harmonics increases with the harmonic number, k. Clearly, the dc component ($k = 0$) of the output current encounters the lowest impedance, equal to the load resistance only, while the load inductance attenuates only the ac component. In other words, the RL load acts as a low-pass filter. In the next section we provide a detailed explanation of terms related to the components and harmonic spectra of waveforms.

Interestingly, if an ac output voltage is to be produced and the generic converter is supplied from a dc source, so that the input voltage is $v_i = V_i = $ const, the switches are operated in the same manner as in the preceding case. Specifically, for every half-period of the desired output frequency, states 1 and 2 are interchanged. In this way, the input terminals are alternately connected and cross-connected with the output terminals, and the output voltage acquires the ac (although not sinusoidal) waveform shown in Figure 1.5. The output current is composed of growth- and decay-function segments, typical for transient conditions of an RL circuit subjected to dc excitation. Again, thanks to the attenuating effects of the load inductance, the current waveform is closer than the voltage waveform to the sinusoid desired. In practice, the dc-to-ac power conversion is performed by power electronic *inverters*. In the case described, the generic inverter is said to operate in the *square-wave mode*.

If the input or output voltage is to be a three-phase ac voltage, the topology of the generic power converter portrayed here would have to be expanded, but it still would be a network of switches. Real power electronic converters are also *networks of semiconductor power switches*. For various purposes, other elements,

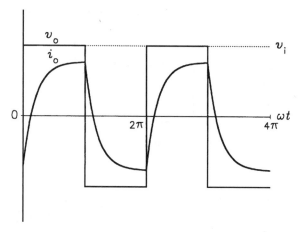

Figure 1.5 Output voltage and current waveforms in a generic inverter.

such as inductors, capacitors, fuses, and auxiliary circuits, are employed besides the switches in power circuits of practical power electronic converters. Yet in most of these converters, the fundamental operating principle is the same as in the generic converter; that is, the input and output terminals are being connected, cross-connected, and disconnected in a specific manner and sequence required for the given type of power conversion. Typically, as in the generic rectifier and inverter presented, the load inductance inhibits the switching-related undesirable high-frequency components of the output current.

Although a voltage source has been assumed for the generic power converter, some power electronic converters are supplied from current sources. In such converters, a large inductor is connected in series with the input terminals to prevent rapid changes in the input current. Analogously, voltage-source converters usually have a large capacitor connected across the input terminals to stabilize the input voltage. Inductors or capacitors are also used at the output of some converters to smooth the output current or voltage, respectively.

According to one of the principles of circuit theory, two ideal current sources may not be connected in series, and two ideal voltage sources may not be connected in parallel. Consequently, the load of a current-source converter may not appear as a current source, while that of a voltage-source converter may not appear as a voltage source. As illustrated in Figure 1.6, this means that in a current-source power electronic converter a capacitor should be placed in parallel with the load. Apart from smoothing the output voltage, the capacitor prevents the potential hazards of connecting the input inductance conducting a certain current with a load inductance conducting a different current. In contrast, in voltage-source converters, no capacitor may be connected across the output terminals, and it is the load inductance (or an extra inductor between the converter and the load) that smoothes the output current.

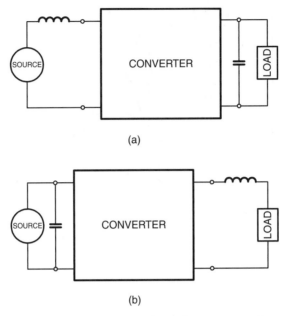

Figure 1.6 Basic configurations of power electronic converters: (a) current-source; (b) voltage-source.

1.3 WAVEFORM COMPONENTS AND FIGURES OF MERIT

Terms such as the *dc component*, *ac component*, and *harmonics* used in the preceding section deserve closer examination. Knowledge of the basic components of voltage and current waveforms allows evaluation of the performance of a converter. Certain relations of these components are commonly used as performance indicators, or *figures of merit*.

A time function $\psi(t)$, here a waveform of voltage or current, is said to be *periodic* with a period T if

$$\psi(t) = \psi(t + T) \tag{1.5}$$

that is, if the pattern (shape) of the waveform is repeated every T seconds. In the realm of power electronics, it is often convenient to analyze voltages and currents in the *angle domain* instead of the *time domain*. The *fundamental frequency*, f_1, in hertz, is defined as

$$f_1 = \frac{1}{T} \tag{1.6}$$

and the corresponding *fundamental radian frequency*, ω_1, in rad/s, as

$$\omega_1 = 2\pi f_1 = \frac{2\pi}{T}. \tag{1.7}$$

Now, a periodic function $\psi(\omega_1 t)$ can be defined such that

$$\psi(\omega_1 t) = \psi(\omega_1 t + 2\pi). \tag{1.8}$$

The *rms value*, Ψ, of waveform $\psi(\omega_1 t)$ is defined as

$$\Psi \equiv \sqrt{\frac{1}{2\pi} \int_0^{2\pi} \psi^2(\omega_1 t)\, d\omega_1 t} \tag{1.9}$$

and the *average value*, or *dc component*, Ψ_{dc}, of the waveform as

$$\Psi_{dc} \equiv \frac{1}{2\pi} \int_0^{2\pi} \psi(\omega_1 t)\, d\omega_1 t \tag{1.10}$$

When the dc component is subtracted from the waveform, the remaining waveform, $\psi_{ac}(\omega_1 t)$, is called the *ac component*, or *ripple*, that is,

$$\psi_{ac}(\omega_1 t) = \psi_{ac}(\omega_1 t) - \psi_{dc}. \tag{1.11}$$

Clearly, the ac component has an average value of zero and a fundamental frequency of f_1.

The rms value, Ψ_{ac}, of $\psi_{ac}(\omega_1 t)$ is defined as

$$\Psi_{ac} \equiv \sqrt{\frac{1}{2\pi} \int_0^{2\pi} \psi_{ac}^2(\omega_1 t)\, d\omega_1 t} \tag{1.12}$$

and it is easy to show that

$$\Psi^2 = \Psi_{dc}^2 + \Psi_{ac}^2. \tag{1.13}$$

For waveforms of the desirable ideal dc quality, such as the load current of a rectifier, a figure of merit called a *ripple factor*, RF, is defined as

$$RF = \frac{\Psi_{ac}}{\Psi_{dc}}. \tag{1.14}$$

A low value of the ripple factor indicates the high quality of a waveform.

Before proceeding to other waveform components and figures of merit, the terms and formulas introduced so far will be illustrated using the waveform of output voltage, v_o, of the generic rectifier, shown in Figure 1.4. The waveform pattern is repeating itself every π radians and, within the 0-to-π interval, $v_o = v_i$. Therefore, the average value, $V_{o,dc}$, of the output voltage can most conveniently be determined by calculating the area under the waveform from $\omega t = 0$ to $\omega t = \pi$ and dividing it by the length, π, of the interval considered. Thus,

$$V_{o,\text{dc}} = \frac{1}{\pi} \int_0^\pi V_{i,p} \sin \omega t \, d\omega t = \frac{2}{\pi} V_{i,p} = 0.64 V_{i,p}. \tag{1.15}$$

Note that formula (1.15) differs from (1.10). Since $\omega_1 = \omega_0 = 2\omega$, the integration is performed in the 0-to-π interval of ωt instead of the 0-to-2π interval of $\omega_1 t$.

Similarly, the rms value, V_o, of the output voltage can be calculated as

$$V_0 = \sqrt{\frac{1}{\pi} \int_0^\pi (V_{i,p} \sin \omega t)^2 \, d\omega t} = \frac{V_{i,p}}{\sqrt{2}} = 0.71 V_{i,p}. \tag{1.16}$$

This result agrees with the well-known relation for a sine wave as $v_o^2 = v_i^2$.

Based on Eqs. (1.13) and (1.14), the rms value, $V_{o,\text{ac}}$, of the ac component of the voltage in question can be calculated as

$$V_{o,\text{ac}} = \sqrt{V_0^2 - V_{0,\text{dc}}^2} = \sqrt{\left(\frac{V_{i,p}}{\sqrt{2}}\right)^2 - \left(\frac{2}{\pi} V_{i,p}\right)^2} = 0.31 V_{i,p} \tag{1.17}$$

and the ripple factor, RF_V, of the voltage as

$$\text{RF}_V = \frac{V_{0,\text{ac}}}{V_{0,\text{dc}}} = \frac{0.31 V_{i,p}}{0.64 V_{i,p}} = 0.48. \tag{1.18}$$

Decomposition into the dc and ac components of the waveform being analyzed is shown in Figure 1.7.

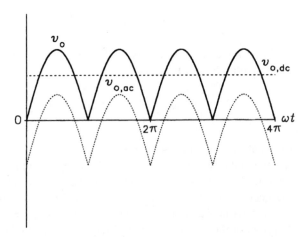

Figure 1.7   Decomposition of an output voltage waveform in a generic rectifier.

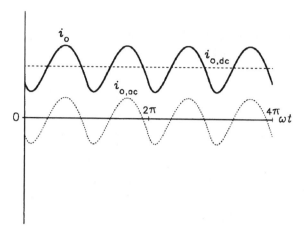

Figure 1.8 Decomposition of the output current waveform in a generic rectifier.

To determine the ripple factor, RF_I, of the output current analytically, the output current waveform, $i_o(\omega t)$, would have to be expressed in a closed form. Instead, numerical computations were performed on the waveform in Figure 1.4 and RF_I was found to equal 0.31. This value is 36% lower than that of the output voltage. This is an example only, but output currents in power electronic converters are indeed of higher quality than output voltages. It is worth mentioning that the value of RF_I obtained is poor. Practical high-quality dc current waveforms have a ripple factor on the order of a few percentage points, and below the 5% level, the current is considered to be of practically ideal dc quality. The current ripple factor depends on the type of converter, and it decreases with an increase in the inductive component of a load. Components of the current waveform evaluated are shown in Figure 1.8.

The ripple factor is of no use for quality evaluation of ac waveforms such as the output current of an inverter, which ideally should be a pure sinusoid. However, as already mentioned and exemplified by the waveforms in Figure 1.5, purely sinusoidal voltages and currents cannot be produced by switching power converters. Therefore, an appropriate figure of merit must be defined as a measure of deviation of a practical ac waveform from its ideal counterpart.

Following the theory of Fourier series (see Appendix B), the ac component, $\psi_{ac}(t)$, of a periodic function, $\psi(t)$, can be expressed as an infinite sum of the harmonics, that is, sine waves whose frequencies are multiples of the fundamental frequency, f_1, of $\psi(t)$. In the angle domain,

$$\psi_{ac}(\omega_1 t) = \sum_{k=1}^{\infty} \psi_k k\omega_1 t = \sum_{k=1}^{\infty} \Psi_{k,p} \cos(k\omega_1 t + \varphi_k) \qquad (1.19)$$

where k is the *harmonic number* and $\Psi_{k,p}$ and φ_k denote the peak value and phase angle of the kth harmonic, respectively. The first harmonic, $\psi_1(\omega_1 t)$, is usually called a *fundamental*. The terms *fundamental voltage* and *fundamental current* are used throughout the book to denote the fundamental of a given voltage or current.

The peak value, $\Psi_{1,p}$, of the fundamental of a periodic function, $\psi(\omega_1 t)$, is calculated as

$$\Psi_{1,p} = \sqrt{\Psi_{1,c}^2 + \Psi_{1,s}^2} \tag{1.20}$$

where

$$\Psi_{1,c} = \frac{1}{\pi} \int_0^{2\pi} \psi(\omega_1 t) \cos \omega_1 t \, d\omega_1 t \tag{1.21}$$

$$\Psi_{1,s} = \frac{1}{\pi} \int_0^{2\pi} \psi(\omega_1 t) \sin \omega_1 t \, d\omega_1 t \tag{1.22}$$

and the rms value, Ψ_1, of the fundamental is

$$\Psi_1 = \frac{\Psi_{1,p}}{\sqrt{2}}. \tag{1.23}$$

Since the fundamental of a function does not depend on the dc component of the function, the ac component, $\psi_{ac}(\omega_1 t)$, can be used in Eqs. (1.21) and (1.22) in place of $\psi(\omega_1 t)$.

When the fundamental is subtracted from the ac component, the *harmonic component*, $\psi_h(\omega_1 t)$, is obtained as

$$\psi_h(\omega_1 t) = \psi_{ac}(\omega_1 t) - \psi_1(\omega_1 t). \tag{1.24}$$

The rms value, Ψ_h, of $\psi_h(\omega_1 t)$, called a *harmonic content* of function $\psi(\omega_1 t)$, can be calculated as

$$\Psi_h = \sqrt{\Psi_{ac}^2 - \Psi_1^2} = \sqrt{\Psi^2 - \Psi_{dc}^2 - \Psi_1^2} \tag{1.25}$$

and used for calculation of the *total harmonic distortion*, THD, defined as

$$\text{THD} \equiv \frac{\Psi_h}{\Psi_1}. \tag{1.26}$$

The concept of total harmonic distortion is widely employed in practice, also outside power electronics, as, for example, in characterization of the quality of audio equipment. Conceptually, the total harmonic distortion constitutes an ac counterpart of the ripple factor.

Using as an example the generic inverter whose output waveforms are shown in Figure 1.6, the rms value, V_o, of the output voltage is equal to the dc input voltage, V_i. Since v_o is either V_i or $-V_i$, then $v_0^2 = V_i^2$. The peak value, $v_{o,1,p}$, of the fundamental output voltage is

$$V_{o,1,p} = V_{0,1,s} \tag{1.27}$$

because the waveform in question has odd symmetry (see Appendix B). Consequently,

$$V_{o,1,p} = \frac{2}{\pi} \int_0^\pi V_i \sin \omega t \; d\omega t = \frac{4}{\pi} V_i = 1.27 V_i. \tag{1.28}$$

In this case, ω denotes the fundamental output frequency. Now, the fundamental output voltage, $v_{o,1}(\omega t)$, can be expressed as

$$v_{o,1}(\omega t) = V_{o,1,p} \sin \omega t = \frac{4}{\pi} V_i \sin \omega t. \tag{1.29}$$

The rms value, $V_{o,1}$, of the fundamental output voltage is

$$V_{o,1} = \frac{V_{0,1,p}}{\sqrt{2}} = \frac{2\sqrt{2}}{\pi} V_i = 0.9 V_i \tag{1.30}$$

and the harmonic content, $v_{o,h}$, is

$$V_{o,h} = \sqrt{V_0^2 - V_{0,1}^2} = \sqrt{V_i^2 - \left(\frac{2\sqrt{2}}{\pi} V_i\right)^2} = 0.44 V_i. \tag{1.31}$$

Thus, the total harmonic distortion of the output voltage, THD_V, is

$$\text{THD}_V = \frac{V_{o,h}}{V_{o,1}} = \frac{0.44 V_i}{0.9 V_i} = 0.49. \tag{1.32}$$

The high value of THD_V is not surprising since the output voltage waveform of a generic inverter operating in the square-wave mode differs strongly from a sine wave. Decomposition of the waveform analyzed is illustrated in Figure 1.9.

The numerically determined total harmonic distortion, THD_I, of the output current, i_o, is 0.216 in this example, that is, less than that of the output voltage by as much as 55%. Indeed, as shown in Figure 1.10, which shows decomposition of the current waveform, the harmonic component is quite small compared with the fundamental. As in a generic rectifier, it shows the attenuating influence of the load inductance on the output current. In practical inverters, the output current is considered to be of high quality if THD_I does not exceed 0.05 (5%).

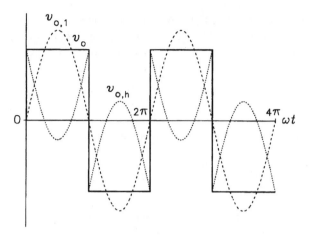

Figure 1.9   Decomposition of the output voltage waveform in a generic inverter.

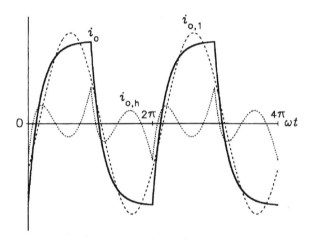

Figure 1.10   Decomposition of the output current waveform in a generic inverter.

Other figures of merit often employed for performance evaluation of power electronic converters are:

1. *The power efficiency*, η, of the converter, defined as

$$\eta \equiv \frac{P_0}{P_i} \tag{1.33}$$

where P_o and P_i denote the output and input powers of the converter, respectively.

2. *The conversion efficiency*, η_c, of the converter, defined as

$$\eta_c \equiv \frac{P_{o,\text{dc}}}{P_i} \tag{1.34}$$

for dc output converters, and

$$\eta_c \equiv \frac{P_{o,1}}{P_i} \tag{1.35}$$

for ac output converters. The symbol $P_{o,\text{dc}}$ denotes the dc output power, that is, the product of the dc components of the output voltage and current, while $P_{o,1}$ is the ac output power carried by the fundamental components of the output voltage and current.

3. *The input power factor*, PF, of the converter, defined as

$$\text{PF} = \frac{P_i}{S_i} \tag{1.36}$$

where S_i is the apparent input power. The power factor can also be expressed as

$$\text{PF} = K_d K_\Theta. \tag{1.37}$$

Here K_d denotes the *distortion factor* (not to be confused with the total harmonic distortion, THD), defined as the ratio of the rms fundamental input current, $I_{i,1}$, to the rms input current, I_i, and K_Θ is the *displacement factor*, that is, the cosine of the phase shift, Θ, between the fundamentals of input voltage and current.

The power efficiency, η, of a converter simply indicates what portion of the power supplied to the converter reaches the load. In contrast, the conversion efficiency, η_c, expresses the relative amount of *useful* output power and, therefore constitutes a more valuable figure of merit than the power efficiency. Since the input voltage to a converter is usually constant, the power factor serves mainly as a measure of utilization of the input current, drawn from the source that supplies the converter. With a constant power consumed by the converter, a high power factor implies a low current and, consequently, low power losses in the source. The reader is probably already familiar with the term *power factor* as the cosine of the phase shift between the voltage and current waveforms, as used in the theory of ac circuits. However, it must be stressed that it is true for purely sinusoidal waveforms only, and the general definition of the power factor is that given by Eq. (1.36).

In an ideal power converter, all three figures of merit defined above would equal unity. To illustrate the relevant calculations, the generic rectifier will again be employed. Since ideal switches have been assumed, no losses are incurred in the generic

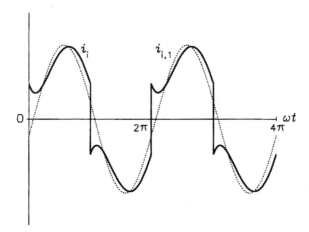

Figure 1.11   Decomposition of the output current waveform in a generic inverter.

converter, so that the input power, P_i, and output power, P_o, are equal and the power efficiency, η, of the converter is always unity.

For the RL load assumed for the generic rectifier, it can be shown that the conversion efficiency, η_c, is a function of the current ripple factor, RF_I. Specifically,

$$\eta_c = \frac{P_{o,dc}}{P_i} = \frac{P_{o,dc}}{P_0/\eta} = \eta \frac{RI_{o,dc}^2}{RI_o^2}$$

$$= \eta \frac{I_{o,dc}^2}{I_{o,dc}^2 + I_{o,ac}^2} = \frac{\eta}{1 + (I_{o,ac}/I_{o,dc})^2} = \frac{\eta}{1 + (RF_1)^2} \tag{1.38}$$

where R is the load resistance. The ripple factor for the current in Figure 1.4 has already been found to be 0.307. Hence, with $\eta = 1$, the conversion efficiency is $1/(1 + 0.307^2) = 0.914$.

The input current waveform, $i_i(t)$, of the rectifier is shown in Figure 1.11 with the fundamental found numerically, $i_{i,1}(t)$. The converter either passes the input voltage and current directly to the output or inverts them. Therefore, the rms values of input voltage, V_i, and current, I_i, are equal to those of the output voltage, V_o, and current, I_o. The apparent input power, S_i, is a product of the rms values of input voltage and current. Hence,

$$PF = \frac{P_i}{S_i} = \frac{P_o/\eta}{V_i I_i} = \frac{RI_o^2}{\eta V_o I_o} = \frac{RI_o}{\eta V_o} \tag{1.39}$$

and specific values of R, I_o, and V_o are needed to determine the power factor. As mentioned in the preceding section, the peak input voltage, $V_{i,p}$, in the example rectifier described is 100 V and the load resistance, R, is 1.3 Ω. Thus, according to Eq. (1.16), the dc output voltage, V_o, is 70.7 V. The rms output current, I_o, computed numerically is 51.3 A. Consequently, PF = $(1.3 \times 51.3)/(1 \times 70.7) = 0.943$ (lag), "lag" indicating that the fundamental input current lags the fundamental input voltage (compare Figures 1.1 and 1.11).

1.4 PHASE CONTROL

Based on the idea of a generic converter whose switches connect, cross-connect, or disconnect the input and output terminals, the principles of ac-to-dc and dc-to-ac power conversion were explained in Section 1.2. However, the question of how to control the magnitude of the output voltage and, consequently, that of the output current, has not yet been answered.

The reader is likely to be familiar with electric transformers and autotransformers that allow magnitude regulation of ac voltage and current. They are heavy and bulky, designed for a fixed frequency and impractical for wide-range magnitude control. Moreover, their principle of operation inherently excludes transformation of dc quantities. In the early days of electrical engineering, adjustable resistors were employed predominantly for voltage and current control. Today, the *resistive control* can still be encountered in relay-based starters for electric motors and obsolete adjustable-speed drive systems. Small rheostats and potentiometers are, of course, still used widely in low-power electric and electronic circuits, in which the issue of power efficiency is not of major importance.

Resistive control does not have to involve real resistors. Actually, any of the existing transistor-type power switches could serve this purpose. Between the state of saturation, in which a transistor offers minimum resistance in the collector–emitter path, and the blocking state, resulting in practically zero collector and emitter currents, a wide range of intermediate states are available. Therefore, such a switch can be viewed as a controlled resistor, and one may wonder if, for example, the transistor switches used in power electronic converters could be operated in the same way as are transistors in low-power analog electronic circuits.

To show why the resistive control *should not* be used in high-power applications, two basic schemes, depicted in Figure 1.12, will be considered. For simplicity it is assumed that the circuits shown are to provide control of a dc voltage supplied to a resistive load. The dc input voltage, V_i, is constant, while the output voltage, V_o, is to be adjustable within the zero-to-V_i range. Generally, for power converters with a controlled output quantity (voltage or current), the *magnitude control ratio*, M, can be defined as

$$M \equiv \frac{\Psi_{o,\text{adj}}}{\Psi_{o,\text{adj(max)}}} \tag{1.40}$$

where $\Psi_{o,\text{adj}}$ denotes the value of the adjustable component of the output quantity: for example, the dc component of the output voltage in a controlled rectifier or the fundamental output voltage in an inverter, while $\Psi_{o,\text{adj(max)}}$ is the maximum available

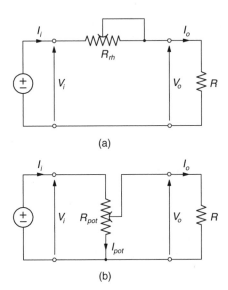

Figure 1.12   Resistive control schemes: (a) rheostatic control; (b) potentiometric control.

value of this component. Usually, it is required that M be adjustable from a certain minimum value to unity.

The magnitude control ratio should not be confused with the *voltage gain*, K_V, which generally represents the ratio of the output voltage to the input voltage. Specifically, the average value is used for dc voltages and the peak value of the fundamental for ac voltages. For example, the voltage gain of a rectifier is defined as the ratio of the dc output voltage, $V_{o,dc}$, to the peak input voltage, $V_{i,p}$. In the resistive control schemes considered, $K_V = V_o/V_i$, and since the maximum available value of the output voltage equals V_i, the voltage gain equals the magnitude control ratio, M.

Figure 1.12a illustrates *rheostatic control*. The active part, R_{rh}, of the controlling rheostat forms a voltage divider with the load resistance, R. Here,

$$M = \frac{V_o}{V_i} = \frac{R}{R_{rh} + R} \qquad (1.41)$$

and since the input current, I_i, equals the output current, I_o, the efficiency, η, of the power transfer from the source to the load is

$$\eta = \frac{R I_o^2}{(R_{rh} + R)I_i^2} = \frac{R I_i^2}{(R_{rh} + R)I_i^2} = \frac{R}{R_{rh} + R} = M. \qquad (1.42)$$

The identity relation between η and M is a serious drawback of rheostatic control, since decreasing the output voltage causes an equal reduction in efficiency.

Potentiometric control, shown in Figure 1.12b and based on the principle of current division, fares even worse. Note that the input current, I_i, is greater than the output current, I_o, by the amount of the potentiometer current, I_{pot}. The power efficiency, η, is

$$\eta = \frac{V_o I_o}{V_i I_i} = M \frac{I_o}{I_i} \tag{1.43}$$

that is, less than M.

It can be seen that the principal trouble with resistive control is that the output current flows through the controlling resistance. As a result, power is lost in that resistance, and the power efficiency is reduced to a value equal to, or less than, the magnitude control ratio. In practical power electronic systems, this is totally unacceptable. Imagine, for example, a 100-kVA converter (not excessively large against today's standards) that at $M = 0.5$ loses as much power in the form of heat as do 25 typical 2-kW domestic heaters! Efficiencies of power electronic converters are seldom lower than 90% in low-power converters and exceed 95% in high-power converters. Apart from economical considerations, large power losses in a converter would require an extensive cooling system. Even in contemporary high-efficiency power conversion schemes, cooling is often quite a problem since the semiconductor power switches are of relatively small size and, consequently, of limited thermal capacity. Therefore, they tend to get overheated quickly if the cooling is inadequate.

The resistive control allows adjustment of the *instantaneous values* of voltage and current, which is important in many applications: for example, those requiring amplification of analog signals, such as those of radio, TV, and tape recorders. There, transistors and operational amplifiers operate on the principle of resistive control, and because of the low levels of power involved, the low efficiency is of minor concern. As illustrated later, in power electronic converters it is sufficient to control the *average value* of dc waveforms and the *rms value* of ac waveforms. This can be accomplished by periodic application of state 0 of the converter (see Section 1.2), in which the connection between the input and output is broken and the output terminals are shorted. In this way the output voltage is made zero within specified intervals of time and, depending on the length of zero intervals, its average or rms value is more or less reduced compared with that of the full waveform.

Clearly, the mode of operation described can be implemented by appropriate use of switches of the converter. Note that there are no power losses in ideal switches, because when a switch is on (closed), there is no voltage drop across it, whereas when it is off (open), there is no current through it. For this reason, both the power conversion and control in power electronic converters are accomplished by means of switching. Analogously to the switches in a generic power converter, the semiconductor devices used in practical converters are allowed to assume two states only. The device is either fully conducting, with a minimum voltage drop between its main electrodes (on-state), or fully blocking, with a minimum current passing between these electrodes (off-state). That is why the term semiconductor power *switches* is used for the devices employed in power electronic converters.

The only major difference between the ideal switches in a hypothetical generic converter and practical semiconductor switches lies in the unidirectionality of the latter devices. In the on-state, a current in the switch can flow in only one direction: for example, from the anode to the cathode in an SCR. Therefore, an idealized semiconductor power switch can be thought of as the connection of an ideal switch and an ideal diode in series.

Historically, for a major part of the past century, only semicontrolled power switches such as mercury-arc rectifiers, gas tubes (thyratrons), and SCRs were available for power conditioning purposes. As mentioned in Section 1.1, once turned on ("fired"), a semicontrolled switch cannot be turned off ("extinguished") as long as the current conducted drops below a certain minimum level for a sufficient amount of time. This condition is required for extinguishing the arc in thyratrons and mercury-arc rectifiers, or for SCRs to recover their blocking capability. If a switch operates in an ac circuit, the turn-off occurs naturally when the current changes polarity from positive to negative. After turn-off, a semicontrolled switch must be refired in every cycle when the anode–cathode voltage becomes positive, that is, when the switch becomes *forward biased*.

Because the forward bias of a switch in an ac-input power electronic converter lasts a half-cycle of the input voltage, firing can be delayed by up to a half-cycle from the instant when the bias changes from reverse to forward. This creates an opportunity for control of the average or rms value of the output voltage of the converter. To demonstrate this control method, the generic power converter will again be employed. For simplicity, the firing delay of the converter switches, previously assumed zero, is now made to equal a quarter of the period of the ac input voltage, that is, $90°$ in the angle domain, ωt.

Controlled ac-to-dc power conversion is illustrated in Figure 1.13. As explained in Section 1.2, when switches S1 through S4 of a generic converter are open, switch S5 must close to provide a path for the load current, which because of the inductance

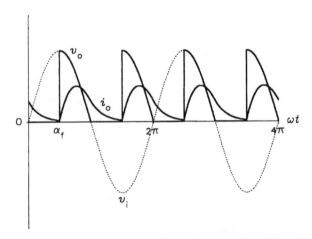

Figure 1.13   Output voltage and current waveforms in a generic rectifier with a firing angle of $90°$.

of the load may not be interrupted. Thus, states 1 and 2 of the converter are separated by state 0. It can be seen that the sinusoidal half-waves of the output voltage in Figure 1.4 have been replaced by quarter-waves. As a result, the dc component of the output voltage has been reduced by 50%. Clearly, a longer delay in closing switches S1–S2 and S3–S4 would reduce this component further, until, with a delay of 180°, it would drop to zero. The generic power converter now operates as a *controlled rectifier*. However, as shown in Chapter 4, practical controlled rectifiers based on SCRs do not need to employ an equivalent of switch S5. As already explained, an SCR cannot be turned off when conducting a current. Therefore, state 1 can be terminated only by switching to state 2, and the other way around, so that one pair of switches takes over the current from another pair. Both these states provide a closed path for the output current. State 0, if any, occurs only when the output current has already died out.

In the angle domain, the firing delay is referred to as a *delay angle*, or *firing angle*, and the method of output voltage control described is called *phase control*, since the firing occurs at a specified phase of the input voltage waveform. In practice, phase control is limited to power electronic converters based on SCRs. Fully controlled semiconductor switches allow more effective control by means of *pulse width modulation*, described in the next section.

The voltage control characteristic, $V_{o,\text{dc}}(\alpha_f)$, of a generic controlled rectifier can be determined as

$$V_{o,\text{dc}}(\alpha_f) = \frac{1}{\pi} \int_0^\pi v_0(\omega t)\, d\omega t = \frac{1}{\pi} \int_{\alpha_f}^\pi V_{i,p} \sin \omega t\, d\omega t = \frac{V_{i,p}}{\pi}(1 + \cos \alpha_f).$$

$$(1.44)$$

If $\alpha_f = 0$, Eq. (1.44) becomes identical to Eq. (1.15). The control characteristic, which is nonlinear, is shown in Figure 1.14.

Ac-to-ac conversion performed by the generic converter for adjustment of the rms value of an ac output voltage is illustrated in Figure 1.15. Power electronic converters used for this type of power conditioning are called *ac voltage controllers*. Practical ac voltage controllers are based primarily on *triacs*, whose internal structure is equivalent to two SCRs connected antiparallel. Phase-controlled ac voltage controllers do not require a counterpart of switch S5.

The voltage control characteristic, $V_o(\alpha_f)$, of a generic ac voltage controller, given by

$$V_o(\alpha_f) = \sqrt{\frac{1}{\pi} \int_0^\pi v_0^2(\omega t)\, d\omega t}$$

$$= \sqrt{\frac{1}{\pi} \int_{\alpha_f}^\pi (V_{i,p} \sin \omega t)^2\, d\omega t} = V_{i,p} \sqrt{\frac{1}{2\pi}\left(\pi - \alpha_f + \frac{\sin 2\alpha_f}{2}\right)}$$

$$(1.45)$$

is shown in Figure 1.16. Again, the characteristic is nonlinear. The fundamental output voltage, $V_{o,1}$, can also be shown to depend nonlinearly on the firing angle.

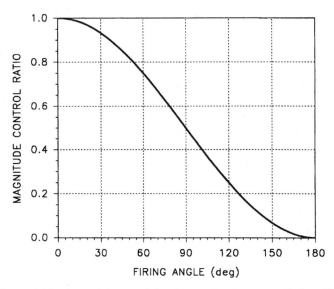

Figure 1.14   Control characteristic of a generic phase-controlled rectifier.

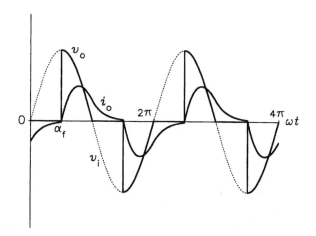

Figure 1.15   Output voltage and current waveforms in a generic ac voltage controller with a firing angle of 90°.

1.5   PULSE WIDTH MODULATION

As explained in the preceding section, control of the output voltage of a power electronic converter by means of an adjustable firing delay has been dictated primarily by the operating properties of semicontrolled power switches. Phase control, although conceptually simple, results in serious distortion of the output current of a converter, which is greater the longer that delay is used. Clearly, the distorted current is a

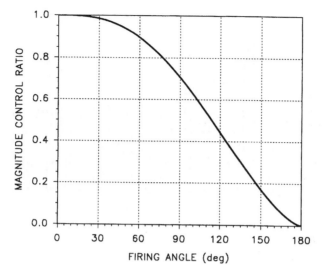

Figure 1.16   Control characteristic of a phase-controlled generic ac voltage controller.

consequence of the distorted voltage. However, the shape of the output voltage waveforms in power electronic converters is of much less practical concern than that of the output current waveforms. Plainly speaking, *it is the current that does the job*, whether it is the torque production in an electric motor or the electrolytic process in an electrochemical plant.

As mentioned earlier, most practical loads contain inductance. Since such a load constitutes a low-pass filter, high-order harmonics of the output voltage of a converter have a weaker impact on the output current waveform than do low-order harmonics. Therefore, the quality of the current depends strongly on the amplitudes of low-order harmonics in the frequency spectrum of the output voltage. Such spectra for a phase-controlled generic rectifier and an ac voltage controller are illustrated in Figure 1.17. Amplitudes of the harmonics are expressed in the per-unit format, with the peak value, $V_{i,p}$, of input voltage taken as the base voltage. Both spectra display several high-amplitude low-order harmonics.

In ac-input converters, distortion of the *input* current is of equal importance. Distorted currents drawn from the power system cause *harmonic pollution* of the system, resulting in faulty operation of system protection relays and electromagnetic interference (EMI) with communication systems. To alleviate these problems, utility companies require that input filters be installed, which raises the total cost of power conversion. The lower the frequency of harmonics to be attenuated, the larger and more expensive the filters that are needed. The alternative method of voltage and current control, by *pulse width modulation* (PWM), results in better spectral characteristics of converters and smaller filters. Therefore, pulse width modulation schemes are used increasingly in modern power electronic converters.

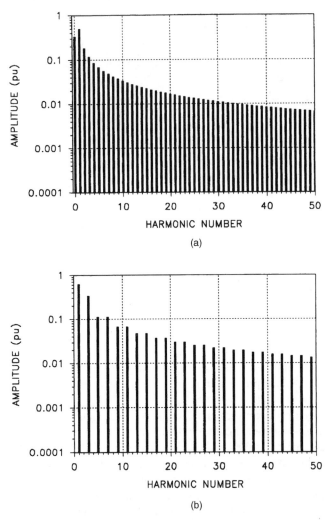

(a)

(b)

Figure 1.17   Harmonic spectra of the output voltage with a firing angle of 90° in (a) a phase-controlled generic rectifier, (b) a phase-controlled generic ac voltage controller.

The principle of pulse width modulation can best be explained by considering the dc-to-dc power conversion performed by a generic converter supplied with fixed dc voltage. The converter controls the dc component of the output voltage. As shown in Figure 1.18, this is accomplished by using the converter switches such that the output voltage consists of a train of pulses (state 1 of the generic converter) interspersed with notches (state 0). Fittingly, the respective practical power electronic converters are called *choppers*. Low-power converters used in power supplies for electronic equipment are usually referred to as *dc voltage regulators* or, simply, *dc-to-dc converters*.

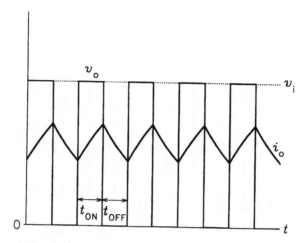

Figure 1.18 Output voltage and current waveforms in a generic chopper.

In the case illustrated, the pulses and notches are of equal duration; that is, switches S1 and S2 operate with a *duty ratio* of 0.5. A duty ratio, d, of a switch is defined as

$$d \equiv \frac{t_{ON}}{t_{ON} + t_{OFF}} \tag{1.46}$$

where t_{ON} denotes the on-time, that is, the interval within which the switch is closed, and t_{OFF} is the off-time, that is, the interval within which the switch is open. Here, switch S5 also operates with a duty ratio of 0.5. However, if the duty ratio of switches S1 and S2 were, for example, 0.6, the duty ratio of switch S5 would have to be 0.4. Switches S3 and S4 of the generic converter are not utilized (unless a reversal of polarity of the output voltage is required), so their duty ratio is zero.

It is easy to see that the average value (dc component), $V_{o,dc}$, of the output voltage is proportional to the fixed value, V_i, of the input voltage and to the duty ratio, d_{12}, of switches S1 and S2; that is,

$$V_{o,dc} = d_{12} V_i. \tag{1.47}$$

Since the range of a duty ratio is zero (switch open all the time) to unity (switch closed all the time), adjusting the duty ratio of appropriate switches allows setting $V_{o,dc}$ at any level between zero and V_i. As follows from Eq. (1.47), the voltage control characteristic, $V_{o,dc} = f(d_{12})$, of the generic chopper is linear.

Although the *switching frequency*, f_{sw}, defined as

$$f_{sw} \equiv \frac{1}{t_{ON} + t_{OFF}} \tag{1.48}$$

does not affect the dc component of the output voltage, the quality of the output current depends strongly on f_{sw}. As illustrated in Figure 1.19, if the number of pulses

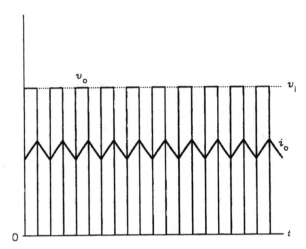

Figure 1.19 Output voltage and current waveforms in a generic chopper with a switching frequency twice as high as that in Figure 1.18.

per second is doubled compared with that in Figure 1.18, the increased switching frequency results in reduction of the current ripple by about 50%. The time between consecutive state changes of the converter is simply short enough to prevent significant current changes between consecutive "jumps" of the output voltage.

The reduction of current ripple can also be explained by the harmonic analysis of the output voltage. Note that the fundamental output frequency equals the switching frequency. As a result, the harmonics of the ac component of the voltage appear around frequencies that are integer multiples of f_{sw}. The corresponding inductive reactances of the load are proportional to these frequencies. Therefore, if the switching frequency is sufficiently high, the ac component of the output current is so strongly attenuated that the current is practically of ideal dc quality.

Voltage control using PWM can be employed in all the other types of electric power conversion described in this chapter. Instead of taking out solid chunks of the output voltage waveforms by the firing delay as shown in Figures 1.13 and 1.15, a large number of narrow segments can be removed in the PWM operating mode illustrated in Figure 1.20. Here, the ratio, N, of the switching and input frequencies is 12. This is also the number of pulses of output voltage per cycle. The output current waveforms are of significantly higher quality than those in Figures 1.13 and 1.15, exemplifying the operational superiority of pulse width–modulated converters over phase-controlled converters.

It should be mentioned that, for clarity, in most examples of PWM converters throughout the book, the switching frequencies employed in preparing the figures are lower than practical frequencies. Typically, depending on the type of power switches, switching frequencies are on the order of a few kilohertz, seldom less than 1 kHz and, in *supersonic converters*, higher than 20 kHz. Therefore, if, for example, a switching frequency in a 60-Hz PWM ac voltage controller is 3.6 kHz, the output voltage is "sliced" into 60 segments per cycle instead of the 12 segments shown in Figure 1.20b.

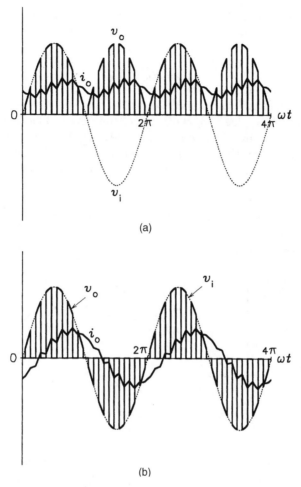

(a)

(b)

Figure 1.20 Output voltage and current waveforms in (a) a generic PWM rectifier; (b) a generic PWM ac voltage controller ($N = 12$).

Generally, a switching frequency should be several times higher than the reciprocal of the dominant (the longest) time constant of the load.

It can be shown that the linear relation (1.46) pertaining to the chopper can be extended to other types of PWM converters operating with fixed duty ratios of switches, such as rectifiers and ac voltage controllers. In all these converters,

$$V_{o,\text{adj}} = d V_{o,\text{adj(max)}} \tag{1.49}$$

that is, the magnitude control ratio, M, equals the duty ratio, d, of switches connecting the input and output terminals. Depending on the type of converter, the symbol $V_{o,\text{adj}}$ represents the adjustable dc component or fundamental ac component of the output voltage, while $V_{o,\text{adj(max)}}$ is the maximum available value of this component. However, concerning the rms value, V_o, of the output voltage, the dependence on the duty ratio is radical, that is,

$$V_o = \sqrt{d}\, V_{o,\text{max}} \tag{1.50}$$

where $V_{o\text{(max)}}$ is the maximum available rms value of output voltage. Specific values of $V_{o,\text{adj(max)}}$ and $V_{o\text{(max)}}$ depend on the type of converter and the magnitude of the input voltage. The control characteristics of a generic PWM rectifier [Eq. (1.47)] and an ac voltage controller [Eq. (1.50)] are shown in Figure 1.21 for comparison with those of phase-controlled converters depicted in Figures 1.14 and 1.16.

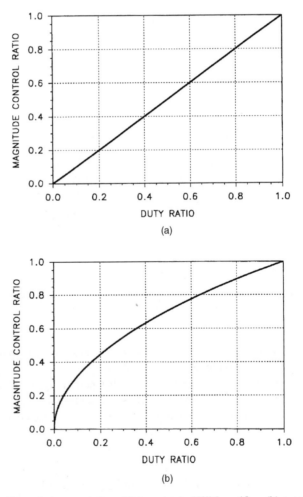

Figure 1.21 Control characteristics of (a) a generic PWM rectifier; (b) a generic PWM ac voltage controller.

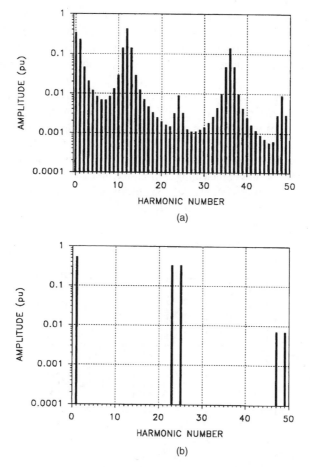

Figure 1.22 Harmonic spectra of output voltage in (a) a generic PWM rectifier; (b) a generic PWM ac voltage controller ($N = 24$).

Harmonic spectra of the output voltage of a generic PWM rectifier and an ac voltage controller are shown in Figure 1.22, again in the per-unit format. Here the switching frequency is 24 times higher than the input frequency; that is, the output voltage has 24 pulses per cycle. The duty ratio of converter switches is 0.5. Comparing these spectra with those in Figure 1.17, essential differences can be observed. The amplitudes of the low-order harmonics have been suppressed, especially in the spectrum for the ac voltage controller. Higher harmonics appear in clusters centered about integer multiples of 12 for the rectifier and 24 for the ac voltage controller. The reduced amplitudes of low-order harmonics of the output voltage translate into enhanced quality of the output current waveforms.

It should be pointed out that the simple pulse width modulation described, with a constant duty ratio of switches, has only been used to illustrate the basic principles of PWM converters. In practice, constant duty ratios are typical for choppers and common for ac voltage controllers. PWM rectifiers and inverters employ more sophisticated PWM techniques, in which the duty ratios of switches change throughout the cycle of the output voltage. Such techniques can also be used in PWM ac voltage controllers.

PWM techniques characterized by variable duty ratios of converter switches are explained in detail in Chapters 4 and 7. Waveforms of the output voltage and current of a generic inverter with $N = 10$ voltage pulses per cycle and with variable duty ratios are shown in Figure 1.23. The magnitude control ratio, M, pertaining to the amplitude of the fundamental output voltage, is 1 in Figure 1.23a and 0.5 in Figure 1.23b. The varying widths of the voltage pulses and the proportionality of these widths to the magnitude control ratio can easily be observed. The output current waveforms, although rippled, are much closer to ideal sine waves than those in the square-wave

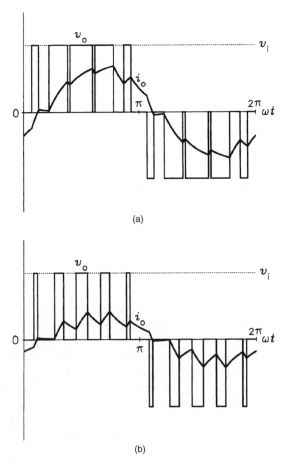

Figure 1.23   Output voltage and current waveforms in a generic PWM inverter: (a) $M = 1$; (b) $M = 0.5$ ($N = 10$).

operating mode of the inverter (see Figure 1.5). The harmonic content of the current would decrease further with an increase in the switching frequency.

All the examples of PWM converters presented in this section show that the higher the switching frequency, the better the quality of the output current. However, the allowable switching frequency in practical power electronic converters is limited by several factors. First, any power switch requires a certain amount of time for transitions from the on-state to the off-state, and vice versa. Therefore, the operating frequency of a switch is restricted, the maximum value depending on the type and ratings of the switch. Second, the control system of a converter has a limited operating speed. Finally, as explained in Chapter 2, the *switching losses* in practical switches increase with the switching frequency, reducing the efficiency of power conversion. Therefore, the switching frequency employed in a PWM converter should represent a sensible trade-off between the quality and efficiency of operation of the converter.

1.6   CALCULATION OF CURRENT WAVEFORMS

Practical sources of raw electric power are predominantly voltage sources. Such a source can be converted into a current source only by providing closed-loop control, maintaining the load current at a steady level. This is a rather complicated arrangement, so most power electronic converters are supplied from voltage sources. Since, as already demonstrated, a static power converter is simply a network of switches, waveforms of voltages are easy to determine from the converter's principle of operation. It is not so with currents, which depend on both the voltages and the load. Typically, however, once one current, usually the output current, is known, the operating algorithm of the converter can again be employed to find the other currents.

As exemplified by the generic converter, power electronic converters operate most of the time in *quasi-steady state*, which is a sequence of transient states. The converters are variable-topology circuits, and each change of topology initiates a new transient state. Final values of currents and voltages in one topology become initial values in the next topology. Since linear electric circuits can be described by ordinary linear differential equations, the task of finding a current waveform becomes a classic initial value problem. As shown later, in place of differential equations, difference equations can be applied conveniently to converters operating in the PWM mode.

If the load of a converter is linear and the output voltage can be expressed in closed form, the output current waveform corresponding to individual states of the converter can also be expressed in closed form. However, when a computer is used for converter analysis, numerical algorithms can be employed to compute consecutive points of the current waveform. Both the analytical and numerical approaches are illustrated in subsequent sections.

1.6.1   Analytical Solution

To show the typical way of finding closed-form expressions for the output current in a power electronic converter, the generic rectifier, generic inverter, and generic PWM

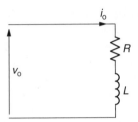

Figure 1.24   Resistive–inductive (RL) load circuit.

ac voltage controller are considered. In all three cases, a resistive–inductive (RL) load is assumed.

The RL load circuit of a generic converter is shown in Figure 1.24. If the converter operates as a rectifier, the input and output terminals are cross-connected when the ac input voltage, given by Eq. (1.1), is negative. Consequently, the output voltage, $v_o(t)$, and current, $i_o(t)$, are periodic with a period of $T/2$, where T is the period of the input voltage, equal to $2\pi/\omega$. Therefore, all subsequent consideration of the generic rectifier is limited to the interval 0 to $T/2$.

Kirchhoff's voltage law for the circuit in Figure 1.24 can be written as

$$Ri_o(t) + L\frac{di_o(t)}{dt} = V_{i,p}\sin\omega t. \tag{1.51}$$

Equation (1.51) can be solved for $i_o(t)$ using the Laplace transformation or, less tediously, taking advantage of the known property of linear differential equations, allowing $i_o(t)$ to be expressed as

$$i_o(t) = i_{o,F}(t) + i_{o,N}(t) \tag{1.52}$$

where $i_{o,F}(t)$ and $i_{o,N}(t)$ denote the *forced* and *natural components* of $i_o(t)$, respectively. The convenience of this approach lies in the fact that the forced component constitutes the steady-state solution of Eq. (1.51), that is, the steady-state current excited in the circuit in Figure 1.25 by the sinusoidal voltage $v_i(t)$. As is well known to every electrical engineer,

$$i_{o,F}(t) = \frac{V_{i,p}}{Z}\sin(\omega t - \varphi) \tag{1.53}$$

where

$$Z = \sqrt{R^2 + (\omega L)^2} \tag{1.54}$$

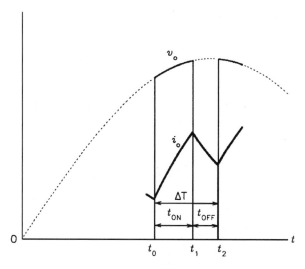

Figure 1.25 Fragments of output voltage and current waveforms in a generic PWM ac voltage controller.

and

$$\varphi = \tan^{-1} \frac{\omega L}{R} \tag{1.55}$$

are the impedance and phase angle of the RL load, respectively. The shortcut from Eq. (1.51) to the forced solution (1.53) saves a lot of work.

The natural component, $i_{o,N}(t)$, represents a solution of the homogeneous equation

$$R i_{o,N}(t) + L \frac{d i_{o,N}(t)}{dt} = 0 \tag{1.56}$$

obtained from Eq. (1.51) by equating the right-hand side (the excitation) to zero. It can be seen that $i_{o,N}(t)$ must be such that the linear combination of it and its derivative is zero. Clearly, an exponential function

$$i_{o,N}(t) = A e^{BT} \tag{1.57}$$

is the most likely candidate for the solution. Substituting $i_{o,N}(t)$ in Eq. (1.56) yields

$$R A e^{BT} + L A B e^{BT} = 0 \tag{1.58}$$

from which $B = -R/L$ and

$$i_{o,N}(t) = A e^{-(R/L)t} \tag{1.59}$$

Thus, based on Eqs. (1.52) and (1.53),

$$t_o(t) = \frac{V_{i,p}}{Z} \sin(\omega t - \varphi) + A e^{-(R/L)t}. \tag{1.60}$$

To determine the constant A, note that in the quasi-steady state of the rectifier, $i_o(0) = i_o(T/2) = i_o(\pi/\omega)$ (see Figure 1.4). Hence,

$$\frac{V_{i,p}}{Z} \sin(-\varphi) + A = \frac{V_{i,p}}{Z} \sin(\pi - \varphi) + A e^{-R\pi/L\omega} \tag{1.61}$$

where

$$-\frac{R\pi}{L\omega} = -\frac{\pi}{\tan \varphi}. \tag{1.62}$$

Equation (1.61) can now be solved for A, yielding

$$A = \frac{2V_{i,p} \sin \varphi}{Z(1 - e^{-\pi/\tan \varphi})} \tag{1.63}$$

and

$$i_o(t) = \frac{V_{i,p}}{Z} \left[\sin(\omega t - \varphi) + \frac{2 \sin \varphi}{1 - e^{-\pi/\tan \varphi}} e^{-(R/L)t} \right]. \tag{1.64}$$

The approach that we have shown to derivation of the rather complex relation (1.64) takes much less time and effort than that required when the Laplace transformation is employed. The reader is encouraged to confirm this observation by his or her own computations.

When the generic power converter works as an inverter, the output voltage equals V_i when the converter is in state 1 and $-V_i$ in state 2. If T now denotes the period of the output voltage, it can be assumed that state 1 lasts from 0 to $T/2$ and state 2 from $T/2$ to T. Because it is only the polarity of output voltage that changes every half-cycle, the output current, $i_{o(2)}(t)$, in the second half-cycle equals $-i_{o(1)}(t-T/2)$, where $i_{o(1)}(t)$ is the output current in the first half-cycle. Consequently, it is sufficient to determine $i_{o(1)}(t)$ only.

The forced component, $i_{o,F(1)}(t)$, of current $i_{o(1)}(t)$, excited in the RL load by the dc voltage $v_o = V_i$, is given by Ohm's law as

$$i_{o,F(1)} = \frac{V_i}{R}. \tag{1.65}$$

The natural component, which depends solely on the load, is the same as that of the output current of the generic rectifier. Thus,

$$i_{o(1)}(t) = \frac{V_i}{R} + Ae^{-R/L}. \tag{1.66}$$

Constant A can be found from the condition $i_{o(1)}(0) = -i_{o(1)}(T/2)$ (see Figure 1.5), that is,

$$\frac{V_i}{R} + A = -\left(\frac{V_i}{R} + Ae^{-\pi/\tan\varphi} \right) \tag{1.67}$$

which yields

$$A = \frac{2}{1 + e^{-\pi/\tan\varphi}} \frac{V_i}{R}. \tag{1.68}$$

This gives

$$i_o(t) = \begin{cases} \dfrac{V_i}{R} \left[1 - \dfrac{2}{1 + e^{-\pi/\tan\varphi}} e^{-(R/L)t} \right] & \text{for } 0 < t \leq \dfrac{T}{2} \\[4mm] -\dfrac{V_i}{R} \left[1 - \dfrac{2}{1 + e^{-\pi/\tan\varphi}} e^{-(R/L)(t-T/2)} \right] & \text{for } \dfrac{T}{2} < t \leq T \end{cases}. \tag{1.69}$$

In the final example, a PWM converter, specifically the PWM ac voltage controller, is dealt with. Within a single cycle of output voltage, the converter undergoes a large number of state changes, and within each corresponding interval of time, the waveform of output current is described by a different equation. Therefore, to express the waveform in a useful format, iterative formulas are employed. The general form of such a formula is $i_o(t + \Delta t) = f[i_o(t), t]$, where Δt is a small increment of time, meaning that consecutive values of the output current are calculated from the previous values.

In developing the formulas, two simplifying assumptions were made; first, that the current waveform is piecewise linear; second, that the initial value, $i_o(0)$, of the output current is equal to that of a hypothetical sinusoidal current that would be generated in the same load by the fundamental of the actual output voltage of the converter. Thus, if the input voltage of the generic ac voltage controller is given by Eq. (1.1), the initial value of the output current is

$$i_o(0) = M\frac{V_{i,p}}{Z} \sin(\omega t - \varphi) \Big|_{t=0} = -M\frac{V_{i,p}}{Z} \sin\varphi. \tag{1.70}$$

The output voltage waveform of the controller, shown in Figure 1.20b, is a chopped sinusoid. A single switching cycle of that waveform is depicted in Figure 1.25, which also shows the coinciding fragment of the output current waveform. The time interval,

t_0 to t_2, corresponding to the switching cycle is called a *switching interval*. From t_0 to t_1, the controller is in state 1 and the output voltage equals the input voltage, while from t_1 to t_2, the output voltage is zero, the controller assuming state 0. Denoting the length of the switching interval, $t_2 - t_0$, by ΔT, the on-time, t_{ON}, is $M \Delta T$, and the off-time, t_{OFF}, is $(1 - M) \Delta T$.

The differential equation of the load circuit is

$$Ri_o(t) + L\frac{di_o(t)}{dt} = v_o(t) \tag{1.71}$$

which for the instant $t = t_0$ can be rewritten as

$$Ri_o(t_0) + L\frac{\Delta i_o}{\Delta t} = V_{i,p} \sin \omega t_0 \tag{1.72}$$

where $\Delta i_o = i_o(t_1) - i_o(t_0)$ and $\Delta t = t_1 - t_0$, that is, as a difference equation. Taking into account that $\Delta t = M \Delta T$, Eq. (1.72) can be solved for $i_o(t_0)$ to yield

$$t_o(t_1) = i_o(t_0) + \frac{M}{L}[V_{i,p} \sin \omega t_0 - Ri_o(t_0)] \Delta T. \tag{1.73}$$

Similarly, at $t = t_1$, the load circuit equation

$$Ri_o(t_1) + L\frac{\Delta i_o}{\Delta t} = 0 \tag{1.74}$$

where $\Delta i_o = i_o(t_2) - i_o(t_1)$ and $\Delta t = t_2 - t_1$, can be rearranged to

$$i_o(t_2) = i_o(t_1)\left[1 - \frac{R}{L}(1 - M) \Delta T\right]. \tag{1.75}$$

Formulas (1.70), (1.73), and (1.75) allow computation of consecutive segments of the piecewise linear waveform of the output current. If, as is typical in practical PWM converters, the switching frequency, $f_{sw} = 1/\Delta T$, is at least one order of magnitude higher than the input or output frequency, the accuracy of those approximate relations is sufficient for all practical purposes.

1.6.2 Numerical Solution

When a computer program is used to simulate a converter, the output voltage is calculated as a series of values, $v_{o,0}, v_{o,1}, v_{o,2}, \ldots$, for the consecutive instants, t_0, $t_1, t_2, \ldots$. These instants should be close, so that $t_{n+1} - t_n < \tau$, where τ denotes the shortest time constant of the system simulated. Then the respective values of the output current, $i_{o,1}, i_{o,2}, i_{o,3}, \ldots$, can be computed as responses of the load circuit to step excitations $v_{o,0}u(t - t_0), v_{o,1}u(t - t_1), v_{o,2}u(t - t_2), \ldots$, where $u(t - t_n)$ denotes a unit step function at $t = t_n$.

For example, for the RL load considered in the preceding section, the differential equation of the load circuit for $t \geq t_n$ can be written as

$$Ri_o(t) + L\frac{di_o(t)}{dt} = v_{o,n}u(t - t_n) \tag{1.76}$$

where $u(t - t_n) = 1$. The forced component, $i_{o,F}(t)$, of the solution, $i_o(t)$, of Eq. (1.76), is given by

$$i_{o,F}(t) = \frac{v_{o,n}}{R} \tag{1.77}$$

and the natural component, $i_{o,N}(t)$, by

$$i_{o,N}(t) = Ae^{-(R/L)t}. \tag{1.78}$$

Hence,

$$i_o(t) = \frac{v_{o,n}}{R} + Ae^{-(R/L)t} \tag{1.79}$$

where the constant A can be determined from

$$i_o(t_n) = i_{o,n} = \frac{v_{o,n}}{R} + Ae^{-(R/L)t_n} \tag{1.80}$$

as

$$A = \left(i_{o,n} - \frac{v_{o,n}}{R}\right)e^{-(R/L)t_n}. \tag{1.81}$$

Substituting $t = t_{n+1}$ and Eq. (1.81) in Eq. (1.79) and denoting $i_o(t_n + 1)$ by $i_{o,n+1}$ yields

$$i_{o,n+1} = \frac{v_{o,n}}{R} + \left(i_{o,n} - \frac{v_{o,n}}{R}\right)e^{-(R/L)(t_{n+1}-t_n)} \tag{1.82}$$

which is an iterative formula, $i_{o,n+1} = f(i_{o,n}, t_{n+1} - t_n)$, that allows easy computation of consecutive points of the output current waveform.

The other common loads and corresponding numerical formulas for the output current are:

Resistive load (R load):

$$i_{o,n} = \frac{v_{o,n}}{R}. \tag{1.83}$$

Inductive load (L load):

$$i_{o,n+1} = i_{o,n} + \frac{v_{o,n}}{L}(t_{n+1} - t_n). \qquad (1.84)$$

Resistive–EMF load (RE load):

$$i_{o,n} = \frac{v_{o,n} - E_o}{R} \qquad (1.85)$$

where E_n denotes the value of load EMF at $t = t_n$. The EMF, in series with resistance R, is assumed to oppose the output current.

Inductive–EMF load (LE load):

$$i_{o,n+1} = i_{o,n} + \frac{v_{o,n} - E_n}{L}(t_{n+1} - t_n) \qquad (1.86)$$

where the load EMF is connected in series with inductance L.

Resistive–inductive–EMF load (RLE load):

$$i_{o,n+1} = \frac{v_{o,n} - E_n}{R} + \left(i_{o,n} - \frac{v_{o,n} - E_n}{R}\right) e^{-(R/L)(t_{n+1}-t_n)} \qquad (1.87)$$

where the load is represented by a series connection of resistance R, inductance L, and EMF E.

The considerations presented can be adapted for analysis of the less common current-source converters. There, those are currents that are easily determinable from the operating principles of a converter, whereas voltage waveforms require analytical or numerical solution of appropriate differential equations.

In engineering practice, specialized computer programs are used for modeling and analysis of power electronic converters. The popular software package PSpice for simulation of electronic circuits is called for in most of the computer assignments in this book (see Appendix A). Other commercial versions of UC Berkeley's Spice, such as the Electronic Bench, are also available on the software market. Many advanced simulation programs, of which Saber is the best known example, have been developed in several countries specifically for modeling of switching power converters. Program EMTP, used primarily by utilities for power system analysis, allows simulation of power electronic converters, which is particularly useful for studies of the impact of converters on power system operation. General-purpose dynamic simulators, such as Simulink, Simplorer, or ACSL, can be used successfully for analysis not only of converters themselves but also of entire systems that include converters, such as electric motor drives.

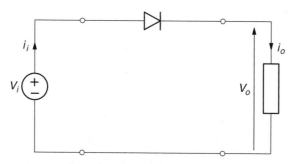

Figure 1.26   Single-pulse diode rectifier.

1.6.3   Practical Examples: Single-Phase Diode Rectifiers

To illustrate the considerations presented in the preceding two sections and to start introducing practical power electronic converters, single-phase diode rectifiers are described. They are the simplest static converters of electrical power. The power diodes employed in the rectifiers can be thought of as uncontrolled semiconductor power switches. A diode in a closed circuit begins conducting (turns on) when its anode voltage becomes higher than the cathode voltage, that is, when the diode is *forward biased*. The diode ceases to conduct (turns off) when its current changes polarity.

A *single-pulse* (single-phase half-wave) diode rectifier is shown in Figure 1.26. Functionally, it is analogous to a generic power converter with only switches S1 and S2 and the provision that they stay closed as long as the current flows from terminal I1 to terminal O1. If the load is purely resistive (R load), and the sinusoidal input voltage is given by Eq. (1.1), the output current waveform resembles that of the output voltage, as shown in Figure 1.27. Comparing Figures 1.27 and 1.4, it can easily be seen that the average output voltage, $V_{o,\mathrm{dc}}$, in a single-pulse rectifier is only half as large as that in the generic rectifier described in Section 1.2, that is, $V_{o,\mathrm{dc}} = V_{i,p}/\pi \approx 0.32 V_{i,p}$ [see Eq. (1.15)]. The single "pulse" of output current per cycle of the input voltage explains the name of the rectifier.

The voltage gain is even lower when the load contains an inductance (RL load). Equation (1.59) can be used to describe the output current, but only until it reaches zero at $\omega t = \alpha_e$, where α_e is called an *extinction angle*. Thus,

$$i_o(0) = \frac{V_{i,p}}{Z}\sin(-\varphi) + A = 0 \tag{1.88}$$

from which

$$A = \frac{V_{i,p}}{Z}\sin\varphi \tag{1.89}$$

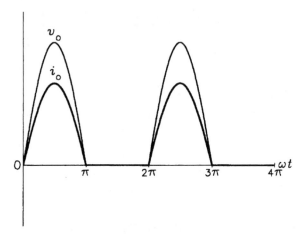

Figure 1.27   Output voltage and current waveforms in a single-pulse diode rectifier with an R load.

and

$$i_o(t) = \frac{V_{i,p}}{Z}\left[\sin(\omega t - \varphi) + e^{-(R/L)t}\sin\varphi\right] \qquad (1.90)$$

when $i_o > 0$, that is, when $0 < \omega t \le \alpha_e$. Substituting α_e/ω for t and zero for i_o, the extinction angle can be found from

$$\frac{V_{i,p}}{Z}\left[\sin(\alpha_e - \varphi) + e^{-\alpha_e/\tan\varphi}\sin\varphi\right] = 0. \qquad (1.91)$$

Clearly, no closed-form expression exists for α_e, and the extinction angle, which is a function of the load angle φ, can only by found numerically. This problem is analyzed in greater detail in Chapter 4.

The output voltage and current waveforms of a single-pulse rectifier with an RL load are shown in Figure 1.28. As the extinction angle is greater than 180°, the output voltage is negative in the interval $\omega t = \pi$ to $\omega t = \alpha_e$, and its average value is lower than that with a resistive load. This can be remedied by connecting the *freewheeling diode*, DF, across the load, as shown in Figure 1.29.

The freewheeling diode, which corresponds to switch S5 in the generic power converter, shorts the output terminals when the output voltage reaches zero and provides a path for the output current in the interval π to α_e. Until the voltage reaches zero, the waveform of output voltage is the same as that in Figure 1.28, and that of output current is described by Eq. (1.90). Later, the current dies out while freewheeled

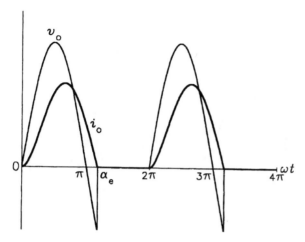

Figure 1.28   Output voltage and current waveforms in a single-pulse diode rectifier with an RL load.

by diode DF. As $v_o = 0$ now, the equation of the current is simply

$$i_o(t) = Ae^{-(R/L)t} \tag{1.92}$$

where constant A can be found by equaling currents given by Eqs. (1.90) and (1.92) at $\omega t = \pi$, that is,

$$\frac{V_{i,p}}{Z}(1 - e^{-\pi/\tan\varphi})\sin\varphi = Ae^{-\pi/\tan\varphi}. \tag{1.93}$$

From Eq. (1.93),

$$A = \frac{V_{i,p}}{Z}(e^{\pi/\tan\varphi} - 1)\sin\varphi \tag{1.94}$$

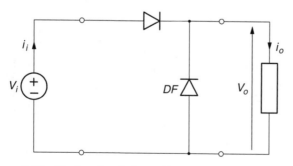

Figure 1.29   Single-pulse diode rectifier with a freewheeling diode.

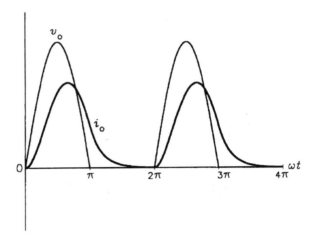

Figure 1.30 Output voltage and current waveforms in a single-pulse diode rectifier with a freewheeling diode and an RL load.

and

$$i_o(t) = \frac{V_{i,p}}{Z}(e^{\pi/\tan\varphi} - 1)e^{-(R/L)t}\sin\varphi. \tag{1.95}$$

Waveforms of the output voltage and current of a single-pulse rectifier with the freewheeling diode are shown in Figure 1.30.

More radical enhancement of the single-pulse rectifier shown in Figure 1.31 consists of connecting a large capacitor across the output terminals. The capacitor charges up when the input voltage is high and discharges through the load when the input voltage drops below a specific level that depends on the capacitor and load. Typical waveforms of the output voltage, v_o, and capacitor current, i_C, with a resistive load are shown in Figure 1.32. Advanced readers are encouraged to try and obtain analytical expressions for these waveforms using the approach sketched in this chapter.

Practical single-pulse rectifiers do not belong in the realm of true power electronics, as the output capacitors would have to be excessively large. The average output

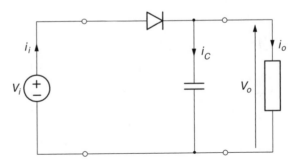

Figure 1.31 Single-pulse diode rectifier with an output capacitor.

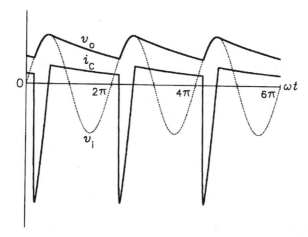

Figure 1.32 Output voltage and current waveforms in a single-pulse diode rectifier with an output capacitor and an RL load.

voltage, $V_{o,\text{dc}}$, can be increased to $2v_{i,p}/\pi \approx 0.64v_{i,p}$ in the two-pulse (single-phase full-wave) diode rectifier in Figure 1.33. Diodes D1 through D4 correspond to the respective switches in the generic converter, connecting directly and cross-connecting the input terminals with the output terminals in dependence on the polarity of the input voltage. As a result, the output voltage and current waveforms are the same as in the generic converter (see Figure 1.4). In low-power rectifiers, a capacitor can be placed across the load, similar to the single-pulse rectifier in Figure 1.31, to smooth out the output voltage and increase its dc component. Generally, however, single-phase rectifiers have operating parameters distinctly inferior to those of three-phase rectifiers, presented in Chapter 4.

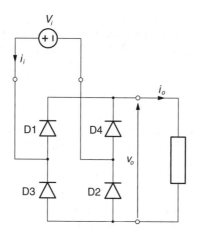

Figure 1.33 Two-pulse diode rectifier.

1.7   SUMMARY

Power conversion and control are performed in power electronic converters that are networks of semiconductor power switches. Most practical converters are supplied from dc or ac voltage sources. Such sources have also been assumed for a generic converter, introduced here for instructional purposes. Three basic states of a voltage-source converter can be distinguished (although in some converters only two states suffice): the input and output terminals are directly connected, cross-connected, or separated. In the last state, the output terminals are shorted to maintain a closed path for the output current. An appropriate sequence of states results in conversion of the given input (supply) voltage to the output (load) voltage desired.

Current-source converters are also feasible, although less common than voltage-source converters. A load of a current-source converter must appear as a voltage source, and that of a voltage-source converter as a current source. In practice, the current- and voltage-source requirements imply a series-connected inductor and a parallel-connected capacitor, respectively.

Control of the output voltage in voltage-source power electronic converters is realized by periodic use of state 0 of the converter. This results in removal of certain portions of the output voltage waveform. Two approaches are employed: phase control and pulse width modulation. The latter technique, promoted by the availability of fast fully controlled semiconductor power switches, is employed increasingly in modern power electronics. The PWM results in higher power conversion and control quality than that of phase control.

The output voltage waveforms in voltage-source converters are usually easy to determine. Current waveforms, however, require analytical or numerical so-lution of differential or, for PWM converters, difference equations describing quasi-steady-state operation of the converters. A dual approach is applicable to current-source converters. Simulation software is used extensively in engineering practice.

Single-phase diode rectifiers are the simplest ac-to-dc power converters. However, their voltage gain, especially that of a single-pulse rectifier, is low, and ripple factors of the output voltage and current are high. These values can be improved by installing an output capacitor, but that solution is feasible only in low-power rectifiers.

EXAMPLES

Example 1.1   A generic power converter is supplied from a 120-V 50-Hz ac voltage source and controlled so that its output voltage has the waveform shown in Figure 1.34. It can be seen that the converter performs ac-to-ac conversion, resulting in the output fundamental frequency reduced with respect to the input frequency. This type of power conversion is characteristic for power electronic converters called *cycloconverters*. The generic cycloconverter in this example operates in a simple *trapezoidal mode*, in which the ratio of the input frequency to output fundamental frequency is an integer.

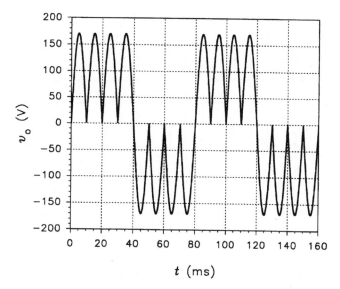

Figure 1.34 Output voltage waveform in the generic cycloconverter in Example 1.1.

What is the fundamental frequency of the voltage in Figure 1.34? Sketch the timing diagram of the converter switches.

Solution: The output fundamental frequency in question is four times lower than the input frequency of 50 Hz, that is, 12.5 Hz. The timing diagram of the generic cycloconverter switches is shown in Figure 1.35. Switch S5 is open all the time since either switches S1 and S2 or S3 and S4 are closed.

Example 1.2 For the output voltage waveform in Figure 1.34, find:

    a. The rms value, V_o
    b. The peak value, $V_{o,1,p}$, and rms value, $V_{o,1}$, of the fundamental
    c. The harmonic content, $V_{o,h}$
    d. The total harmonic distortion, $\mathrm{THD_V}$.

Solution: The peak value of the input voltage is $\sqrt{2} \times 120 = 170$ V. Denoting the fundamental output radian frequency by ω, the output voltage waveform can be expressed as

$$
v_o(\omega t) =
\begin{cases}
|170| \sin 4\omega t & \text{for } 2n\dfrac{\pi}{4} \le \omega t < (2n+1)\dfrac{\pi}{4} \\[3mm]
-|170| \sin 4\omega t & \text{for } (2n+1)\dfrac{\pi}{4} \le \omega t < (2n+2)\dfrac{\pi}{4}
\end{cases}
$$

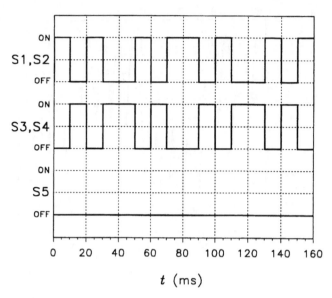

Figure 1.35   Timing diagram of switches in the generic cycloconverter in Example 1.1.

where $n = 0, 1, 2, \ldots,$ and the rms value, V_o, of the voltage is the same as that of the input voltage, that is, 120 V. The waveform has the odd and half-wave symmetries, so the peak value, $V_{o,1,p}$, of the fundamental is

$$V_{o,1,p} = \frac{4}{\pi} \int_0^{\pi/2} v_o(\omega t) \sin \omega t \; d\omega t$$

$$= \frac{4}{\pi} \int_0^{\pi/4} 170 \sin 4\omega t \; d\omega t - \frac{4}{\pi} \int_{\pi/4}^{\pi/2} 170 \sin 4\omega t \; d\omega t$$

$$= \frac{4 \times 170}{\pi} \left\{ \left[\frac{\sin 3\omega t}{6} - \frac{\sin 5\omega t}{10} \right]_0^{\pi/4} \right.$$

$$\left. = - \left[\frac{\sin 3\omega t}{6} - \frac{\sin 5\omega t}{10} \right]_{\pi/4}^{\pi/2} \right\} = 139 \text{ V}.$$

The rms value of the fundamental is

$$V_{o,1} = \frac{V_{o,1,p}}{\sqrt{2}} = \frac{139}{\sqrt{2}} = 98 \text{ V}$$

and since there is no dc component, the harmonic content is

$$V_{o,k} = \sqrt{V_o^2 - V_{o,1}^2} = \sqrt{120^2 - 98^2} = 69 \text{ V}.$$

Thus, the total harmonic distortion is

$$\mathrm{THD}_V = \frac{V_{o,h}}{V_{o,1}} = \frac{69}{98} = 0.7$$

Example 1.3 For a generic phase-controlled rectifier, find the relation between the ripple factor of the output voltage, RF_V, and the firing angle, α_f.

Solution: Equation (1.44) gives the relation between the dc component, $V_{o,\mathrm{dc}}$, of the output voltage in question and the firing angle, while Eq. (1.45), concerning the rms value, V_o, of the output voltage of a generic ac voltage controller, can be applied directly to the generic rectifier (why?). Consequently,

$$\begin{aligned} \mathrm{RF}_V(\alpha_f) &= \frac{V_{o,\mathrm{ac}}(\alpha_f)}{V_{o,\mathrm{dc}}(\alpha_f)} = \frac{\sqrt{V_o^2(\alpha_f) - V_{o,\mathrm{dc}}^2(\alpha_f)}}{V_{o,\mathrm{dc}}^2(\alpha_f)} \\ &= \sqrt{\frac{V_o^2(\alpha_f)}{V_{o,\mathrm{dc}}^2(\alpha_f)} - 1} = \sqrt{\frac{(1/2\pi)\left[\pi - \alpha_f + (\sin 2\alpha_f/2)\right]}{(1/\pi^2)(1 + \cos\alpha_f)^2} - 1} \\ &= \sqrt{\frac{\pi}{2}\frac{\pi - \alpha_f + (\sin 2\alpha_f/2)}{(1 + \cos\alpha_f)^2} - 1}. \end{aligned}$$

Graphical representation of the relation derived is shown in Figure 1.36. It can be seen that when the firing angle exceeds 150°, the voltage ripple increases rapidly because the dc component approaches zero.

Example 1.4 A generic converter is supplied from a 100-V dc source and operates as a chopper with a switching frequency, f_{sw}, of 2 kHz. The average output voltage, $V_{o,\mathrm{dc}}$, is −60 V. Determine the duty ratios of all switches of the converter and the corresponding on- and off-times, t_{ON} and t_{OFF}.

Solution: The negative polarity of the output voltage implies that switches S3, S4, and S5 perform the modulation, and switches S1 and S2 are permanently open (turned off). Therefore, the duty ratio, d_{12}, of the latter switches is zero. Adapting Eq. (1.47) to switches S3 and S4, the duty ratio, d_{34}, of these switches is

$$d_{34} = \frac{V_{o,\mathrm{dc}}}{V_1} = \frac{60}{100} = 0.6.$$

The negative sign at 60 V is omitted since the very use of switches S3 and S4 causes reversal of the output voltage (obviously, a duty ratio can only assume values between zero and unity). Switch S5 is turned on when the other switches are off. Hence, its

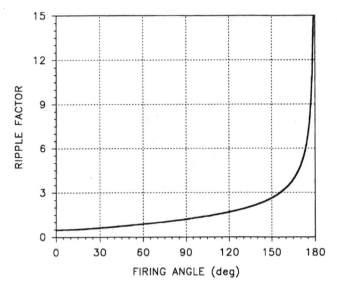

Figure 1.36   Voltage ripple factor versus firing angle in the generic rectifier in Example 1.3.

duty ratio, d_5, is

$$d_5 = 1 - d_{34} = 1 - 0.6 = 0.4.$$

From Eq. (1.48),

$$t_{ON} + t_{OFF} = \frac{1}{f_{sw}} = \frac{1}{2 \times 10^3} = 0.0005 \text{ s} = 0.5 \text{ ms}$$

and from Eq. (1.45),

$$t_{ON,34} = d_{34}(t_{ON} + t_{OFF}) = 0.6 \times 0.5 = 0.3 \text{ ms}$$

while

$$t_{ON,5} = d_5(t_{ON} + t_{OFF}) = 0.4 \times 0.5 = 0.2 \text{ ms}.$$

Consequently,

$$t_{OFF,34} = 0.5 - 0.3 = 0.2 \text{ ms}$$

and

$$t_{OFF,5} = 0.5 - 0.2 = 0.3 \text{ ms}.$$

Example 1.5   A generic converter is supplied from a 120-V 60-Hz ac voltage source and operates as a PWM rectifier with an RL load, where $R = 2\ \Omega$ and $L = 5$ mH. The switching frequency is 720 Hz and the magnitude control ratio is 0.6. Develop iterative formulas for the output current and calculate values of that current at the switching instants within one cycle of the output voltage.

Solution: Inspecting similar formulas, (1.73) and (1.75), for the generic PWM ac voltage controller, it can be seen that they are easy to adapt to the PWM rectifier by replacing the $\sin \omega t_0$ term in Eq. (1.73) with $|\sin \omega t_0|$. Then the output current of the rectifier can be computed from the equations

$$i_o(t_1) = i_o(t_0) + \frac{M}{L}\left[V_{i,p}\,|\sin \omega t_0| - Ri_o(t_0)\right]\,\Delta T$$

and

$$i_o(t_2) = i_o(t_1)\left[1 - \frac{R}{L}(1 - M)\,\Delta T\right].$$

The input frequency, ω, is $2\pi \times 60 = 377$ rad/s, and the length, ΔT, of a switching interval is 1/720 s. Thus, the general iterative formulas above give

$$i_o(t_1) = i_o(t_0) + \frac{0.6}{5 \times 10^{-3}}\left[\sqrt{2} \times 120\,|\sin(377t_0)| - 2i_o(t_0)\right] \times \frac{1}{729}$$

$$= 0.667i_o(t_0) + 28.3\,|\sin(377t_0)|$$

and

$$i_o(t_2) = i_o(t_1)\left[1 - \frac{2}{5 \times 10^{-3}}(1 - 0.6)\frac{1}{720}\right] = 0.778i_o(t_1).$$

As shown in Figure 1.20a, the period of the output voltage of the rectifier equals half of that of the input voltage, that is, 1/120 s = 8.333 ms. Comparing it with the switching frequency, the number of switching intervals per cycle of the output voltage turns out to be six. With $M = 0.6$, the on-time, t_{ON}, equal to $t_1 - t_0$, is 0.6/720 s = 0.833 ms, and the off-time, t_{OFF}, equal to $t_2 - t_1$, is 0.4/720 s = 0.556 ms.

To start the computations, the initial value, $i_o(0)$, is required as $i_o(t_0)$ in the first switching interval. It can be estimated as the dc component, $I_{o,dc}$, of the output current, given by

$$I_{o,dc} = \frac{M(2/\pi)V_{i,p}}{R} = \frac{0.6 \times 2/\pi \times \sqrt{2} \times 120}{2} = 32.4\ \text{A}.$$

Now, the computations of $i_o(t)$ can be performed for consecutive switching intervals, the t_2 instant for a given interval being the t_0 instant for the next interval.

First interval ($t_0 = 0$, $t_1 = 0.833$ ms, $t_2 = 1.389$ ms):

$$i_o(0.833 \text{ ms}) = 0.667 \times 32.4 + 28.3 \left|\sin(377 \times 0)\right| = 21.6 \text{ A}$$

and

$$i_o(1.389 \text{ ms}) = 0.778 \times 21.6 = 16.8 \text{ A.}$$

Second interval ($t_0 = 1.389$ ms, $t_1 = 2.222$ ms, $t_2 = 2.778$ ms):

$$i_o(2.222 \text{ ms}) = 0.667 \times 16.8 + 28.3 \left|\sin(377 \times 1.389 \times 10^{-3})\right| = 25.4 \text{ A}$$

and

$$i_o(2.778 \text{ ms}) = 0.778 \times 25.4 = 19.8 \text{ A.}$$

Third interval ($t_0 = 2.778$ ms, $t_1 = 3.611$ ms, $t_2 = 4.167$ ms):

$$i_o(3.611 \text{ ms}) = 0.667 \times 19.8 + 28.3 \left|\sin(377 \times 2.778 \times 10^{-3})\right| = 37.7 \text{ A}$$

and

$$i_o(4.167 \text{ ms}) = 0.778 \times 37.7 = 29.3 \text{ A.}$$

Fourth interval ($t_0 = 4.167$ ms, $t_1 = 5$ ms, $t_2 = 5.556$ ms):

$$i_o(4.167 \text{ ms}) = 0.667 \times 29.3 + 28.3 \left|\sin(377 \times 4.167 \times 10^{-3})\right| = 47.8 \text{ A}$$

and

$$i_o(5.556 \text{ ms}) = 0.778 \times 47.8 = 37.2 \text{ A.}$$

Fifth interval ($t_0 = 2.778$ ms, $t_1 = 3.611$ ms, $t_2 = 4.167$ ms):

$$i_o(6.389 \text{ ms}) = 0.667 \times 37.2 + 28.3 \left|\sin(377 \times 5.556 \times 10^{-3})\right| = 49.3 \text{ A}$$

and

$$i_o(6.944 \text{ ms}) = 0.778 \times 49.3 = 38.4 \text{ A.}$$

Sixth interval ($t_0 = 6.944$ ms, $t_1 = 7.778$ ms, $t_2 = 8.333$ ms):

$$i_o(7.778 \text{ ms}) = 0.667 \times 38.4 + 28.3 \left|\sin(377 \times 6.944 \times 10^{-3})\right| = 39.8 \text{ A}$$

and

$$i_o(8.333 \text{ ms}) = 0.778 \times 39.8 = 31.0 \text{ A}.$$

As the rectifier is assumed to operate in the quasi-steady state, the last, final value, $i_o(8.333 \text{ ms})$, should equal the initial value, $i_o(0)$. It is not exactly so, which implies an incorrect assumption of the initial value of 32.4 A. Note that the impact of the initial value on the final value is minimal, since at each step of the computations the previous value of the output current is multiplied by either 0.667 or 0.778. Thus, after the 12 steps resulting from the six switching intervals, the initial value error, $\Delta i_o(0)$, is translated into the respective final value error, $\Delta i_o(8.333 \text{ ms})$, of $(0.667 \times 0.778)^6 \Delta i_o(0) \approx 0.02 \Delta i_o(0)$ only. Therefore, it is a safe guess that the final value of 31 A obtained is only slightly greater than the actual initial value. Assuming $i_o(0)$ to be 30.9 A and repeating the iterative calculations yields the following points of the output current waveform:

0	30.9 A
0.833 ms	20.6 A
1.389 ms	16.0 A
⋮	⋮
6.944 ms	38.3 A
7.778 ms	39.7 A
8.333 ms	30.9 A

Waveforms of the output voltage and current are shown in Figure 1.37. Clearly, the calculations presented can easily be computerized, greatly reducing the amount of work involved.

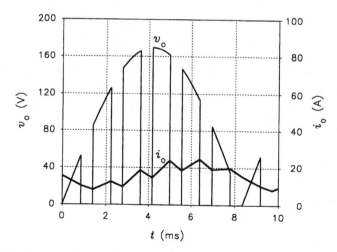

Figure 1.37 Output voltage and current waveforms in the generic PWM rectifier in Example 1.5.

PROBLEMS

P1.1   Refer to Figure 1.13 and sketch the waveforms of the output voltage of the generic rectifier with a firing angle of:

(a) 36°

(b) 72°

(c) 108°

(d) 144°

P1.2   Refer to Figure 1.15 and sketch the waveforms of the output voltage of the generic ac voltage controller with a firing angle of:

(a) 36°

(b) 72°

(c) 108°

(d) 144°

P1.3   From the harmonic spectrum in Figure 1.17a, find:

(a) The per-unit value of the dc component of the output voltage of a generic rectifier with a firing angle of 90° [use Eq. (1.44) to verify the result]

(b) The peak and rms per-unit values of the fundamental output voltage with the same firing angle as in part (a)

P1.4   From the harmonic spectrum in Figure 1.17b, read the peak per-unit value of the fundamental output voltage of the generic ac voltage controller with a firing angle of 90°, and find:

(a) The rms per-unit value of the output voltage [use Eq. (1.44)]

(b) The per-unit harmonic content of the output voltage

(c) The total harmonic distortion of the output voltage

P1.5   For a generic inverter in the square-wave operating mode, derive an equation for the peak per-unit value of the kth harmonic of the output voltage (take the dc input voltage, V_i, as the base voltage). Calculate the peak per-unit values of the first 10 harmonics.

P1.6   A generic rectifier is supplied from a 115-V 60-Hz ac source and operates with a firing angle of 30°. For the output voltage of the rectifier, find:

(a) The dc component

(b) The rms value of the ac component

(c) The ripple factor

P1.7   For the generic rectifier in Problem 1.6, find the fundamental frequency of the output voltage.

P1.8   For a generic ac voltage controller, derive an equation for the peak per-unit value of the kth harmonic of the output voltage as a function of the firing angle (take the peak value, $V_{i,p}$, of the input voltage as the base voltage). Use the spectrum in Figure 1.17b to verify the equation for a firing angle of $90°$ and the five lowest-order harmonics.

P1.9   Review the output voltage waveforms in Figures 1.13 and 1.15 and the corresponding harmonic spectra in Figure 1.17. Which harmonics present in the spectrum for the generic phase-controlled rectifier are absent in the spectrum for the generic ac voltage controller? Why?

P1.10   A generic converter operates as a chopper and is supplied from a 100-V dc source. What is the magnitude control ratio of the chopper, and what are the duty ratios of individual switches if the average output voltage is 70 V?

P1.11   Repeat Problem 1.10 for an output voltage of -35 V.

P1.12   A generic converter operates as a chopper with a magnitude control ratio of 0.4. The duration of a pulse of the output voltage is 0.2 ms. Find the switching frequency.

P1.13   A generic converter operates as a chopper with 1250 pulses of output voltage per second and with a magnitude control ratio of 0.3. Find the duration of a pulse and a notch of the output voltage.

P1.14   A generic PWM rectifier operates with the average output voltage reduced by 20% with respect to the maximum available value of this voltage. Find the magnitude control ratio and the ratio of the width of the pulse of the output voltage to the notch width.

P1.15   A generic PWM rectifier operates as in Problem 1.14 with 50 pulses of the output voltage per period of the 60-Hz input voltage. Find the width of a pulse and the switching frequency.

P1.16   A generic PWM ac voltage controller is supplied from a 60-Hz ac voltage source. The fundamental output voltage is reduced by two-thirds with respect to the supply voltage, and the switching frequency is 1.5 kHz. Find:

(a) The number of pulses of the output voltage per cycle of this voltage

(b) The duty ratios of switches S1, S2, and S5

(c) The pulse width

P1.17   Refer to Figure 1.23a and determine the states (on or off) of all the switches of the generic PWM inverter at:

(a) $\omega t = \pi/2$ rad

(b) $\omega t = \pi$ rad

(c) $\omega t = 3\pi/2$ rad

P1.18   Waveforms in Figure 1.23 for the generic PWM inverter, with 10 pulses of the output voltage per cycle, were obtained in the following way:

1. The 360° period of the voltage was divided into 10 equal switching intervals of 36° each.

2. Denoting by α_n the central angle of the nth switching interval ($\alpha_1 = 18°$, $\alpha_2 = 54°$, etc.), the duty ratio, D_n, of operating switches S1–S2 or S3–S4 for this interval was calculated as

$$D_n = M |\sin \alpha_n|$$

where M denotes the magnitude control ratio.

For 8 pulses per cycle and $M = 0.75$, find the widths (in degrees) of individual pulses of the output voltage and sketch to scale the resulting voltage waveform, $v_o(\omega t)$. The voltage pulses should be located in the middle of switching intervals.

P1.19   Refer to Figure 1.34 and sketch the output voltage waveform of a generic phase-controlled cycloconverter with a magnitude control ratio of 0.5 and an input/output frequency ratio of 2. Identify the states of the converter.

P1.20   Refer to Figure 1.34 and sketch the output voltage waveform of a generic PWM cycloconverter with an input/output frequency ratio of 3 and a magnitude control ratio of 0.5 (for convenience, assume a low number of pulses of the output voltage).

P1.21   A generic PWM rectifier is supplied from a 240-V 60-Hz ac line and operates with a magnitude control ratio of 0.5 and a switching frequency of 720 Hz. The rectifier feeds a dc motor, which under the given operating conditions can be represented as an RLE load with $R = 0.5\ \Omega$, $L = 12\ \text{mH}$, and $E = 90\ \text{V}$. Determine the piecewise linear waveform of the output current for one cycle of output voltage.

P1.22   A generic PWM ac voltage controller with an RL load is supplied from a 115-V 50-Hz source and operates with 10 switching intervals per cycle and a magnitude control ratio of 0.75. The resistance and inductance of the load are 10 Ω and 25 mH, respectively. Determine the piecewise linear waveform of the output current for one cycle of the output voltage.

COMPUTER ASSIGNMENTS

A generic power converter, which represents an idealized theoretical concept, cannot be modeled precisely using PSpice, which is designed for simulation of practical circuits. In contrast to the ideal, infinitely fast switches assumed for the generic converter, PSpice switch models have finite time of transition from one state to another. To avoid interruptions of the output current, the output terminals of the converter must therefore be shorted by switch S5 just before the output is separated from the input by opening switches S1–S2 or S3–S4. As a result, the supply source is temporarily shorted, too, albeit very briefly, and large impulses ("spikes") of the input current appear in the Probe oscillograms. Such short-circuit currents are not present in practical correctly designed and controlled power electronic converters. However, the output voltage and current in the generic converter can be simulated quite accurately.

To calculate the figures of merit, make use of the avg(x) (average value of x) and rms(x) (rms value of x) functions. Also, employ the *Fourier* option for the X-axis to obtain harmonic spectra of voltages and currents. Refer to Appendix A and the book by Rashid and Rashid [4] for instructions on PSpice simulations.

Assignments involving PSpice files accompanying the book are identified by an asterisk.

*CA1.1 Run PSpice file *Gen_Ph-Contr_Rect.cir* for a generic phase-controlled rectifier. Perform simulations for firing angles of 0° and 90° and find for the output voltage of the converter:

(a) The dc component

(b) The rms value

(c) The rms value of the ac component

(d) The ripple factor

Observe oscillograms of the input and output voltages and currents. Explain what causes the spikes in the oscillograms of the input quantities and what limits the amplitude of the current spikes.

CA1.2 Develop a PSpice circuit file for a generic inverter operating in the square-wave mode with an adjustable fundamental output frequency. Perform a simulation for an output frequency of 50 Hz and find for the output voltage of the inverter:

(a) The rms value

(b) The rms value of the fundamental (from the harmonic spectrum)

(c) The harmonic content

(d) The total harmonic distortion

Observe oscillograms of the input and output voltages and currents.

CA1.3 Develop a PSpice circuit file for a generic ac voltage controller. Perform simulations for firing angles of 0° and 90° and find for the output voltage of the converter:

(a) The rms value

(b) The rms value of the fundamental (from the harmonic spectrum)

(c) The harmonic content

(d) The total harmonic distortion.

Observe oscillograms of the input and output voltages and currents.

CA1.4 Refer to Example 1.1 and Problem P1.19 and develop a PSpice circuit file for a generic phase-controlled cycloconverter with an input frequency of

50 Hz and a fundamental output frequency of 25 Hz. Perform simulations for firing angles of $0°$ and $90°$ and find for the output voltage of the converter:

(a) The rms value

(b) The rms value of the fundamental (from the harmonic spectrum)

(c) The harmonic content

(d) The total harmonic distortion

Observe oscillograms of the input and output voltages and currents.

CA1.5 Develop a PSpice circuit file for a generic chopper. Perform simulations for a switching frequency of 1 kHz and magnitude control ratios of 0.6 and -0.3, and find for the output voltage of the converter:

(a) The dc component

(b) The rms value

(c) The rms value of the ac component

(d) The ripple factor

Observe oscillograms of the input and output voltages and currents.

*CA1.6 Run PSpice program *Gen_PWM_Rect.cir* for a generic PWM rectifier. Perform simulations for 12 pulses of the output voltage per cycle of the input voltage and a magnitude control ratio of 0.5, and find for the output voltage of the converter:

(a) The dc component

(b) The rms value

(c) The rms value of the ac component

(d) The ripple factor

Observe oscillograms of the input and output voltages and currents. Explain what causes the spikes in the oscillograms of the input quantities and what limits the amplitude of the current spikes.

CA1.7 Develop a PSpice circuit file for a generic PWM ac voltage controller. Perform a simulation for 12 pulses of the output voltage per cycle and a magnitude control ratio of 0.5, and find for the output voltage of the converter:

(a) The rms value

(b) The rms value of the fundamental (from the harmonic spectrum)

(c) The harmonic content

(d) The total harmonic distortion

Observe oscillograms of the input and output voltages and currents.

CA1.8 Refer to Example 1.1 and Problem P1.20 and develop a PSpice circuit file for a generic PWM cycloconverter. Perform a simulation for an input frequency of 50 Hz, a fundamental output frequency of 25 Hz, a magnitude control ratio of 0.5, and 10 pulses of output voltage per cycle of the input voltage. Find for the output voltage of the converter:

(a) The rms value

(b) The rms value of the fundamental (from the harmonic spectrum)

(c) The harmonic content

(d) The total harmonic distortion

Observe oscillograms of the input and output voltages and currents.

CA1.9 Develop a computer program for calculation of the harmonics of a given periodic waveform, $\psi(\omega t)$. Data points that represent one cycle of the waveform are stored in an ASCII file in the ωt, $\psi(\omega t)$ format. Generate and store the voltage waveform in Figure 1.13 (generic phase-controlled rectifier) and use your program to obtain the harmonic spectrum of the waveform. Compare the results with those in Figure 1.17a.

CA1.10 Develop a computer program for calculation of the following parameters of a given periodic waveform, $\psi(\omega t)$:

(a) The rms value

(b) The dc component

(c) The rms value of the ac component

(d) The rms value of the fundamental

(e) The harmonic content

(f) The total harmonic distortion

Data points that represent one cycle of the waveform are stored in an ASCII file in the ωt, $\psi(\omega t)$ format. Generate and store the voltage waveform in Figure 1.15 (generic phase-controlled ac voltage controller) and apply the program to compute parameters (a) through (f).

CA1.11 Develop a computer program for the determination of a current waveform generated in a given load by a given voltage. The load can be of the R, RL, RE, LE, or RLE type, and the voltage waveform, $v(t)$, is given as either a closed-form function of time, $v = f(t)$, or as an ASCII file of (t, v) pairs.

*CA1.12 Run PSpice program *Diode_Rect_1P.cir* for a single-pulse diode rectifier. Repeat the simulation for the rectifier with a freewheeling diode and with an output capacitor ("comment out" the unused components). In each case, determine the average output voltage and ripple factor of that voltage.

LITERATURE

[1] Agrawal, J. P., *Power Electronic Systems: Theory and Design*, Prentice Hall, Upper Saddle River, NJ, 2001, Chap. 1.

[2] Rashid, M. H., *Power Electronics Handbook*, 2nd ed., Academic Press, San Deigo, CA, 2007, Chap. 1.

[3] Rashid, M. H., *Power Electronics: Circuits, Devices, and Applications*, 3rd ed., Prentice Hall, Upper Saddle River, NJ, 2003, Chap. 1.

[4] Rashid, M. H., and Rashid H. M., *SPICE for Power Electronics and Electric Power*, 2nd ed., CRC Press, Boca Raton, FL, 2006.

2 Semiconductor Power Switches

Semiconductor power switches used in power electronic converters are presented in this chapter. A behavioral classification of switches as uncontrolled, semicontrolled, and fully controlled is then introduced. Parameters and characteristics of power diodes, SCRs, triacs, GTOs, IGCTs, power BJTs and MOSFETs, and IGBTs are described and compared. Power modules combining several semiconductor devices are shown.

2.1 GENERAL PROPERTIES OF SEMICONDUCTOR POWER SWITCHES

The hypothetical generic power converter in Chapter 1 was a simple two-port network of five switches capable of various types of electric power conversion and control. Practical power electronic converters are based on semiconductor power switches, which differ from the ideal switches of the generic converter in two aspects. The chief difference is that of the unidirectionality of semiconductor switches, which can conduct substantial currents in one direction only. The other difference lies in the control properties of the switches. An ideal switch is assumed to react instantly to the turn-on (close) or turn-off (open) signals. In contrast, certain semiconductor power switches are *uncontrolled* (i.e., they lack the control electrode) or *semicontrolled* (i.e., they lack the controlled turn-off capability). Even in *fully controlled* switches, the speed of response to control signals is limited, depending on the type and size of the switch. As a result, the maximum allowable switching frequencies of practical power switches vary from about 1 kHz for large SCRs and GTOs to more than 1 MHz for low-power MOSFETs.

Also, in contrast to the ideal lossless switches of the generic converter, practical semiconductor power switches a dissipate certain amount of energy when conducting and switching. Voltage drop across a conducting semiconductor device is at least on the order of 1 V, often higher, up to a few volts. The *conduction power loss*, P_c, is given by

$$P_c = \frac{1}{T} \int_0^T p \, dt \tag{2.1}$$

Introduction to Modern Power Electronics, Second Edition, by Andrzej M. Trzynadlowski
Copyright © 2010 John Wiley & Sons, Inc.

where T denotes the conduction period and p is the instantaneous power loss, that is, the product of the voltage drop across the switch and the current conducted. Power losses increase during turn-on and turn-off of the switch, because during the transition from one conduction state to another, both the voltage and current are temporarily large. The resulting *switching power loss*, P_{sw}, is given by

$$P_{sw} = \left(\int_0^{t_{ON}} p\, dt + \int_0^{t_{OFF}} p\, dt \right) f_{sw} \tag{2.2}$$

where f_{sw} denotes the switching frequency, that is, the number of switching cycles (on–off) per second, and t_{ON} and t_{OFF} are the turn-on and turn-off times, whose precise meaning is explained later. Losses in a typical semiconductor switch during the turn-on, conduction, and turn-off are illustrated in Figure 2.1, which shows waveforms of voltage, current, and instantaneous power loss.

An *intrinsic semiconductor* is defined as a material whose resistivity is too low for an insulator and too high for a conductor. Semiconductor power switches are based on silicon, a member of group IV of the periodic table of elements, which has four electrons in its outer orbit. If pure silicon is doped with a small amount of a group V element, such as phosphorus, arsenic, or antimony, each atom of the dopant forms a covalent bond within the silicon lattice, leaving a loose electron. These loose electrons greatly increase the conductivity of the material, which is referred to as *n-type* (for "negative") *semiconductor*. Vice versa, if the dopant is a group III element, such as boron, gallium, or indium, a vacant location called a *hole* is introduced into the silicon lattice. Analogously to an electron, a hole can be considered a mobile charge

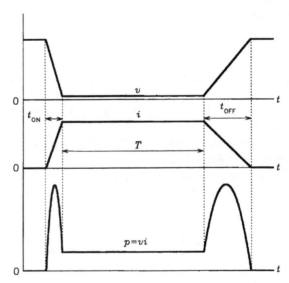

Figure 2.1   Waveforms of voltage, current, and power loss in a semiconductor power switch at turn-on, conduction, and turn-off.

carrier, as it can be filled by an adjacent electron, which in this way leaves a hole behind. Such material, also more conductive than pure silicon, is known as *p-type* (for "positive") *semiconductor*. The doping involves a single atom of the added impurity in over a million silicon atoms. Semiconductor power switches are based on various structures of n- and p-type semiconductor layers.

Because of the introductory nature of this book, no further solid-state physics considerations are presented in the subsequent coverage of semiconductor power switches. Instead, practical parameters and characteristics are described to give the reader a general idea of the capabilities of individual types of those devices. It should be pointed out that the specific parameter values provided in the coverage of individual switches, particularly the maximum voltage and current ratings, are those for the typical, easily available devices. The fast pace of technology and a large number of manufacturers make it difficult to keep precise track of the most recent developments and extreme ratings.

2.2 POWER DIODES

Power diodes, which are uncontrolled semiconductor power switches, are widely used in power electronic converters. The semiconductor structure and circuit symbol of a diode are shown in Figure 2.2. A high current in a diode can only flow from *anode* (A) to *cathode* (C). The voltage–current characteristic of a power diode is illustrated in Figure 2.3. Note that different scales are used for positive and negative half-axes. The *maximum reverse leakage current*, I_{RM}, is lower by many orders of magnitude than the current that can safely be conducted in the forward direction. Similarly, the *reverse breakdown voltage*, V_{RB}, which causes a dangerous *avalanche*

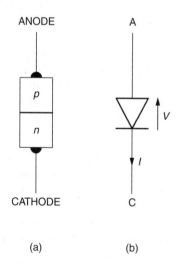

(a)   (b)

Figure 2.2 Power diode: (a) semiconductor structure; (b) circuit symbol.

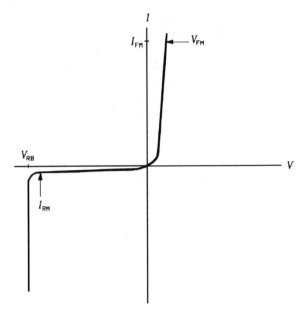

Figure 2.3 Voltage–current characteristic of a power diode.

breakdown, is much higher than the *maximum forward voltage drop*, V_{FM}, across the diode. Typically, V_{FM}, specified for a given forward current, I_{FM}, is on the order of 1 to 2 V and can be neglected in most practical considerations of power converters. However, the voltage drop across a diode must be taken into account in calculations of converter losses and the design of a cooling system. Otherwise, a diode can be considered as an almost ideal switch. It turns on when *forward biased*, that is, with a positive anode–cathode voltage, and turns off when a forward current reverses its polarity.

Main parameters of power diodes that are listed in catalogs of semiconductor devices and form a base for proper selection of diodes for a given application are described below. The symbols used are the most common ones, but they may vary somewhat with individual manufacturers. The same observation applies to the number and choice of parameters provided in various catalogs.

1. The maximum allowable *reverse repetitive peak voltage*, V_{RRM} (V). Note the peak value employed, since it is the instantaneous voltage that if equal to or greater than the reverse breakdown voltage, V_{BR}, may damage a semiconductor device. High voltage across a diode can occur only when the diode is *reverse biased* and $V_{RRM} < V_{RB}$. This parameter will subsequently be called a *rated voltage* since it appears in price lists of diodes as one of only two parameters specified, the other being the rated current. Therefore, besides V_{RRM}, the symbol V_{rat} is also used in this book.

2. The maximum allowable *full-cycle average forward current*, $I_{F(av)}$ (A). This parameter will subsequently be called a *rated current* and, alternatively, denoted by I_{rat}. Note that overcurrent damages a semiconductor device by excessive heat. Therefore, the average value is more appropriate here than the peak value. Because of the typical use of diodes as rectifiers, the average current for standard diodes is assumed as that in a single-pulse rectifier operating with an R load and a frequency of 60 Hz. That frequency is 1 kHz for *fast recovery* diodes. The rated current is less than the value, I_{FM}, of forward current for which the maximum forward voltage drop is defined.

3. The maximum allowable *full-cycle rms forward current*, $I_{F(rms)}$ (A). This is the rms value of current in a diode operating as a single-pulse rectifier with an R load and conducting an average current of $I_{F(av)}$. Thus, if the diode selected on the basis of $I_{F(av)}$ actually works as a rectifier, there is no danger that the rms forward current will be excessive. Only in applications characterized by high-ripple currents, the rms forward current in a diode must be checked against the $I_{F(rms)}$ value. If not listed in a catalog, this can easily be determined as

$$I_{F(rms)} = \frac{\pi}{2} I_{F(av)}. \tag{2.3}$$

4. The maximum allowable *nonrepetitive surge current*, I_{FSM} (A). This is a maximum one-half cycle (60 Hz) peak surge current in excess of the rated current that the diode can survive. In practice, it pertains to possible short circuits in a converter. Duration of the surge current must be limited to a short period of time, specifically 8.3 ms (the 60-Hz half-period), within which the overcurrent protection system is expected to interrupt the current.

5. The maximum allowable junction and case temperatures, Θ_{JM} and Θ_{CM} (°C). These temperatures usually vary from 150 to 200°C. The range of allowable storage temperatures can also be specified: for example, from −65 to 175°C.

6. The junction-to-case and case-to-sink *thermal resistances*, $R_{\Theta JC}$ and $R_{\Theta CS}$ (°C/W). These are used in designing the cooling system, proper selection of the *heat sink* (radiator) in particular. It must be stressed that current-related parameters of semiconductor power switches pertain to switches with heat sinks. Because of the small size of contemporary switches (high-power stud-type diodes are not larger than a man's fist), the thermal capacity of a bare device is too low to safely dissipate the heat generated by internal losses. Heat-flow systems can be represented by equivalent quasi-electric circuits in which low thermal resistances indicate high heat-dissipation properties of the physical components being modeled.

7. *The fuse coordination*, I^2t (A² · s). This parameter is used for fuse selection for a given diode. If the I^2t value of the fuse is less than that of the diode, an overcurrent will cause the fuse to burn sooner than the diode gets damaged. The I^2t value for a diode is obtained by squaring the rms value of current that can be withstood for a period of 8.3 ms, then multiplying the result by the same 8.3ms.

Taking into account the definition of maximum allowable nonrepetitive surge current, $I^2 t$ is given by

$$I^2 t = \frac{I_{\text{FSM}}^2}{240} \qquad (2.4)$$

(In 50-Hz countries, the 240 is replaced by 200.) Vice versa, if I_{FSM} is not given in a catalog, it can be calculated as

$$I_{\text{FSM}} = \sqrt{240 I^2 t}. \qquad (2.5)$$

Parameters 1 through 5 can be called *restrictive*, since their values may not be exceeded without endangering the diode's integrity. In contrast, parameters 6 and 7, as well as the maximum leakage current, I_{RM}, and maximum forward voltage drop, V_{FM}, introduced previously, are *descriptive* in nature, providing information about certain properties of the device. Another important descriptive parameter, called a *reverse recovery time*, t_{rr}, requires a more detailed explanation.

When a conducting diode is abruptly reverse-biased, the device does not regain its reverse blocking capability instantly. As a result, for a short time the diode passes a high current in the reverse direction. This phenomenon is illustrated in Figure 2.4, which shows the voltage and current waveforms when a reverse voltage, V_R, is applied to a conducting diode at $t = 0$. It can be seen that the reverse recovery time, t_{rr}, can be defined loosely as the duration of the period when a diode recovers its blocking capability. Within this period, a negative current overshoot, I_{rrM}, occurs, followed closely by a voltage overshoot, V_{RM}. The latter overshoot is proportional to the slope, di_{rr}/dt, of the *current tail*, that is, the decaying portion of the current waveform. The value of di_{rr}/dt, often listed in catalogs as another descriptive parameter, affects the reverse recovery time, which for standard diodes varies from several microseconds to over 20 μs. In comparison with standard diodes, t_{rr} is shorter by one order of magnitude in the fast recovery diodes already mentioned, which, however, produce higher-voltage

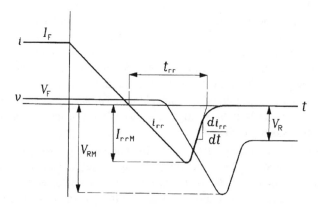

Figure 2.4   Voltage and current waveforms during reverse recovery in a power diode.

TABLE 2.1   Examples of High-Power Diodes

Designation:	RDK86040	R7014405	R9G23615	R6031635
	(Powerex)	(Powerex)	(Powerex)	(Powerex)
Type:	General purpose	General purpose	Fast recovery	Fast recovery
Case:	Disk	Stud	Disk	Stud
V_{RRM}:	6 kV	4.4 kV	3.6 kV	1.6 kV
$I_{F(av)}$:	4 kA	0.55 kA	1.5 kA	0.35 kA
$I_{F(rms)}$:	6.3 kA	0.86 kA	2.35 kA	0.55 kA
I_{FSM}:	60 kA	10 kA	18 kA	6 kA
I^2t:	$1.5 \times 10^7 A^2 \cdot s$	$4.5 \times 10^5 A^2 \cdot s$	$1.35 \times 10^6 A^2 \cdot s$	$1.5 \times 10^5 A^2 \cdot s$
V_{FM}:	1.65 V	1.2 V	1.65 V	1.5 V
I_{RRM}:	300 mA	50 mA	75 mA	50 mA
t_{rr}:	25 μs	15 μs	5 μs	2 μs
Diameter:	132 mm	38 mm	74 mm	27 mm
Height:	38 mm	96 mm	28 mm	79 mm

overshoots. Small "ultra fast" recovery diodes have the reverse recovery times, on the order of hundreds of nanoseconds only. Generally, for a given class of diodes, the length of reverse recovery time depends mostly on the size of the device.

It must be stressed that the descriptive parameters are not strictly constant but dependent on operating conditions of a diode. The anode current and junction temperature are particularly influential. This observation applies not only to power diodes but to all semiconductor devices. Power diodes are among the largest semiconductor devices. Their voltage and current ratings extend to 6.5 kV and 10 kA, respectively. However, it must be stressed that the highest-voltage diodes are not the highest-current diodes, and vice versa. The same is true for other semiconductor power switches. Parameters of some high-power diodes are listed in Table 2.1. It can be seen that the disk ("hockey puck") case, in which the device constitutes a flat cylinder to be sandwiched between two heat sinks (radiators), allows for higher ratings than those of stud-type diodes, whose single end is screwed into a radiator. General-purpose diodes are used primarily in uncontrolled rectifiers; fast recovery diodes are used in other converters. The maximum voltage and current ratings of the latter diodes are 6 kV and 2 kA.

Much less powerful uncontrolled semiconductor switches, *Shottky diodes*, based on a metal–semiconductor junction, deserve mention because of their use in switching dc power supplies. Fast, characterized by a very low forward voltage drop (less than 0.5 V), and capable of conducting currents of up to 0.5 kA, Shottky diodes have inherently low breakdown voltages that do not exceed 200 V. This weakness makes the diodes unsuitable for application in most power electronic converters.

2.3   SEMICONTROLLED SWITCHES

Members of the *thyristor family* of semiconductor devices can be considered semi-controlled power switches. They can be turned on by an appropriate gate signal, but they turn off by themselves similar to a diode, that is, when the current conducted changes its polarity to negative. Two subsequently presented prominent thyristor-type devices are the *SCR* (silicon-controlled rectifier), mentioned in Chapter 1, and the *triac*, an electric equivalent of two SCRs connected antiparallel.

2.3.1 SCRs

An SCR (also known as a thyristor), a four-layer three-electrode semiconductor device shown in Figure 2.5, can be thought of as a controlled diode that when off blocks currents of either polarity. When forward biased and turned on ("fired"), an SCR operates as an ordinary diode as long as the current conducted remains above the *holding current*, I_H, level. In multi-SCR converters, such as the phase-controlled rectifiers described in Chapter 4, *commutation* occurs; that is, one SCR turns off when another SCR takes over conduction of the current.

In power electronic converters, an SCR is turned on by a gate current, i_G, supplied by an external source connected between the *gate* (G) and the cathode. A high forward voltage between the anode and cathode, and even a rapid change, dv/dt, of such voltage can also cause turn-on, albeit unwanted. The voltage–current characteristic of the SCR is shown in Figure 2.6 (for clarity, the forward and reverse leakage currents are exaggerated). With no gate current, when a forward voltage applied to the SCR exceeds the *forward breakover voltage*, V_{BF}, the forward leakage current increases to a *latching current*, I_L, level and the SCR starts conducting. Injecting a current through the gate into the central p-type layer reduces the forward breakover voltage to a value less than the voltage actually applied, providing controlled turn-on.

Most of the diode parameters listed in the preceding section are shared by SCRs, although somewhat different symbols are generally used to account for the fact that

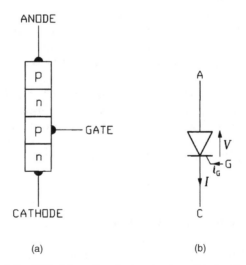

(a)   (b)

Figure 2.5   SCR: (a) semiconductor structure; (b) circuit symbol.

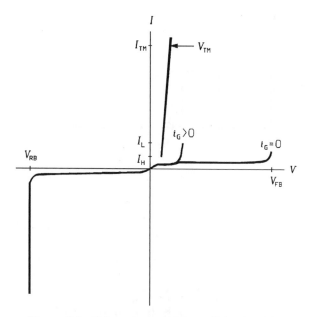

Figure 2.6   Voltage–current characteristic of an SCR.

SCRs can block voltages of either polarity. Consequently, for the on-state, a subscript "T" is used instead of an "F." For instance, as shown in Figure 2.6, the maximum forward voltage drop is denoted by V_{TM} instead of V_{FM}. Quantities pertaining to the forward blocking state carry subscripts beginning with a "D." For example, V_{DRM} represents the maximum allowable forward repetitive peak voltage across the blocking SCR that will not cause turn-on without a firing signal. Typically, $V_{DRM} = V_{RRM}$.

In addition to parameters related to the cathode–anode path, catalogs of SCRs provide information on the dc gate current and voltage (precisely, gate–cathode voltage) required for successful firing and designated I_{GT} and V_{GT}, respectively. Data sheets of SCRs, which are more detailed than general catalogs, contain diagrams of firing areas with the instantaneous gate current and voltage as coordinates. Depending on the application, the gate current in an SCR to be fired is produced by employing a single pulse of the gate voltage, v_G, or a *multipulse*, both shown in Figure 2.7. A multipulse is obtained by rectifying and clipping a high-frequency sinusoidal voltage. Typically, frequencies on the order of several kilohertz are used, so that a multipulse, which can be as long as a quarter of the 60-Hz cycle, contains tens of individual half-wave sine pulses. Multipulse firing is employed in situations when it is not certain that a single pulse would accomplish the required turn-on. A typical gate current for large SCRs is on the order of 0.1 to 0.3 A, resulting in current gains (ratio of the anode current to the gate current) of several thousands.

The *critical dv/dt* parameter expresses the minimum rate of change of anode voltage that causes turn-on without a gate current. Also listed in catalogs and data

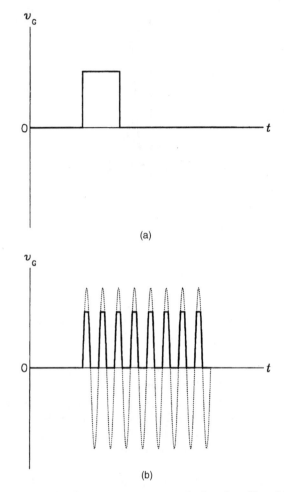

Figure 2.7   SCR gate voltage signals: (a) single pulse; (b) multipulse.

sheets is the *maximum allowable repetitive di/dt* value. The limit on the allowable rate of rise of the anode current is required to allow the conduction area to spread over the entire cross section of the SCR before the current reaches a high level. Otherwise, an excessive current density in the small initial area of conduction would cause spot overheating and permanent damage to semiconductor material.

Two time parameters are usually listed in SCR catalogs. The turn-on time, t_{ON}, is counted from the instant of application of the gate signal to the instant when the anode voltage drops to 10% of its initial, full value. The turn-on time period consists of two subperiods: the delay time, from the firing instant to the instant when the anode voltage has decreased by the first 10% (from 100% to 90% of the initial value), and the voltage fall time (from 90% to 10%). Alternatively, the turn-on time can be defined in terms of the anode current, with the delay time counted from the firing

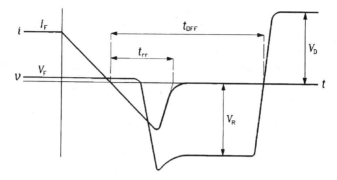

Figure 2.8   Anode voltage and current waveforms during forced commutation of an SCR.

instant to the instant when the current has risen to 10% of its final value, and the current rise time, during which the current increases to 90% of the final value. Typical turn-on times are on the order of few microseconds, and the firing pulses should be one order of magnitude longer.

The meaning of the other time parameter, the turn-off time, t_{OFF}, is illustrated in Figure 2.8, which shows the anode voltage and current waveforms during the *forced commutation*, that is, forced turn-off, followed by reapplication of the forward voltage. The forced commutation is realized by applying a reverse-biasing voltage across a conducting SCR to reverse the anode current and extinguish the SCR. Even when the reverse recovery process has been finished, a negative anode voltage must be maintained for some time, so that the SCR recovers its forward blocking capability. Therefore, $t_{OFF} > t_{rr}$, and typical turn-off times lie within the range 10 to 100 μs. SCRs are usually classified as either phase control or inverter grade. Inverter-grade SCRs, used in power inverters, are faster than phase-control SCRs, designated for 60-Hz applications such as rectifiers and ac voltage controllers.

Besides power diodes, SCRs are by far the largest existing semiconductor power switches. Standard phase-control SCRs are available with ratings as high as 8 kV and 6 kA. Fast-switching, or inverter-grade, SCRs are rated up to 2.5 kV and 3 kA. Special light-activated thyristors used in high-voltage dc transmission have voltage and current ratings similar to those of regular phase-control SCRs. Table 2.2 shows typical parameters of large SCRs.

2.3.2   Triacs

The triac is a semiconductor device that is electrically equivalent to two SCRs connected antiparallel, although, as shown in Figure 2.9, its internal structure is not exactly the same as that of two SCRs. Because of the resulting capability of bidirectional current conduction, the power electrodes are simply called *main terminal 1* (T1) and *main terminal 2* (T2), instead of anode and cathode. The gate signal is applied between the gate and terminal 1. The triac can be turned on by a positive or negative gate current, and the direction of the current that is depends conducted on the polarity of the supply voltage.

TABLE 2.2   Examples of High-Power SCRs

Designation:	5STP 12N8500 (ABB)	5STP 50Q1800 (ABB)	C770L (Powerex)	T7071430 (Powerex)
Type:	Phase control	Phase control	Fast switching	Fast switching
Case:	Disk	Disk	Disk	Stud
V_{RRM}/V_{DRM}:	8 kV	1.8 kV	2 kV	1.4 kV
$I_{T(av)}$:	1.2 kA	6.1 kA	2.1 kA	0.3 kA
$I_{T(rms)}$:	1.88 kA	9.6 kA	3.3 kA	0.475 kA
I_{TSM}:	35 kA	94 kA	38 kA	8 kA
I^2t:	6×10^6 A$^2 \cdot$ s	4.3×10^7 A$^2 \cdot$ s	6×10^6 A$^2 \cdot$ s	2.65×10^5 A$^2 \cdot$ s
V_{TM}:	2 V	1.04 V	1.55 V	1.45 V
I_{RRM}/I_{DRM}:	400 mA/1 A	300 mA/300 mA	100 mA/100 mA	30 mA/30 mA
t_{ON}:	3 μs	3 μs	2 μs	3 μs
t_{OFF}:	600 μs	500 μs	100 μs	60 μs
I_{GT}:	400 mA	400 mA	300 mA	150 mA
V_{GT}:	2.6 V	2.6 V	3 V	3 V
Diameter:	150 mm	150 mm	110 mm	38 mm
Height:	35 mm	35 mm	37 mm	106 mm

In comparison with a pair of equivalent SCRs, a triac has a longer turn-off time, lower critical dv/dt, and lower current gain. However, the compact construction is advantageous in certain applications, such as lighting and heating control, solid-state relays, and control of small motors. The voltage and current ratings available are limited to 1.4 kV and 0.1 kA, respectively.

Bidirectionally controlled thyristors (BCTs) are worth mentioning in this context, as they are functionally similar to triacs. A BCT consists of two integrated antiparallel

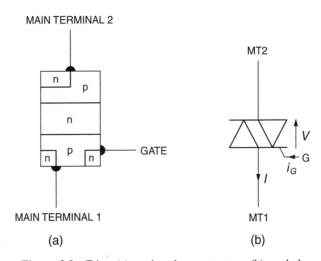

Figure 2.9   Triac: (a) semiconductor structure; (b) symbol.

SCR-like devices on one silicon wafer. Unlike in the triac, the constituent thyristors are triggered individually. BCTs are much larger devices than triacs, with ratings of up to 6.5 kV and 5.5 kA.

2.4 FULLY CONTROLLED SWITCHES

Although SCRs initiated the era of semiconductor power electronics, several types of fully controlled switches have gradually phased them out from many applications, particularly dc-input converters. The feasibility of controlled turn-on and turn-off makes fully controlled semiconductor power switches very attractive for use in modern power electronic converters, especially those with pulse width modulation. Common representatives of this class of semiconductor devices are described in subsequent sections.

2.4.1 GTOs

GTO is an acronym for *gate turn-off thyristor*, whose structure and circuit symbol are shown in Figure 2.10. It is a thyristor-type semiconductor switch that similarly to an SCR is turned on by a low positive gate current. However, in contrast to an SCR, a GTO can also be turned off, using a large negative pulse of the gate current, on the order of up to 30% of the anode current. Thus, the turn-off current gain is poor, but the turn-off gate pulse is only on the order of 50 μs, so that the associated energy of the gate signal is low.

Interestingly, the current gain depends on the rate of change, di_G/dt, of the reverse gate current. With di_G/dt high, the turn-off time is short, but the current gain approaches unity. Vice versa, a slowly applied reverse gate current can result in a turn-off gain of over 20, but the turn-off time and associated switching losses are high. GTOs have been the first fully controlled truly high-power semiconductor switches,

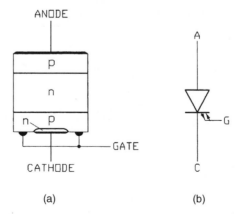

Figure 2.10 GTO: (a) semiconductor structure; (b) circuit symbol.

but they are slow and their switching and conduction losses are high. They also require *snubbers*, which suppress ("snub") voltage transients during turn-off. Ratings of GTOs are comparable with those of SCRs, reaching 6 kV and 6 kA.

2.4.2 IGCTs

An *integrated gate commutated thyristor* (IGCT), whose circuit symbol is shown in Figure 2.11, is similar to the GTO, as it can be turned on and off by the gate signal. The gate turn-off current is actually greater than the anode current, which results in short turn-off times. The *gate driver*, that is, the circuitry generating the gate current, is integrated into the package of the device and surrounds the IGCT. As a result, IGCTs differ from the typical stud or disk packages of other high-power switches by their boxy shape. The impedance of the connection between the driver and the device is low, thanks to the large contact area and short length of the connection. This is a crucial feature considering the high and fast-changing gate current, which precludes the use of wire leads. Snubbers are not required.

Most available IGCTs cannot block reverse voltages and are referred to as asymmetrical. Asymmetrical IGCTs integrated in a single package with a reverse-conducting diode are called *reverse-conducting IGCTs*. Symmetrical, reverse-blocking IGCTs are under development.

IGCTs can switch much faster than GTOs, so they can operate at high switching frequencies. However, high switching losses limit the sustained switching frequencies to less than 1 kHz. The voltage and current ratings reach 6.5 kV and 6 kA. Parameters of IGCTs are listed in Table 2.3.

2.4.3 Power BJTs

An npn *bipolar junction transistor*, (BJT) is shown in Figure 2.12. The collector (C)-to-emitter (E) path serves as the switch, conducting or interrupting the main current, while the base (B) is the control electrode. In contrast to thyristors, the collector

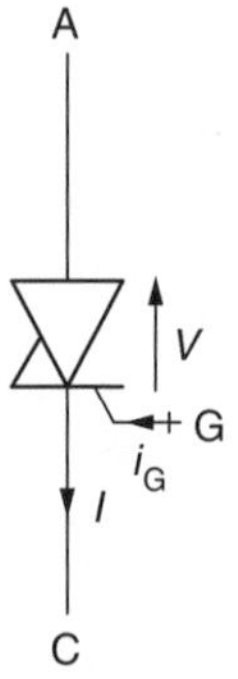

Figure 2.11 Circuit symbol of an IGCT.

TABLE 2.3   Examples of IGCTs

Designation:	5SHY42L6500 (ABB)	5SHY55L4500 (ABB)	5SHX19L6010 (ABB)	5SHX26L4510 (ABB)
Type:	Asymmetric	Asymmetric	Reverse cond.	Reverse cond.
V_{DRM}:	6.5 kV	4.5 kV	5.5 kV	4.5 kV
$I_{T(av)}$:	1.27 kA	1.86 kA	0.84 kA	1.01 kA
$I_{T(rms)}$:	2.00 kA	2.92 kA	1.32 kA	1.59 kA
I_{TSM}:	26 kA	33 kA	18 kA	17 kA
$I^2 t$:	3.38×10^6 A$^2 \cdot$ s	5.45×10^5 A$^2 \cdot$ s	1.62×10^6 A$^2 \cdot$ s	1.45×10^5 A$^2 \cdot$ s
V_{TM}:	2.0 V	1.15 V	1.9 V	1.8 V
I_{DRM}:	50 mA	50 mA	50 mA	50 mA
t_{ON}:	40 μs	12 μs	11.5 μs	11.5 μs
t_{OFF}:	40 μs	15 μs	14 μs	14 μs
$I_{GQM}{}^a$:	4.2 kA	5.5 kA	1.8 kA	2.2 kA
$E_{off}{}^a$:	44 J	31.5 J	11 J	12 J
Length:	429 mm	429 mm	429 mm	429 mm
Height:	40 mm	40 mm	40 mm	40 mm
Width:	173 mm	173 mm	173 mm	173 mm

$^a I_{GQM}$, maximum controllable turn-off gate current; E_{off}, turn-off energy per pulse of gate current.

current, I_C, can be controlled continuously by the base current, I_B, as

$$I_C = \beta I_B \tag{2.6}$$

where β denotes a *dc current gain* of the transistor. In high-power BJTs, the current gain is low, on the order of 10. The emitter current, I_E, is a sum of the collector and base currents.

The voltage–current characteristics of the BJT, specifically the collector current, I_C, versus collector–emitter voltage, V_{CE}, relations for various values of the base

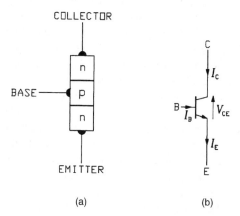

Figure 2.12   BJT: (a) semiconductor structure; (b) circuit symbol.

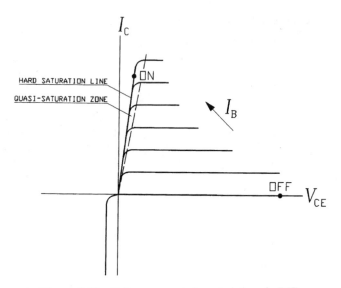

Figure 2.13   Voltage–current characteristics of a BJT.

current, I_B, are shown in Figure 2.13. The conduction power loss, P_c, is given by

$$P_c = V_{CE}I_C. \tag{2.7}$$

Therefore, for the BJT to emulate an ideal lossless switch, the base current in the on-state must be high enough for the operating point to lie on, or close to, the *hard saturation line* associated with the lowest voltage drop across the device. In the off-state, the base current is zero, and the collector current is reduced to the leakage level (assumed zero in Figure 2.13).

The rated current of a BJT represents the maximum allowable dc collector current, usually denoted simply as I_C. The rated voltage, denoted by V_{CEO}, is the maximum collector–emitter voltage that the transistor can safely block with a zero base current. As indicated in Figure 2.13, BJTs cannot block negative collector–emitter voltages. Thus, when used in an ac-input converter, a BJT must be protected from reverse breakdown with a diode connected in series with the collector. The same applies to other types of asymmetrical blocking semiconductor switches.

BJTs are susceptible to *second breakdown*, so named to distinguish it from reverse, avalanche breakdown (first breakdown). In contrast to reverse breakdown caused by excessive voltage across a blocking device, second breakdown occurs when both the collector–emitter voltage and collector current are high, that is, during turn-on or turn-off. As a result of crystal faults or doping fluctuations and high power loss, local hot spots appear in the semiconductor. The positive temperature coefficient of the collector current causes a positive feedback effect, manifested in an increased current density in the hot-spot region. Given sufficient time, this *thermal runaway* can cause irreparable damage. Limiting the power dissipation in a transistor is the best

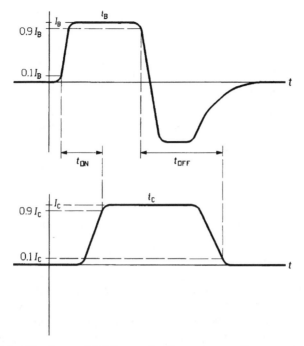

Figure 2.14 Waveforms of BJT base and collector currents for turn-on and turn-off.

way to prevent the second breakdown. To counterbalance the positive temperature coefficient, *emitter ballast resistance* can be incorporated in the transistor structure. However, it increases the on-state voltage drop.

To accelerate the turn-off process, it is recommended that the base current be changed temporarily from positive to negative. The base current and collector current waveforms for turn-on and turn-off are shown in Figure 2.14. The turn-off time can be reduced further by avoiding full saturation in the on-state. Instead, by reducing the on-state base current, the operating point of the transistor is shifted to the *quasi-saturation zone*, in the vicinity of the hard saturation line.

Single BJTs are seldom used in power electronic converters because of their low current gain. Instead, the *Darlington connection* of two or three transistors is employed, as illustrated in Figure 2.15. The current gain of a two-transistor Darlington connection is on the order of 100, and that of a three-transistor connection, of 1000. Power BJTs can operate with switching frequencies on the order of 10 kHz, and their maximum available voltage and current ratings reach 1.5 kV and 1.2 kA.

In recent years, power BJTs have been losing their market share to insulated-gate bipolar-junction transistors (IGBTs), described in Section 2.4.5. Voltage-controlled IGBTs possess all the advantages of BJTs without the weaknesses of the latter devices, such as second breakdown or current-controlled switching. Thus, IGBTs are perfectly suited for most applications where power BJTs have been employed.

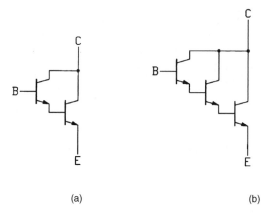

(a)   (b)

Figure 2.15   BJT Darlington connections: (a) two-transistor; (b) three-transistor.

2.4.4   Power MOSFETs

A power MOSFET (*metal-oxide semiconductor field-effect transistor*), whose simplified structure and circuit symbol are shown in Figure 2.16, is a semiconductor
power switch characterized by the highest switching speed. Three electrodes of the
MOSFET, the *drain* (D), *source* (S), and *gate* (G), correspond to the collector, emitter,
and base of a BJT, respectively. However, in contrast to the BJT, the power MOSFET
is voltage controlled, as the dc impedance of the gate–source path is practically infinite (10^9 to 10^{11} Ω). It is worth mentioning that during fast turn-on and turn-off, the
gate circuit does carry a short current pulse, which is associated with the respective
charging and discharging of the gate–source capacitance.

Voltage–current characteristics of the power MOSFET are illustrated in Figure
2.17. The characteristics show the drain current, I_D, as a function of the drain–source

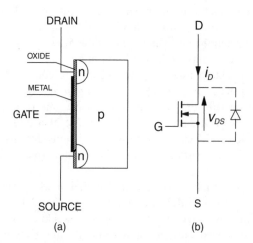

(a)   (b)

Figure 2.16   Power MOSFET: (a) semiconductor structure; (b) circuit symbol.

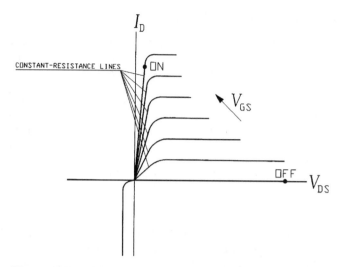

Figure 2.17   Voltage–current characteristics of a power MOSFET.

voltage, V_{DS}, for various values of the gate–source voltage, V_{GS}. Although the characteristics bear a certain resemblance to those of the a BJT (see Figure 2.13), there is no hard saturation line. Instead, each characteristic has a constant-resistance portion in the area of low V_{DS} values, and the control voltage, V_{GS}, should be made sufficiently high for the on-state operating point to lie on that portion. The rated voltage and current represent the maximum allowable values, V_{DSS} and I_{DM}, of the drain-source voltage and drain current, respectively. Similar to BJTs, power MOSFETs may not be exposed to negative drain–source voltages unless protected by a diode connected in series with a transistor.

The technical advantages of power MOSFETs are not limited to high switching speeds. Little power is required to control them, and the control circuitry is simpler than that for BJTs. The typical turn-on gate–source voltage is 20 V, while zero voltage is used to turn the device off. MOSFETs have a negative temperature coefficient on the drain current, which facilitates paralleling several transistors for increased current-handling capability. If the temperature of one of the component MOSFETs increases, the current conducted drops, restoring the thermal balance among the devices connected. This characteristic also makes for uniform current density within the MOSFET, preventing second breakdown from occurring. On-state resistance of high-voltage power MOSFETs used to be quite high, but recent technological advances have resulted in significant reduction of that parameter. The switching losses are low, even with high switching frequencies, thanks to short turn-on and turn-off times. These, usually less than 100 ns, are defined similar to that for BJTs (see Figure 2.14).

The dashed-line diode in parallel with the MOSFET in Figure 2.16b, called a *body diode*, is a by-product of the technological process. It can serve as a freewheeling diode, but being relatively slow it should be bypassed in fast-switching converters

TABLE 2.4   Examples of Power MOSFETs

Designation:	VMO 650-01F (IXYS)	IXFB 100N50P (IXYS)	APT45M100J (Microsemi)	STP4N150 (STMicroel,)
V_{DSS}:	100 V	500 V	1000 V	1500 V
I_D:	690 A	100 A	45 A	4 A
V_{GS}:	20 V	30 V	30 V	30 V
$I_{GS}{}^a$:	0.5 μA	0.2 μA	0.1 μA	0.1 μA
$R_{DS}{}^a$:	1.8 mΩ	49 mΩ	170 mΩ	5 Ω
t_{ON}:	500 ns	36 ns	85 ns	35 ns
t_{OFF}:	800 ns	110 ns	285 ns	45 ns
Size:	110 × 62 × 30 mm	26 × 20 × 5 mm	38 × 25 × 12 mm	16 × 10 × 5 mm

$^a I_{GS}$, gate leakage current; R_{DS}, static drain–source on resistance.

with an external fast recovery diode. Switching frequencies can be as high as hundreds of kilohertz in medium-power converters and on the order of 1 MHz in low-power switching power supplies. Power MOSFETs are available with voltage and current ratings of up to 1.5 kV and 1.8 kA. See Table 2.4 for high-power MOSFETs.

2.4.5   IGBTs

IGBTs, *insulated gate bipolar transistors*, are hybrid semiconductor devices, the combining the advantages of MOSFETs and BJTs. Like MOSFETs, they are voltage controlled but have lower conduction losses and higher voltage and current ratings. The equivalent circuit and circuit symbols of the IGBT are shown in Figure 2.18. To stress the hybrid nature of an IGBT, the control electrode is called a gate (G), while the main current is conducted or interrupted in the collector (C)-to-emitter (E) path.

Voltage–current characteristics of IGBTs are shown in Figure 2.19 as relations between the collector current, I_C, and the collector–emitter voltage, V_{CE}, for various values of the gate–emitter voltage, V_{GE}. The majority of the IGBTs available on the

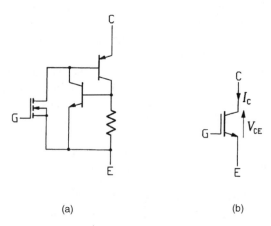

(a)   (b)

Figure 2.18   IGBT: (a) equivalent circuit; (b) circuit symbol.

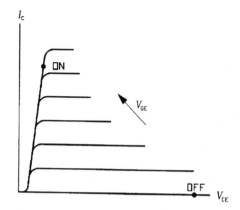

Figure 2.19   Voltage–current characteristics of an IGBT.

market are of the *asymmetrical* (punch-through) type. They do not have reverse-voltage blocking capability and, with low conduction losses, are destined for use in dc-input converters such as choppers and inverters. Usually, they are integrated with an antiparallel freewheeling diode. *Symmetrical* (non-punch-through) IGBTs can block reverse voltages as high as the rated forward-blocked voltage. Although this is an additional advantage of these devices, their conduction losses are higher than those of asymmetrical IGBTs. Symmetrical IGBTs are used primarily in ac-input PWM converters, such as rectifiers and ac voltage controllers.

Regarding the on-state voltage drop, IGBTs are comparable to BJTs but superior to power MOSFETs. Similar to MOSFETs, IGBTs are turned on by a gate–emitter voltage on the order of 20 V and turned off by zero voltage. IGBTs can be switched with "supersonic" (exceeding 20 kHz) frequencies. The maximum available voltage and current ratings are 6.5 kV and 2.4 kA. It can be seen that in addition to their operational superiority, IGBTs have wider ranges of voltage and current ratings than those of power BJTs. Therefore, IGBTs are now the most popular semiconductor power switches. Selected parameters of high-power IGBTs are listed in Table 2.5.

TABLE 2.5   Examples of IGBTs

Designation:	5SNA 2400E170100 (ABB)	5SNA 1500E330300 (ABB)	5SNA 0600G650100 (ABB)
V_{CE}:	1.7 kV	3.3 kV	6.5 kV
I_C:	2.4 kA	1.5 kA	0.6 kA
I_{FSM}:	20 kA	14 kA	6 kA
$V_{CE(sat)}{}^a$:	2.6 V	3 V	5.4 V
$I_{GES}{}^a$:	0.5 μA	0.5 μA	0.5 μA
t_{ON}:	0.32 μs	0.57 μs	0.57 μs
t_{OFF}:	1.1 μs	1.68 μs	1.86 μs
Size:	190 × 140 × 38 mm	190 × 140 × 38 mm	190 × 140 × 48 mm

[a] $V_{CE(sat)}$, collector–emitter saturation voltage, I_{GES}, gate leakage current.

2.5   COMPARISON OF SEMICONDUCTOR POWER SWITCHES

Designers of modern power electronics converters have the choice of many types of semiconductor power switches. The variety of devices available allows optimal selection of switches for a given converter. As each switch type has its advantages and disadvantages, it would be difficult to judge one of them unequivocally as generally better than the others.

A hypothetical perfect switch that can be used as a reference in evaluating real devices would have the following characteristics:

1. Highly rated voltage and current, allowing application of single switches in high-power converters.
2. Low, possibly zero, leakage current in the off-state and a voltage drop across the switch in the on-state. These would result in minimum conduction and off-state power losses in the switch.
3. Short turn-on and turn-off times, allowing the device to be switched with high frequencies and minimum switching losses.
4. Low power requirements to turn the switch on and off. This would simplify the design of control circuits and improve the efficiency and reliability of the entire converter.
5. Negative temperature coefficient of the current, resulting in equal current sharing by paralleled devices.
6. Large allowable dv/dt and di/dt values, limiting the need for snubbers for protection of the switch from failures and structural damage.
7. Low price: an important consideration in today's highly competitive field of commercial power electronics.

Finally, it is desirable for semiconductor power switches to have large *safe operating areas* (SOAs), both *forward*-and *reverse-bias* SOAs. The concept of SOAs, omitted for conciseness from the description of individual types of switches, requires some elaboration. Notice that the voltage–current characteristics in Figures 2.13, 2.17, and 2.19 do not extend uniformly over the entire range of the voltage. The hyperbolical envelope of these characteristics results from the fact that the amount of heat generated in the device is proportional to the power loss, that is, the voltage–current product. For this product to be constant and equal to the maximum allowable value, the current must be inversely proportional to the voltage. This yields a hyperbolic limit on the allowable steady-state operating points in the voltage–current coordinates. When a switch turns on or off, the instantaneous operating points can fall far beyond that limit. However, this is not necessarily dangerous for the switch, since the turn-on and turn-off may be too short for excessive temperature buildup.

A typical SOA for a power MOSFET is shown in Figure 2.20. It is drawn in the logarithmic scale to replace hyperbolic borderlines with linear ones. The sloped line on the left-hand side results from the on-state resistance of the MOSFET, which

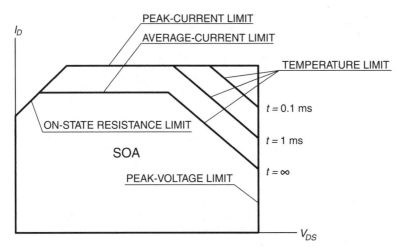

Figure 2.20 Safe operating area for a power MOSFET.

determines the voltage drop across the device for a given drain current. The three sloped lines on the right-hand side are the SOA limits for current pulses of the durations indicated. A *forward-bias SOA* corresponds to the situation when the gate–source voltage is positive, as during turn-on. If that voltage is negative, as for turn-off, the respective SOA is called a *reverse-bias SOA*. Both of them are identical for the power MOSFET, but it is not necessarily so for the other fully controlled switches. Similar to high dv/dt and di/dt ratings, a robust SOA reduces the need for external protection circuits.

In analogy to human athletes, the most powerful semiconductor switches are also the slowest, while the agile, high-frequency switches have much lower power-handling capabilities. Progressing from slow high-power SCRs and GTOs through IGCTs, BJTs, and IGBTs, to fast but relatively low-power MOSFETs, the best balance between voltage–current ratings and high-frequency switching ability seems to be achieved by the IGBT. As already mentioned, it is currently the most popular device in modern PWM converters. The IGBT, although not free of certain weaknesses such as relatively high conduction losses, seems to be closest to the perfection expressed by the wish list of properties.

The future of inverter-grade SCRs looks bleak, although the slower, phase-control SCRs still constitute the best choice for phase-controlled rectifiers and ac voltage controllers. BJTs seem to be undergoing gradual phase-out by hybrid technology devices. In conclusion, it is worth mentioning that the variety of existing semiconductor power switches are by no means limited to those presented in this chapter. In addition to the popular switches described, there is another batch of less common devices, such as, *MOS-controlled thyristors* (MCTs), *static induction transistors* (SITs), and *static induction thyristors* (SITHs).

Basic properties and maximum ratings of most common semiconductor power switches are summarized in Table 2.6. Most often, switches with the highest-rated

TABLE 2.6   Properties and Maximum Ratings of Semiconductor Power Switches

Type	Switching Signal	Switching Characteristic	Switching Frequency	Forward Voltage	Rated Voltage	Rated Current
Diode			20 kHz[a]	1.2–1.7 V	6.5 kV	10 kA
SCR	current	trigger	0.5 kHz	1.5–2.5 V	8 kV	6 kA
Triac	current	trigger	0.5 kHz	1.5–2 V	1.4 kV[b]	0.1 kA[b]
GTO	current	trigger	2 kHz	3–4 V	6 kV	6 kA
IGCT	current	trigger	5 kHz	3-4 V	6.5 kV	6 kA
BJT	current	linear	20 kHz	1.5–3 V	1.5 kV	1.2 kA
MOSFET	voltage	linear	1 MHz	3–4 V	1.5 kV	1.8 kA
IGBT	voltage	linear	20 kHz	3–4 V	6.5 kV	2.4 kA

[a] Fast recovery diodes. General-purpose diodes operate at 50 or 60 Hz.
[b] BCTs (bidirectionally controlled thyristors), whose operating principle is similar to that of the triac, reach 6.5 kV of rated voltage and 5.5 kA of rated current.

voltage offered by manufacturers have only a medium-rated current, and vice versa. The values provided are typical, but it is possible to find devices, usually special-purpose devises, whose ratings are even higher. For example, ABB lists on its Web site 8-kV SCRs and the "housingless" welding diodes, whose rated current exceeds 13.5 kA.

2.6   POWER MODULES

To facilitate the design and simplify the physical layout of power electronic converters, manufacturers of semiconductor devices offer a variety of *power modules*. A power module is a set of semiconductor power switches interconnected into a specific topology and enclosed in a single case. Most popular topologies are single- and three-phase bridges or their subcircuits. Power modules may also contain several switches of the same type connected in series, parallel, or series–parallel, to increase the overall voltage and/or current ratings.

Configurations of the power modules available are illustrated in Figures 2.21 through 2.24. Four two-device assemblies of power diodes and SCRs are shown in Figure 2.21. The series connections of the devices in Figure 2.21a through c are used in rectifiers, while the antiparallel connection of two SCRs in Figure 2.21d, introduced in Section 2.3.2 as the BCT, can form part of an ac voltage controller or a static ac switch.

The Darlington connection of two BJTs is utilized in modules in Figure 2.22, showing a dual-switch module, quad-switch module, and six-switch module. Each switch is equipped with a parallel freewheeling diode, while the resistor–diode circuits connected between the emitter and the base of each component Darlington transistor help to reduce the leakage current and speed up the turn-off. The quad-switch bridge topology can serve, depending on the control algorithm, as a four-quadrant chopper or a single-phase voltage-source inverter. The six-switch bridge configuration constitutes

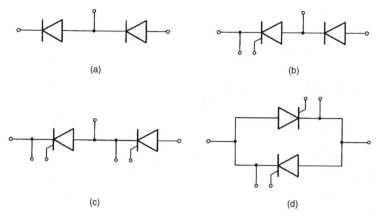

Figure 2.21 Diode and SCR modules: (a) two power diodes; (b) power diode and SCR; (c) two SCRs; (d) antiparallel connection of two SCRs (BCT).

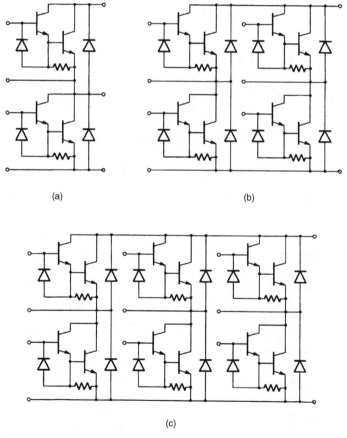

Figure 2.22 Power BJT (Darlington) modules: (a) dual switch; (b) quad switch; (c) six-switch.

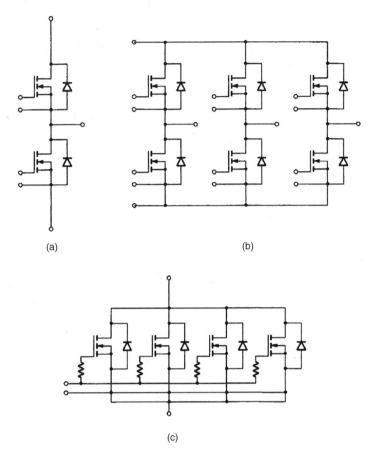

(a)   (b)

(c)

Figure 2.23 Power MOSFET modules: (a) dual switch; (b) six-switch; (c) single four-transistor switch.

a three-phase voltage-source inverter. However, the more devices that are crammed together in a single module, the less heat per device can safely be dissipated and the lower the current ratings are. Hence, the need for the dual-switch module, which can be used as a building block of either of the bridge topologies if the load current required is too high for the quad- or six-switch modules available.

A dual-power MOSFET module is illustrated in Figure 2.23a and a six-switch module in Figure 2.23b. Figure 2.23c shows a single-switch module in which, to increase the overall current rating, four power MOSFETs are connected in parallel and controlled simultaneously using a single gate. The resistors connected to each internal gate of the four transistors equalize the voltage signals applied to the individual MOSFETs. Finally, an arrangement of seven IGBTs and 13 diodes, constituting the power circuit of a three-phase ac-to-dc-to-ac converter called a *frequency changer*, is depicted in Figure 2.24. Additional external components, such as a resistor or capacitor, which are too large to incorporate into the module, are required.

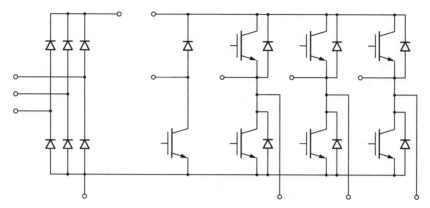

Figure 2.24   IGBT-based modular frequency changer.

The power modules described contain power circuits only. Control circuits are added in *intelligent power modules* (IPMs). These additions include protection circuits and gate drives. For instance, Powerex offers the following IPM configurations based on IGBTs:

- Single transistor
- Two transistors
- Four-transistor bridge
- Four-transistor bridge plus single-transistor chopper
- Four-transistor bridge plus two single-transistor choppers
- Six transistors
- Seven transistors

For example, the high-voltage IPM 1200HCE330-1, a single-IGBT configuration, has voltage and current ratings of 3.3 kV and 1.2 kA, respectively. The module includes a gate drive with undervoltage lock, as well as overtemperature and short-circuit protection circuits.

After finishing Chapters 4 through 7, the reader is encouraged to return to this section briefly to appreciate the practical usefulness of power modules. The topologies described represent only a tip of an iceberg of the great variety of modular circuits available on the market of power electronic devices.

2.7 SUMMARY

Power electronic converters are based on semiconductor power switches that operate in two states only: the on-state and off-state. In the on-state, the voltage drop across a switch is low, resulting in low conduction losses. In the off-state, the current through a switch is practically zero, so no losses are produced. However, during transitions from one state to another, switching losses are generated, because for a short time the transient voltage and current are both substantial. Each switching is thus associated with energy loss, and the more switchings per second are executed (i.e., the higher the switching frequency), the higher the power loss becomes.

Semiconductor power switches can be classified as uncontrolled, semicontrolled, and fully controlled. Power diodes, the most common uncontrolled switches, start conducting when forward biased and cease to conduct when the current changes its polarity. SCRs, triacs, and BCTs are semicontrolled switches that can be triggered into conduction (fired) when forward biased. Once fired, the devices cannot be extinguished by a control signal. Most common fully controlled switches are GTOs, IGCTs, power BJTs and MOSFETs, and IGBTs. The GTO and IGCT can be fired in the same way as the SCR, but they can be extinguished by a strong negative gate current pulse. The current in the BJT, MOSFET, and IGBT can be controlled linearly, but to minimize losses they are operated, like all switches, only in the on–off mode. BJTs are current controlled, whereas power MOSFETs and IGBTs are voltage controlled and require a negligible amount of gate power.

Except for triacs and BCTs, all semiconductor power switches can conduct current in one direction only. It does not mean that all the switches can block voltages of both forward and reverse polarity. Symmetrical blocking capability is typical for SCRs, triacs, and BCTs, some GTOs and IGCTs, and non-punch-through IGBTs. The other devices may not be subjected to reverse voltage unless a series-connected diode is employed to block that voltage.

Catalogs and data sheets provide information on semiconductor power switches in the form of restrictive and descriptive parameters, characteristics, and safe operating areas. Those data, especially the rated voltage and current, help to select the most appropriate devices for a given application. SCRs, GTOs, and IGCTs are the largest and slowest switches, and power MOSFETs are the smallest and fastest. Each type has advantages and disadvantages, but the IGBTs dominate the field of the most common low- and medium-power converters. Manufacturers of semiconductor devices also offer a wide choice of power modules that combine several switches in a single case.

LITERATURE

[1] Baliga, B. J., *Power Semiconductor Devices*, PWS Publishing Co., Boston, 1995.

[2] Bose, B. K., *Modern Power Electronics and AC Drives*, Prentice Hall, Upper Saddle River, NJ, 2001, Chap. 1.

[3] http://www.abb.com/ProductGuide/.

[4] http://www.pwrx.com/.

[5] Rashid, M. H., *Power Electronics Handbook*, 2nd ed., Academic Press, San Diego, CA, 2007, Chapts 2 to 9.

3 Supplementary Components and Systems

Supplementary components and systems for power electronic converters are reviewed in this chapter. Drivers, overcurrent protection schemes, snubbers, filters, cooling methods, and control systems are also described, and example solutions are presented.

3.1 WHAT ARE SUPPLEMENTARY COMPONENTS AND SYSTEMS?

A practical power electronic converter is a complex system comprised of several subsystems and numerous components. Many of them are not shown in converter circuit diagrams, which are usually limited to the power circuit and, sometimes, a block diagram of the control system. The supplementary components and systems for modern power electronic converters include:

1. *Drivers* for individual semiconductor power switches, which provide switching signals, interfacing the switches with the control system.
2. *Overcurrent protection schemes*, which safeguard converter switches and sensitive loads from excessive currents.
3. *Snubbers*, which protect switches from transient overvoltages and overcurrents at turn-on and turn-off, and reduce the switching losses.
4. *Filters*, which improve the quality of the power drawn from the source and that supplied to the load. As parts of the power circuit, filters are usually shown in circuit diagrams of converters.
5. *Cooling systems*, which reduce thermal stresses on switches.
6. *Control systems*, which govern the converter operation, including protection tasks.

Because of its introductory scope, this book is focused on principles of power conversion and control and their realization in power circuits of power electronic converters. However, the reader should be aware of the basic properties of supplementary components and systems, and their impact on the operation of converters.

Introduction to Modern Power Electronics, Second Edition, by Andrzej M. Trzynadlowski
Copyright © 2010 John Wiley & Sons, Inc.

Therefore, subsequent sections of this chapter are devoted to brief coverage of individual types of supplementary equipment. It must be stressed that the great variety of power electronic products offered by hundreds of manufacturers allows for only a sampling of this vast topic.

3.2   DRIVERS

Depending on the types of switches, converter topology, and voltage levels, various driver configurations are employed in power electronic converters. A driver, activated by a logic-gate-level signal from the control system, must be able to provide a sufficiently high voltage or current to the controlling electrode (gate or base) to cause immediate turn-on. The on-state of the switch must be safely maintained until turn-off.

Generally, the low-voltage control system and the high-voltage power circuit must be isolated electrically. This can be realized using pulse transformers or optical coupling. The latter is accomplished by placing a light-emitting diode in the vicinity of a light-activated semiconductor device. Alternatively, instead of transferring light signal through free space, a fiber-optic cable can be used. Because of the fundamentally different driving signal requirements, different solutions are used for semicontrolled thyristors (SCRs, triacs, BCTs), current-controlled switches (GTOs, IGCTs, power BJTs), and voltage-controlled hybrid devices (power MOSFETs and IGBTs).

3.2.1   Drivers for SCRs, Triacs, and BCTs

To fire, that is to trigger an SCR, triac, or BCT into conduction, the pulse of the gate current, i_G, must have sufficient magnitude and duration and a possibly short rise time (i.e., high di_G/dt). Isolation is necessary between the control circuitry and the power circuit, at least for switches with ungrounded cathodes. The isolation can be provided by either an optocoupler or a transformer. Both solutions have advantages and disadvantages. An optocoupler requires a power supply and an amplifier on the thyristor side, which are not needed if a transformer is used. However, extra circuitry must be employed to avoid saturation of the transformer core.

A driver for an SCR, based on a pulse transformer, PTR, and transistor amplifier, TRA, is shown in Figure 3.1. Diode D1 and zener diode DZ connected across the primary winding provide a freewheeling path for the primary current at turn-off and prevent saturation of the transformer core. Diode D2 in the gate circuit rectifies the secondary current of the transformer.

A simple optically isolated driver for an SCR is shown in Figure 3.2. The optocoupler is comprised of a light-emitting diode (LED) and a small light-activated thyristor (LAT). The energy for the gate signal is obtained directly from the power circuit, as it is the voltage across the SCR that produces the gate current when the LAT is activated by the LED. The LAT must withstand the same voltage as the driven SCR. This is not a serious problem, though, as light-activated thyristors, also used

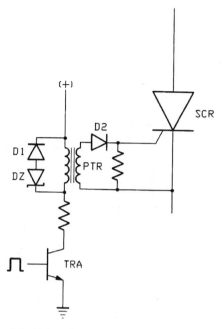

Figure 3.1   Driver for an SCR with transformer isolation.

in high-voltage transmission lines, are among the semiconductor devices with the highest voltage ratings.

Figure 3.3 shows a nonisolated driver for a triac. A transistor amplifier (TRA) provides the gate current for the triac. An optically isolated driver using a light-activated triac (LATR) is illustrated in Figure 3.4.

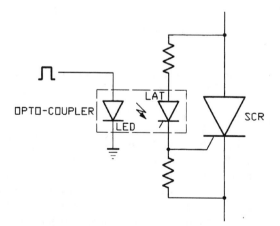

Figure 3.2   Optically isolated driver for an SCR.

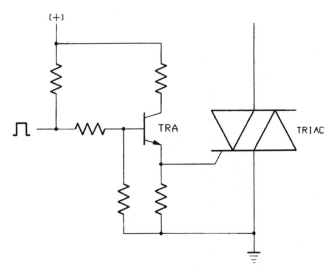

Figure 3.3 Nonisolated driver for a triac.

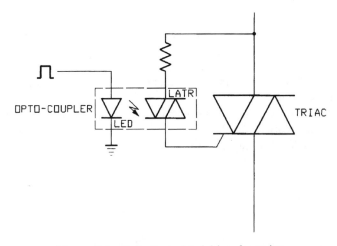

Figure 3.4 Optically isolated driver for a triac.

3.2.2 Drivers for GTOs and IGCTs

Although GTOs and IGCTs are turned on similarly to SCRs, drivers for these switches are more complex than those for the semicontrolled thyristors, because of the very high gate current pulse magnitude required for turn-off. A gate drive circuit is shown in Figure 3.5. To turn the GTO on, the pulse transformer, PTR, transmits high-frequency current pulses generated by alternately switched MOSFETs M1 and M2. The firing current is supplied to the gate through the zener diode, DZ, and inductor, L, which controls the rate of change, di_G/dt, of the current. Simultaneously, capacitor C

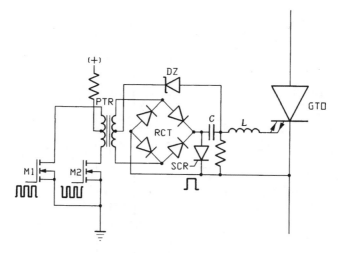

Figure 3.5   Driver for a GTO with transformer isolation.

is charged via the four-diode rectifier, RCT. Cessation of the impulse train indicates that turn-off is to be performed. Turn-off is initiated by the SCR, which causes rapid discharge of the capacitor in the gate–cathode circuit.

Drivers with optical isolation are also known. The GTO side of the driver must have its own power supply to provide the necessary gate current, especially for turn-off. Instead of the SCR in the driver in Figure 3.5, a BJT, a power MOSFET, or a combination can be used to initiate turn-on and turn-off gate pulses.

3.2.3   Drivers for BJTs

Because of the properties of BJTs, base drive circuits must be of the current-source type. A high-quality driver should have the following characteristics:

1. A high current pulse at turn-on, to minimize the turn-on time.
2. An adjustable base current in the on-state, to minimize losses in the base–emitter junction. The initial boost current should be reduced after turn-on.
3. Prevention of hard saturation of the transistor. A saturated BJT has a significantly longer turn-off time than a quasi-saturated BJT.
4. A reverse base current for turn-off, to further minimize the turn-off time.
5. Possibly a low impedance between the base and the emitter in the on-state and a reverse base–emitter voltage in the off-state. These measures increase the collector–emitter voltage blocking capability of the transistor.

Two simple nonisolated drivers are shown in Figure 3.6. The single-ended driver shown in Figure 3.6a requires only one transistor, TR, but its performance is inferior

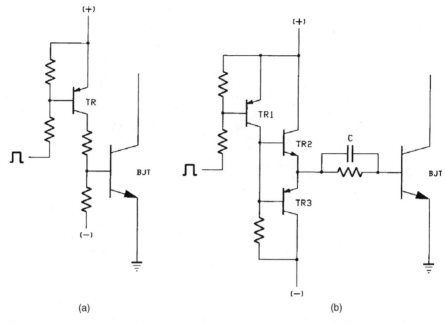

Figure 3.6 Nonisolated drivers for a BJT: (a) single-transistor driver; (b) driver with a class B output stage.

to that of more advanced schemes. Power losses in the driver are reduced in the circuit shown in Figure 3.6b. Input transistor TR1 drives a *class B output stage*, consisting of an npn transistor (TR2) and a pnp transistor (TR3). Capacitor C in both drivers speeds up switching processes, providing a current boost to the base.

To increase the turn-off speed, an antisaturation circuit called a *Baker's clamp* can be used (shown in Figure 3.7). The purpose of the clamp is to shunt the current

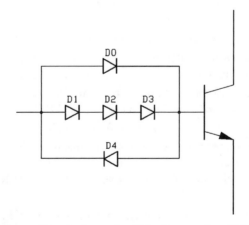

Figure 3.7 Antisaturation Baker's clamp for a BJT.

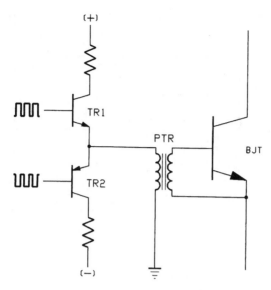

Figure 3.8 Driver for a BJT with transformer isolation.

from the base through diode D0, in dependence on the collector–emitter voltage, to shift the operating point of the transistor from the hard saturation line to the quasi-saturation region (see Figure 2.13). Diodes D1 through D3 produce appropriate bias for the clamping diode D0 (in practice, the number of diodes in the series can be more or less than three). Diode D4 provides the path for negative base current during turn-off.

If a BJT is to be employed in a high-voltage power circuit, an isolation transformer can be used, as shown in Figure 3.8. However, the range of available duty ratios of the driven BJT is then limited to about 0.1 to 0.9. Therefore, isolated drivers for BJTs are usually based on optocouplers. A commercial single-chip driver with optical isolation is shown in Figure 3.9a. Phototransistors are used in the internal optocouplers of the driver, and the BJT is driven by a class B output stage. The waveform of the base current generated by the driver is illustrated on a nonlinear magnitude scale in Figure 3.9b. In reality, the positive and negative peak values of the current are 10 to 20 times higher than those of sustained on-state current.

3.2.4 Drivers for Power MOSFETs and IGBTs

In the steady state, gates of hybrid semiconductor power switches draw almost no current. As such, they can be activated directly from logic gates. However, if high-frequency switching is desired, an electric charge must quickly be transferred to and from the gate capacitance. This requires high pulses of gate current at the beginning of the turn-on and turn-off signals. Standard logic gates by themselves are incapable of supplying (sourcing) or drawing (sinking) such high currents, so that the maximum available switching frequency is seriously limited. Therefore, to fully utilize the

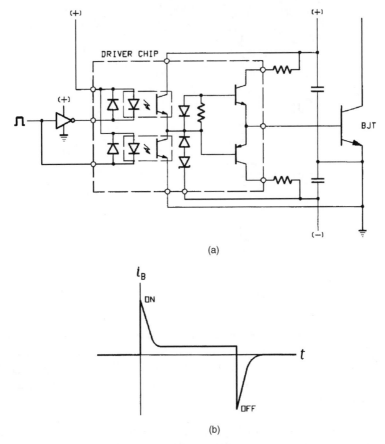

(a)

(b)

Figure 3.9   Driver for a BJT with optical isolation; (a) circuit diagram; (b) waveform of base current.

high-speed potentials of hybrid switches, very fast power MOSFETs in particular, provisions must be made in the drivers to source or sink transient current pulses.

For simplicity, all the subsequent drivers are shown in application to power MOS-FETs, although they can also be used for IGBTs. A simple gate drive circuit with a high-current TTL clock driver (CD) is shown in Figure 3.10. Figure 3.11 illustrates a power MOSFET driven from a pulse transformer, PTR. The internal parasitic diode, D, in the auxiliary MOSFET, AM, provides the path for the charging current of the main MOSFET's gate capacitance. When the pulse transformer saturates, AM blocks the discharge current from the gate until turn-off, which is initiated by a negative pulse from the transformer that turns AM on. The driver is particularly convenient for switches requiring a floating (ungrounded) gate drive.

A driver for power MOSFETs and IGBTs with optical isolation is shown in Figure 3.12. In addition to turning the driven switch on and off, the driver provides overcurrent protection for the switch. The overcurrent condition is detected by sensing the

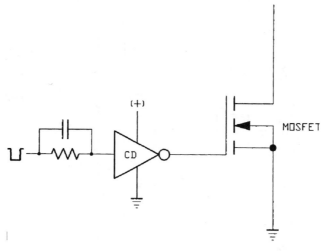

Figure 3.10   Gate drive for a power MOSFET with a high-current TTL clock driver.

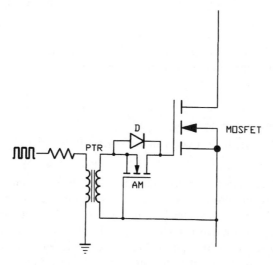

Figure 3.11   Driver for a power MOSFET with transformer isolation.

collector–emitter voltage in the on-state. Since the hybrid devices offer approximately constant resistance in the main circuit, elevated voltage implies an overcurrent. The switch is then turned off and an indicator circuit is activated.

3.3   OVERCURRENT PROTECTION SCHEMES

Semiconductor power switches can easily suffer permanent damage if a short circuit happens somewhere in the converter or the load, or if an overcurrent occurs due to an

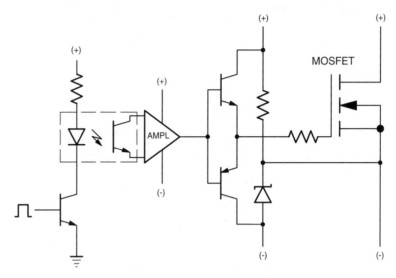

Figure 3.12 Driver for a power MOSFET with optical isolation.

excessive load, that is, a too-low load impedance and/or counter-EMF. Three basic approaches to overcurrent protection are used in power electronic converters:

1. Fuses
2. SCR "crowbar" arrangement
3. Turning the switches off when an overcurrent is detected.

High-power slow-reacting semiconductor power devices such as diodes, SCRs, and GTOs are usually protected by special, fast-melting fuses connected in series with each device. The fuses are made of thin silver bands, no thicker than a tenth of an inch, usually with a number of rows of punched holes. One or more such bands are packed in sand and enclosed in a cylindrical ceramic body with tinned mounting brackets. The I^2t parameter of a fuse must be less than that of the device being protected but not so low as to cause a breakdown under normal operating conditions. A properly selected fuse should melt within a half-cycle of the 60-Hz (or 50-Hz) voltage.

Low-cost power electronic converters can be protected by a single fuse between the supply source and the input to the converter. A more sophisticated solution, illustrated in Figure 3.13, involves an SCR crowbar connected across the input terminals of a converter. The input current to the converter is monitored by sensing the voltage drop across the low-resistance resistor R or employing a current sensor. If overcurrent is detected, the SCR is fired, shorting the supply source and causing a meltdown of the input fuse. Alternatively, a fast circuit breaker can be employed in place of the fuse. In a similar manner, the SCR crowbar can be employed for overvoltage protection.

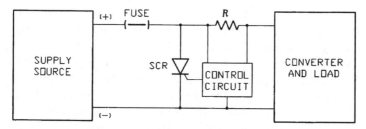

Figure 3.13   SCR crowbar for overcurrent protection of a power electronic converter.

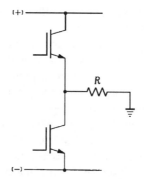

Figure 3.14   Totem-pole arrangement of two switches in a leg of a bridge topology.

Fully controlled semiconductor power switches are best protected by turning them off when overcurrent occurs. The turn-off process is often slowed down to avoid an excessive rate of change, di/dt, of the switch current that could generate a hazardous voltage spike across the device. As exemplified by the system shown in Figure 3.12, dedicated overcurrent protection circuits are often incorporated in modern drivers.

A special case of a potentially dangerous short circuit, called a *shot-through*, is specific for the *bridge topology*, typical for many power converters. In individual legs (branches) of a bridge circuit, two (or more) switches are connected in series in a *totem-pole* arrangement, illustrated in Figure 3.14 by two IGBTs. Normally, their states are mutually exclusive; that is, when one switch is on, the other is off, and the current always flows through the load, R. However, if for any reason (e.g., due to driver failure or incorrect timing of switching signals) one switch is turned on before the other has turned off, a short circuit occurs. To reduce the probability of an overshoot, the beginning of a turn-on signal for one switch is delayed with respect to the end of a turn-off signal for the other switch by *dead time*.

3.4   SNUBBERS

Off all the possible modes of operation of semiconductor devices, switching between two extreme states is certainly the most trying, subjecting switches in power electronic converters to various stresses. For example, if no measures were taken, a rapid change in the switched current at turn-off would produce potentially damaging voltage spikes in stray inductances of the power circuit. At turn-on, a simultaneous high voltage and high current could take the operating point of a switch well beyond the safe operating area (SOA). Therefore, switching-aid circuits called *snubbers* must often accompany semiconductor power switches. Their purpose is to prevent transient overvoltages and overcurrents, attenuate excessive rates of changes of voltage and current, reduce switching losses, and ensure that a switch does not operate outside its SOA. Snubbers also help to maintain uniform distribution of voltages across switches that are connected in series to increase the effective voltage rating, or currents in switches that are connected in parallel to increase the effective current rating. Functions and configurations of snubbers depend on the type of switch and converter topology.

Analysis of snubber circuits is tedious at best, and often quite difficult, due to the nonlinear properties of the semiconductor devices involved and relative complexity of certain schemes. Therefore, subsequent treatment of this topic is mostly qualitative. In practice, besides specialized literature, computer simulations using PSpice or similar software tools are usually employed for snubber development.

To illustrate the need for snubbers, a practical example is considered. A simple BJT-based chopper is shown in Figure 3.15a. The load inductance is assumed to be so high that the output current is practically constant and equal to I_o. Consequently, the load can be modeled by a current source. A stray inductance, L_σ, of the power circuit of the chopper is lumped between the source of input voltage, V_i, and the transistor. The snubber circuit, composed of resistance R_{sn} and capacitance C_{sn}, is connected in parallel with the transistor.

The collector–emitter voltage, v_{CE}, across the BJT is given by

$$v_{CE} = V_i - v_L - v_o \tag{3.1}$$

where v_L and v_o denote the inductor and output voltages, respectively. With the transistor in the on-state, $v_{CE} \approx 0$. At $t = 0$, the BJT is turned off so that its collector current, i_C, decreases linearly from the initial value of I_o, reaching zero at $t = t_0$. As a result, a transient voltage appears across the stray inductance. The voltage waveform has the shape of a pulse with duration (width) t_0 and peak value, $V_{L,p}$, of

$$V_{L,p} = L_\sigma \frac{di_c}{dt} = -L_\sigma \frac{I_o}{t_0}. \tag{3.2}$$

In the meantime, the freewheeling diode, D, has taken over conduction of the output current, thus, $v_o = 0$ and the peak value, $V_{CE,p}$ of the collector–emitter voltage is

$$V_{CE,p} = V_i - V_{L,p} = V_i + L_\sigma \frac{I_o}{t_0}. \tag{3.3}$$

Fast turn-offs are desirable from the point of view of reduction of switching losses, but Eq. (3.3) shows that when t_0 approaches zero, the voltage across the BJT

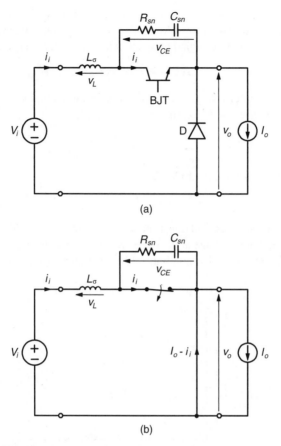

Figure 3.15 BJT-based chopper with an RC snubber; (a) circuit diagram; (b) equivalent circuit in the off-state.

approaches infinity. Even with a finite but short t_0, the voltage can easily be excessive, damaging the transistor.

With the snubber in place, the equivalent circuit of the chopper at turn-off is as shown in Figure 3.15b. When switch S representing the BJT opens, a series RLC circuit is created. As known from the theory of such circuits, if $R_{sn} < 2\sqrt{L_\sigma/C_{sn}}$, the current, i_i, in the circuit is given by

$$i_i = I_p e^{-(R_{sn}/L_\sigma)t} \cos(\omega_d t + \varphi) \tag{3.4}$$

where

$$\omega_d = \sqrt{\frac{1}{L_\sigma C_{sn}} - \left(\frac{R_{sn}}{2L_\sigma}\right)^2}. \tag{3.5}$$

The values of amplitude I_p and angle φ in Eq. (3.4) can be determined from the known initial and final conditions. Since the initial current, $i_i(0)$, equals I_o and the final, steady-state current through the capacitor is zero, then $I_p = I_o$ and $\varphi = 0$, that is,

$$i_i = I_o e^{-(R_{sn}/L_\sigma)t} \cos \omega_d t. \tag{3.6}$$

The collector–emitter voltage across the snubber can be obtained by differentiating i_i and substituting di_i/dt in the equation

$$v_{CE} = V_i - v_L = V_i - L_\sigma \frac{di_i}{dt} \tag{3.7}$$

which after some rearrangement yields

$$v_{CE} = V_i \left[1 - e^{-(R_{sn}/L_\sigma)t} \cos \omega_d t \right]. \tag{3.8}$$

With a properly designed and tuned snubber, the collector–emitter voltage displays only a small overshoot. Voltage and current waveforms for the both cases considered are shown in Figure 3.16. By combining the $i_C(t)$ and $v_{CE}(t)$ waveforms into the

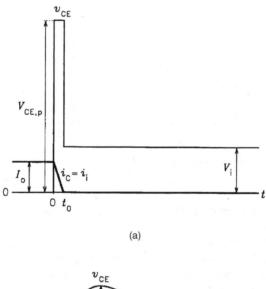

(a)

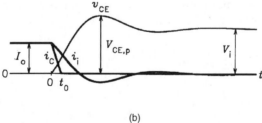

(b)

Figure 3.16 Voltage and current waveforms in the chopper of Figure 3.15: (a) without snubber, (b) with snubber.

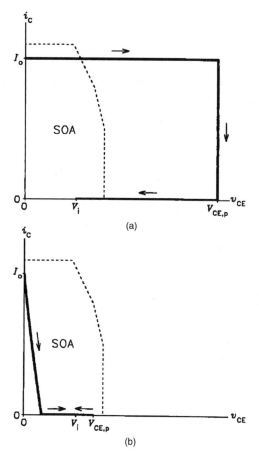

Figure 3.17   Switching trajectories of the BJT of Figure 3.15: (a) without snubber; (b) with snubber.

$i_C = f(v_{CE})$ relation, the *switching trajectory* is obtained, which represents a locus of operating points of the transistor. As shown in Figure 3.17, using the SOA of the BJT as a background, it can easily be checked whether the snubber protects the transistor from unsafe operating conditions. A large portion of the switching trajectory in Figure 3.17a for the snubberless transistor lies outside the SOA. However, that in Figure 3.17b, when a snubber is installed, is contained within the SOA with significant safety margins.

Snubbers are not inherently necessary, as it is often possible to select power switches of such high ratings that even large transient voltages and current would not be dangerous. However, there is always a price to pay for oversized semiconductor devices: not only in the literal sense, but also in terms of the increased weight, bulk, and losses. On the other hand, snubbers also contribute to the cost, weight, and size of a power electronic converter, and they are not free of losses. Therefore, a search for snubbers that are optimal for a given application constitutes a true test of the diligence and expertise of a designer.

Snubbers for individual power switches are disposed of in *resonant converters*, where a single resonant circuit is used to provide safe, low-loss switching conditions for all switches of the converter. Resonant power electronic converters belong mostly in the family of dc-to-dc converters, covered in Section 8.4. Recently, resonant topologies for power inverters have been of interest. The *resonant dc-link inverter* is described in Section 7.5.

3.4.1 Snubbers for Power Diodes, SCRs, and Triacs

The rate of rise of the reverse recovery current in power diodes is high, so that an overvoltage at turn-off is very likely, even with a small amount of stray inductance. Therefore, simple RC snubbers, such as that in Figure 3.15, are connected in parallel with the diode. This type of switching aid circuit is often termed a *turn-off snubber*, since it alleviates the voltage stresses at turn-off.

RC snubbers are also employed in SCR- and triac-based converters, mostly for the purpose of preventing false triggering (firing) from excessive dv/dt. In certain applications, firing an SCR at the wrong time may prove catastrophic. SCRs and triacs should also be protected from extreme di/dt values. This is accomplished by placing an inductor in series with the device. The required inductance of such a *turn-on snubber* is low, so that just the stray inductance of wiring of the power circuit is often sufficient. Snubbers for the power diode and SCR are illustrated in Figure 3.18.

3.4.2 Snubbers for GTOs and IGCTs

Simple turn-on and turn-off snubbers for a GTO are shown in Figure 3.19. A turn-on snubber protects the GTO from overcurrents when it takes over conduction of the

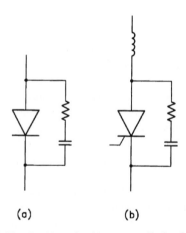

(a)   (b)

Figure 3.18   Snubbers for (a) a power diode; (b) an SCR.

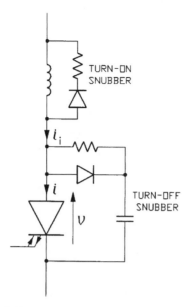

Figure 3.19   GTO with turn-on and turn-off snubbers.

current from another slow device, that is, a high-power freewheeling diode. The di/dt is also attenuated. The diode–resistor circuit allows fast dissipation of the energy stored in the inductor when the GTO turns off.

The RDC (resistor–diode–capacitor) snubber reduces the anode–cathode voltage at turn-off, limiting the switching loss. Similar to that for the SCR, the snubber also prevents the GTO from refiring due to supercritical values of dv/dt. At turn-off, the input current, i_i, is diverted into the snubber capacitor through the diode, while the anode current, i, in the GTO decreases. The anode is clamped to the capacitor, whose initial voltage is zero and rising. Thus, a simultaneous occurrence of high anode voltage, v, and current, i, at turn-off is avoided. The capacitor is discharged quickly in the resistor–GTO circuit at the following turn-on.

3.4.3   Snubbers for Transistors

In the figures in this section, snubbers for transistors (BJTs, MOSFETs, and IGBTs) are shown with the BJT, but the same solutions can be applied to the other devices. Snubbers similar to those for the GTO (see Figure 3.19) are often employed. The inductive turn-on snubber ensures that the collector–emitter voltage, v_{CE}, drops to the saturation level prior to the collector current reaching its full on-state value.

Figure 3.20 shows another, related solution in the form of a combined turn-on and turn-off snubber. At turn-on, the series inductor slows down the rate of increase of the collector current, i_C. Simultaneously, the capacitor discharges through the resistor, inductor, and transistor. At turn-off, the input current, i_i, is diverted from the transistor to the diode–capacitor bypass. When the capacitor is fully charged, the remaining electromagnetic energy of the inductor is dissipated in the resistor.

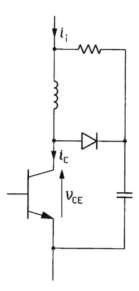

Figure 3.20   Combined on-and-off snubber for a transistor.

It must be pointed out that when selecting a snubber, the entire converter topology should be taken into account, since other components of the power circuit can interfere with proper operation of snubbers. Common snubbers for popular bridge converters based on switches with antiparallel freewheeling diodes are shown in Figure 3.21 for a single leg of the bridge.

In recent years there has been a tendency to design snubberless converters. The accumulated practical experience as to how to minimize stray inductance, and the robust SOA_1 of modern devices, allow turn-off snubbers to be disposed of. Yet to protect internal freewheeling diodes and reduce the electromagnetic interference, simple RC snubbers in parallel with the switches are still recommended. Also, in bridge converters, a single inductor can be placed in series with the input terminals of the bridge. If a converter is supplied directly from a transformer, the secondary leakage inductance can play the role of a turn-on snubber, limiting the rate of change of currents in the converter.

3.4.4   Energy Recovery from Snubbers

The snubbers we have discussed modify the switching trajectory and reduce switching losses. However, the energy stored temporarily in the inductive and capacitive components is dissipated in resistors, that is, lost irretrievably. In high-frequency and high-power converters, the amount of energy lost in snubbers is substantial, straining cooling systems and reducing the efficiency of the converters. Therefore, measures have been developed to recover energy from snubbers and direct it to the load or return it to the supply source.

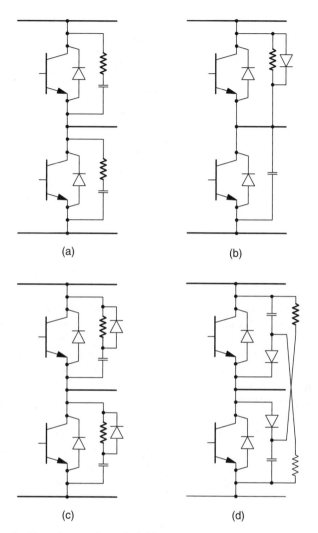

Figure 3.21   Snubbers for transistors in bridge converters: (a) RC; (b) RCD; (c) charge and discharge RCD; (d) discharge-suppressing RCD.

Energy recovery systems can be passive or active, the latter involving an auxiliary power converter. A scheme for the passive recovery of energy from a capacitive snubber is shown in Figure 3.22. Transistor T operates in the PWM mode to control the output voltage, v_o, across a load that is assumed to have an inductive component: hence the need for a freewheeling diode, D_{fw}, that provides a path for the load current when the transistor is off. The simple capacitive turn-off snubber is composed of diode D_{sn} and capacitor C_{sn}. The energy recovery circuit is based on diodes D1 and D2, inductor L, and capacitor C. At turn-off, the snubber capacitor is charged to the input (supply) voltage, V_i. When the transistor turns on, the charge stored in the capacitor

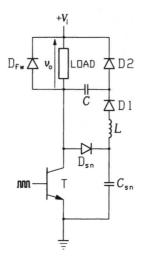

Figure 3.22 Turn-off capacitive snubber with passive energy recovery.

is transferred to capacitor C by means of electric resonance in the C_{sn}–L–D1–C–T circuit. At the subsequent turn-off, capacitor C_{sn} charges again, while capacitor C discharges through diode D2 into the load. Since no resistors are used, most of the energy from the snubber capacitor is recovered and consumed by the load.

In high-power GTO-based converters, auxiliary step-up choppers, described in Section 6.3, are used to transfer energy from snubbers to the power supply. A step-up chopper is a PWM dc-to-dc converter whose pulsed output voltage has an adjustable magnitude which is higher than that of the input voltage. An active energy recovery system is shown in Figure 3.23. When the GTO turns on, the charge of the snubber capacitor, C_{sn}, is resonated into a large storage capacitor C via the GTO, inductor L, and diode D. The storage capacitor acts as a voltage source for the step-up chopper, which boosts the voltage and transfers the energy to the supply line.

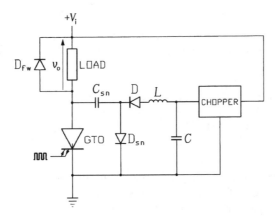

Figure 3.23 Turn-off capacitive snubber for a GTO with active energy recovery.

3.5   FILTERS

Filters are indispensable circuit components of most practical power electronic converters. Generally, depending on their placement, filters can be classified as input, output, and intermediate. Input filters, also called *line* or *front-end filters*, screen the supply source from harmonic currents drawn by converters and improve the input power factor of the converters. Output, or load, filters are employed to improve the quality of the power supplied to the load. Intermediate filters, usually called *links*, interface two converters, forming a cascade, such as that of a rectifier and inverter in the ac-to-dc-to-ac power conversion scheme.

Few simple filter topologies, essentially involving inductive (L) and capacitive (C) components only, can be encountered in practice of in the field of power electronics. Those are covered in the next chapters with converters in which they are applied. Usually, filter inductors carry the entire input or output current. Therefore, their resistance must be keep at a possibly low level, implying a low number of turns, N_t, of a thick-wire winding. Since the coefficient of inductance is proportional to N_t^2 and inversely proportional to the reluctance of the inductor's magnetic circuit, high-permeability cores with large cross-sectional areas are employed. Still, the filter inductances tend to be low, typically on the order of several to tens of millihenries. Inductor cores are made of thin laminations insulated from each other with varnish or shellac to minimize the eddy current losses.

To make up for the usually low inductances, filter capacitors must have large capacitances. Aluminum-foil electrolytic capacitors are therefore used, with available coefficients of capacitance up to 500 mF. However, the highest capacitances are accompanied by low voltage ratings. It is desirable for capacitors in power electronic applications to have very low stray inductance and low equivalent series resistance. Data sheets of capacitors specify maximum allowable values of the working voltage, nonrepetitive surge voltage, voltage and current ripple, and temperature. The highest capacitance-per-volume ratios are obtained in *polarized* electrolytic capacitors, which may not operate under reversed voltage. Nonpolarized capacitors have capacitances about half as high as those of polarized capacitors of the same physical size. In three-phase systems, capacitors are connected in delta to maximize the line-to-line effective capacitance to one and a half times that of a single capacitor.

A filter at the dc side of a power converter is employed to reduce the voltage and current ripple. Depending on specific requirements, the filter consists of a capacitor connected in parallel with the converter terminals, an inductor connected in series, or both of these components. If a dc current I_C is drawn from a capacitor C during a time interval Δt, the voltage across the capacitor drops by

$$\Delta V_C = \frac{I_C}{C} \Delta t. \tag{3.9}$$

It means that the higher the filter capacitance, the more stable the voltage across it. Analogously, if a dc voltage V_L is applied to an inductor L over an interval Δt, the current in the inductor increases by

$$\Delta I_L = \frac{V_L}{L} \Delta t \qquad (3.10)$$

which implies a stabilizing impact of the inductance on the current.

Filters in ac circuits are used to block or shunt ripple currents but pass the fundamental currents. If the harmonic frequencies are much higher than the fundamental frequency, as in PWM converters, filter inductors can be connected in series with converter terminals, carrying the entire input or output current. Since the impedance, Z_L, of an inductor is given by

$$Z_L = 2\pi f L \qquad (3.11)$$

the inductor can serve as a *low-pass filter* for the current. In contrast, a capacitor, whose impedance, Z_C, is

$$Z_C = \frac{1}{2\pi f C} \qquad (3.12)$$

constitutes a *high-pass filter*. Consequently, as in dc filters, capacitors in ac filters are placed in parallel to converter terminals, shunting the high-frequency currents. To avoid a dangerous resonance overvoltage, care should be taken for the resonance frequency of an LC filter to be significantly higher than the supply frequency (60 Hz or 50 Hz).

The series–parallel placement of the inductive and capacitive filter components is not practical if the undesired ac current components have low frequencies, such as those in input currents of phase-controlled converters. In that case, since the dominant ripple frequency is not much higher than the fundamental frequency, the fundamental current would also be suppressed by the filter. Therefore, as described in Sections 4.1.2 and 4.2.1, *resonant filters* are placed between ac lines supplying diode- and SCR-based rectifiers. Each filter is tuned to a specific harmonic frequency, usually the fifth, seventh, and twelfth multiples of fundamental frequency, requiring a total of at least nine inductors and nine capacitors.

Special filters are needed to protect the power system from the conducted *electromagnetic interference* (EMI). If high-frequency currents, particularly those of radio-level frequencies, were allowed to spread in the system, they would severely pollute the electromagnetic environment and disturb the operation of communication systems. *EMI filters*, also called *radio-frequency filters*, are particularly indispensable in power converters characterized by high switching frequencies. High-order EMI filters, involving several resistors, inductors, and capacitors, are common.

Power electronic converters are also sources of the *radiated EMI*, that is, electromagnetic waves generated due to high *di/dt* rates resulting from the switching mode of operation. In practice, two factors help to minimize the environmental impact of this phenomenon: snubbers, which reduce the rate of change of switched currents and voltages, the metal cabinets in which power electronic converter are usually enclosed, and which act as effective electromagnetic shields.

3.6   COOLING

Losses in power electronic converters produce heat that must be transferred away and dissipated to the surroundings. The major losses occur in semiconductor power switches, while their small volume limits their thermal capacity. The high temperatures of the semiconductor structures cause degradation of electrical characteristics of switches, such as the maximum blocked voltage or the turn-off time. Serious overheating can lead to destruction of a semiconductor device in a short time. To maintain a safe temperature, a power switch must be equipped with a heat sink (radiator) and be subjected to at least *natural convection* cooling. The heat generated in the switch is transferred via the heat sink to the ambient air, which then tends to move upward and away from the switch.

Forced air cooling, very common in practice, is more effective than natural cooling. The cooling air is propelled by a fan, typically placed at the bottom of the cabinet that houses the power electronic converter. Slotted openings at the top part of the cabinet allow the heated air to escape to the surroundings. The energy consumption of the low-power fan does not tangibly affect the overall efficiency of the converter.

If the power density of a converter, that is, the rated power-to-weight ratio, is very high, *liquid cooling* may be needed. Water or oil can be used as the cooling medium. The fluid is forced through extended hollow copper or aluminum bars to which semiconductor power switches are bolted. Thanks to the high specific heat of water, water cooling is very effective, although it poses a danger of corrosion. On the other hand, oil, whose specific heat is less than half that of water, has much better insulating and protecting properties.

Thermal equivalent circuits facilitate design and analysis of cooling systems. The concept of those circuits is based on the formal similarity between heat transfer and electrical phenomena. Thermal quantities and their electric equivalents are listed in

TABLE 3.1   Comparison of Thermal and Electrical Quantities

Thermal Quantity	Electrical Quantity
Amount of heat (energy), Q (J)	Electric charge, Q (C)
Heat current (power), P (W)	Electric current, I (A)
Temperature, Θ (K)	Electric voltage, V (V)
Thermal resistance, R_Θ (K/W)	Electric resistance, R (Ω)
Thermal capacity, C_Θ (J/K)	Electric capacitance, C (F)
Thermal time constant, $\tau_\Theta = R_\Theta C_\Theta$(s)	Electrical time constant, $\tau = RC$ (s)

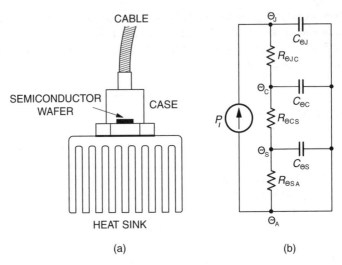

Figure 3.24 Power diode with a heat sink: (a) physical arrangement, (b) thermal equivalent circuit.

Table 3.1. As an example, a power diode with a heat sink and the corresponding thermal equivalent circuit are illustrated in Figure 3.24. The variables indicated on the circuit diagram are:

P_l	power loss in the diode (W)
Θ_J, Θ_C, Θ_S, and Θ_A	absolute temperatures of the pn junction, case, sink, and ambient air, respectively (K)
$R_{\Theta JC}$, $R_{\Theta CS}$, and $R_{\Theta SA}$	junction-to-case, case-to-sink, and sink-to-ambient air thermal resistances, respectively (K/W)
$C_{\Theta J}$, $C_{\Theta J}$, and $C_{\Theta J}$	junction, case, and sink thermal capacities, respectively (J/K).

Values of thermal resistances are listed in data sheets of semiconductor power switches and heat sinks, and the thermal capacity of a component can be determined from its specific heat and mass. Thermal capacities are used for calculation of transient temperatures, while knowledge of thermal resistances is sufficient for the computation of steady-state temperatures. The goal of cooling system design is to ensure that specific maximum temperatures allowed for the converter switches are not exceeded.

3.7   CONTROL

To function efficiently and safely, a power electronic converter must be properly controlled, that is, the process of power conversion must be accompanied by concurrent information processing. Various control technologies have been employed over the last few decades, starting with analog electronic circuits with discrete components, and progressing to contemporary integrated microelectronic digital systems. It is worth mentioning that simple analog control is still employed in many low-power dc-to-dc converters, especially those with very high switching frequencies.

In practical applications, a power electronic converter usually constitutes a part of a larger engineering system, such as an adjustable-speed drive or an active power filter. The control system of the converter is then subordinated to a master controller: for example, that of the speed of a motor fed from the converter. The main task of the converter control system is to generate signals for semiconductor power switches that result in the fundamental output voltage or current desired. Other "housekeeping chores" are also performed, such as control of electromechanical circuit breakers connecting the converter with the supply system. Often, especially in expensive, high-power converters, the control system also monitors operating conditions. When a failure is detected, the system turns the converter off and displays the diagnosis. Generally, the system draws information from the human operator, sensors, and master controllers and converts it into switching signals for the converter switches and external circuit breakers. The control system also supplies information about operation of the converter back to the operator, through displays, indicators, or recorders.

Microcontrollers, *digital signal processors* (DSPs), and *field-programmable logic arrays* (FPGAs) are now firmly established as the tools of choice for control of modern power electronic converters. The awesome and continuously increasing computational power of those devices allows a single processor to control an entire power conversion system, which may include several converters. Prices of digital processors are falling, so that they are economical even in low-cost applications.

Microprocessors employed as microcontrollers come from many manufacturers and vary widely. As the name indicates, microcontrollers are designed for control applications. A microcontroller has a central processing unit (CPU) that executes programs stored in a read-only memory (ROM), a random-access memory (RAM) for data storage, and input and output devices for communication with the outside world. Typically, a microcontroller has a smaller instruction set than that of a comparable microcomputer, but it may have certain additional functional blocks. The number of data and instruction bits (typically, 8, 16, 24, or 32), as well as the memory size, depends on the complexity of tasks required. Microcontrollers are low-power devices, often ruggedized in some way to withstand harsh operating conditions such as high temperatures or vibrations.

The differences between DSPs and microcontrollers are more functional than structural, as both have data-processing units, memories, and input/output circuitry. However, CPUs of DSPs are capable of extremely fast execution of a small set of simple instructions. The most advanced DSPs utilize floating-point arithmetic and can be programmed with higher-level languages instead of the cumbersome assembly language. Hardware multipliers and shift registers dramatically reduce the number of steps required to perform such algebraic operations as multiplication, division, or square-root calculation. Those can now be performed within a fraction of a microsecond with 32-bit accuracy.

In power electronic systems, *DSP controllers* are usually employed. Similar to microcontrollers, in addition to the basic components, they have specialized functional blocks, such as analog-to-digital and digital-to-analog converters, embedded timers, or generators of PWM switching signals. Modern microcontrollers and DSPs can be quite complex, due to the tendency to increase their speed and functionality by placing more memory and peripheral circuitry in a single chip.

In recent years, the distinction between microcontrollers and DSP controllers has blurred. For example, Freescale Semiconductor (spun from Motorola) advertises its *digital signal controllers* (DSCs) as single-chip devices combining the processing power of DSPs with the functionality and user friendliness of microcontrollers.

FPGAs are programmable digital logic chips which contain thousands of small logic blocks with flip-flops (memory elements) and programmable interconnects. After defining and compiling the logic function to be implemented, the function can be downloaded into the FPGA in the form of a binary file. If another logic function is needed, the FPGA can easily be reprogrammed. Thus, an FPGA can be considered a virtual breadboard, which requires no component changes and resoldering and which greatly facilitates fast prototyping. In general, FPGAs are somewhat inferior to application-specific integrated circuits (ASICs), but they are less expensive and more convenient to use.

As an example of a digital control system, a block diagram of an adjustable-speed ac drive governed by a digital control system based on a DSP controller is illustrated in Figure 3.25. The adjustable-frequency and adjustable-magnitude currents for the three-phase induction motor are produced by an inverter, whose three phases are controlled independently by current regulators CR_A through CR_C. Each current regulator receives an analog reference signal from the controller, compares it with the analog current signal obtained from current sensors, and generates appropriate switching signals for the inverter. The control system reacts to the speed control signal from

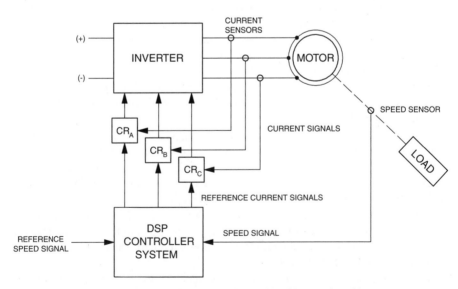

Figure 3.25   Block diagram of an adjustable-speed ac drive.

a digital speed sensor. As an alternative, the currents could be measured by digital sensors, with the current control incorporated in the operating algorithm of the DSP controller. In that case, the input signals to the controller would include those from the current sensors. The controller can be accessed by another processor: for example, one controlling the process of which the drive system shown is a part.

Concluding this brief overview of converter control systems, the growing tendency to integrate semiconductor power switches and control circuits in a common case deserves a note. Intelligent power modules, already mentioned in Section 2.6, and *high-voltage integrated circuits* (HVICs) are fabricated as hybrid assemblies or as fully integrated circuits (ICs). The latter technology constitutes a remarkable achievement because semiconductor power switches operate at high voltage and current levels, which can endanger the integrity and proper operation of low-voltage low-current ICs.

3.8 SUMMARY

A complete power electronic converter includes a number of supplementary components and systems, most of them not shown in regular power circuit diagrams. On command from the control system, drivers generate switching voltages and currents for gates (or bases) of converter switches. Overcurrent protection schemes, based on fast-melting fuses or dedicated circuitry, safeguard the switches from damage caused by short circuits in the converter or load. Snubbers relieve transient voltage and current stresses on switches at turn-on and turn-off, and reduce switching losses.

Power filters, based on inductors and capacitors, are placed at the input and output terminals of converters to improve the quality of power drawn from the power system or supplied to the load. Intermediate filters provide a match between two converters in a cascade arrangement. EMI filters are placed on the supply side of fast-switching converters to reduce the electromagnetic noise conducted.

Cooling systems remove heat from switches to protect them from excessive temperatures. Overall operation of a power electronic converter is governed by a control system. Today's control systems are based on microcontrollers, DSP controllers, and FPGAs, all of which are capable of implementing sophisticated control algorithms. Modern power electronic converters are complex engineering systems which combine efficient energy conversion with advanced information processing. The team effort of engineers of various specialties is required for successful design of this high-quality apparatus.

LITERATURE

[1] Agrawal, J. P., *Power Electronic Systems*, Prentice Hall, Upper Saddle River, NJ, 2001, Chaps. 1 and 2.

[2] Buso, S., and Mattavelli, P., *Digital Control in Power Electronics*, Morgan & Claypool Publishers, Princeton, NJ, 2006 (electronic book).

[3] Habetler, T. G., and Harley, R. G., Power electronic converter and system control, *Proceedings of the IEEE*, vol. 89, no. 6, pp. 913–925, 2001.

[4] Kazmierkowski, M. P. (Ed.), *Control in Power Electronics*, Academic Press, San Diego, CA, 2002.

[5] Rashid, M. H., *Power Electronics Handbook*, 2nd ed., Academic Press, San Diego, CA, 2007, Chaps. 20, 37, 40, and 42.

[6] Rashid, M. H., *Power Electronics: Circuits, Devices, and Applications*, 3rd ed., Prentice Hall, Upper Saddle River, NJ, 2003, Chaps. 17 and 18.

4 AC-to-DC Converters

Three-phase ac-to-dc converters are presented in this chapter. Beginning with un-controlled diode-based rectifiers and progressing to phase-controlled rectifiers, dual converters, and PWM rectifiers, circuit topologies, control principles, and operating characteristics are described. Device selection for ac-to-dc converters is explained and practical applications of rectifiers are outlined.

4.1 DIODE RECTIFIERS

Single-phase diode rectifiers were presented in Chapter 1. Although widely employed in low-power electronic circuits, they are not feasible for ac-to-dc conversion at the medium and high power levels. The output voltage of those rectifiers is of poor quality, due to the low dc component and high ripple factor. Therefore, the realm of medium- and high-power electronics is dominated by three-phase six-pulse rectifiers. For completeness, a three-pulse diode rectifier, although highly impractical, is discussed briefly first.

4.1.1 Three-Pulse Diode Rectifier

The circuit diagram of a three-pulse (three-phase half-wave) diode rectifier is shown in Figure 4.1. Supplied from a three-phase four-wire ac power line, the rectifier consists of three power diodes, DA through DC. The load is connected between the common-cathode node of the diode set and the neutral, N, of the supply line.

It can easily be shown that at a given instant only the diode that is supplied with the highest line-to-neutral voltage is conducting the output current, i_o. A situation when the highest voltage is that of phase B is illustrated in Figure 4.2. As $v_{BN} > v_{AN}$ and $v_{BN} > v_{CN}$, it is diode DB that is conducting, while diodes DA and DC are reverse biased by line-to-line voltages v_{BA} and v_{BC}, respectively.

Each of the three line-to-neutral voltages is higher than the other two for one-third of the cycle of input voltage. Consequently, in the continuous conduction mode, each diode conducts the current within a 120°-wide angle interval. Voltage and current waveforms of the three-pulse rectifier with a resistive load (R load) are shown in Figure 4.3. The waveform pattern of the output voltage is repeated every 120°

Introduction to Modern Power Electronics, Second Edition, by Andrzej M. Trzynadlowski
Copyright © 2010 John Wiley & Sons, Inc.

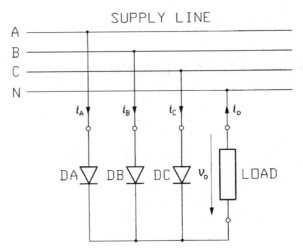

Figure 4.1 Three-pulse diode rectifier.

$(2\pi/3 \text{ rad})$. As in all considerations concerning electronic converters, idealized, loss-less semiconductor devices are assumed, so that the voltage drops across conducting switches are neglected. Since for $0 \leq \omega t \leq 2\pi/3$,

$$v_0 = v_{AN} = V_{LN,p} \; \sin\left(\omega t + \frac{\pi}{6}\right) \tag{4.1}$$

where $V_{LN,p}$ is the peak value of the supply line-to-neutral voltage, the dc component,

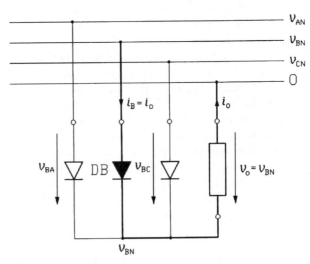

Figure 4.2 Current path and voltage distribution in a three-pulse diode rectifier.

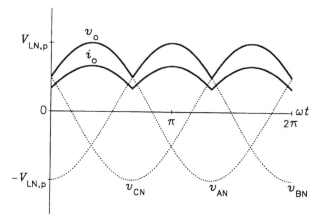

Figure 4.3 Waveforms of output voltage and current in a three-pulse diode rectifier (R load).

$V_{o,dc(C)}$, of the output voltage is

$$V_{o,dc(C)} = \frac{1}{(2/3)\pi} \int_0^{(2/3)\pi} V_{LN,p} \sin\left(\omega t + \frac{\pi}{6}\right) d\omega t = \frac{3}{2\pi} V_{LN,p} \left[\cos\left(\omega t + \frac{\pi}{6}\right)\right]_{(2/3)\pi}^0$$

$$= \frac{3\sqrt{3}}{2\pi} V_{LN,p} \approx 0.827 \, V_{LN,p}. \tag{4.2}$$

The "(C)" subscript indicates the continuous conduction mode.

The output current with the large dc component flows through the neutral wire. However, a dc component equal to one-third of the dc output currents also appears in currents drawn from wires A, B, and C. This property disqualifies the three-pulse rectifier from practical applications. In a power system supplying the rectifier, the dc current would cause saturation of transformer cores, resulting in distortion of voltage waveforms in the system. Another potential difficulty stems from the fact that all components of the power system are designed for ac sinusoidal currents of a specific frequency (60 Hz in the United States). The waveform of the phase A line current, i_A, is shown in Figure 4.4. Clearly, it is completely different from a sine wave. Such currents drawn by the rectifier would disturb the operation of protection systems.

4.1.2 Six-Pulse Diode Rectifier

A six-pulse (three-phase full-wave) diode rectifier, depicted in Figure 4.5 with an RLE load, is the most commonly used ac-to-dc power converter, producing a fixed dc voltage. The power circuit of the rectifier consists of six power diodes in a three-phase bridge configuration. Diodes DA, DB, and DC form a common-cathode group, and diodes DA′, DB′, and DC′ constitute a common-anode group. At a given instant, only one pair of diodes conducts the current, one in the common-cathode group and one in the common-anode group, the two diodes belonging to different phases of the bridge. Consequently, six combinations of conducting diodes are possible, each pair being

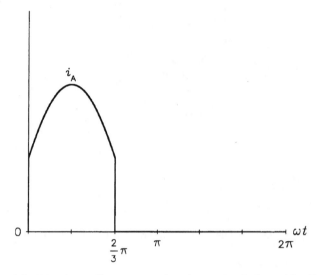

Figure 4.4   Waveform of input current in a three-pulse diode rectifier (R load).

active for one-sixth of the cycle of the supply voltage, that is, for a 60°-wide angle interval.

The conducting diode pair is that supplied with the highest line-to-line voltage. This is illustrated in Figure 4.6 for voltage v_{AC} being higher than the remaining five line-to-line voltages, v_{AB}, v_{BC}, v_{BA}, v_{CA}, and v_{CB}. Note that six line-to-line voltages are distinguished here, as, for instance, v_{AB} is considered separately from v_{BA}. Diodes DA and DB′ form a path for the output current. The other four diodes are subjected to voltages v_{AC} (diode DC), v_{CB} (diode DC′), and v_{AB} (diodes DA′ and DB). The phasor diagram of the ac voltages is shown in Figure 4.7 at the instant when

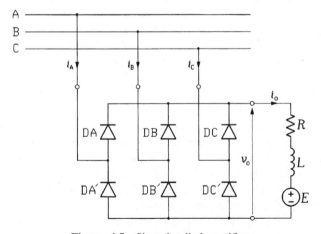

Figure 4.5   Six-pulse diode rectifier.

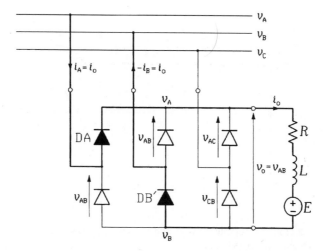

Figure 4.6   Current path and voltage distribution in a six-pulse diode rectifier.

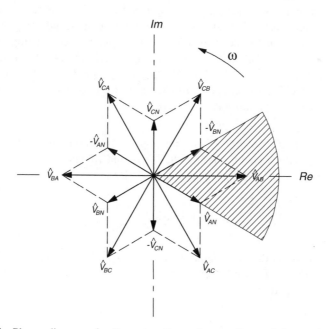

Figure 4.7   Phasor diagram of voltages in a three-phase ac line, and the maximum-voltage area.

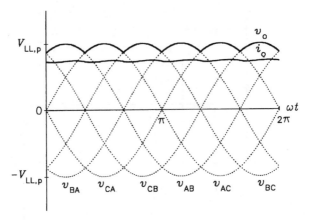

Figure 4.8 Waveforms of output voltage and current in a six-pulse diode rectifier in the continuous conduction mode (RLE load).

voltage v_{AB} (the real part of phasor $\hat{V}_{AB}$) is higher than the other line-to-line voltages, that is, when phasor $\hat{V}_{AB}$ is located in the shaded 60°-wide sector of the complex plane.

It can be seen that voltages v_{AC}, v_{CB}, and v_{AB} are positive (the respective phasors have positive real parts), imposing reverse bias on the nonconducting diodes. The phasor diagram also allows determination of the sequence of conducting diode pairs, which is DA and DB′, DA and DC′, DB and DC′, DB and DA′, DC and DA′, DC and DB′, and so on. Thus, each diode conducts the current for one-third of the cycle of supply voltage. The process of a diode taking over conduction of current from another diode is called *natural commutation*.

Output Voltage and Current   Under most operating conditions, the output current is continuous, as illustrated in Figure 4.8. The output voltage within the zero-to-$\pi/3$ interval equals the line-to-line voltage v_{AB} given by

$$v_{AB} = V_{LL,p} \sin\left(\omega t + \frac{\pi}{3}\right) \tag{4.3}$$

where $V_{LL,p}$ denotes the peak value of the supply line-to-line voltage. The output voltage waveform repeats itself every $\pi/3$ radians, so the average output voltage, $V_{o,dc(C)}$, can be found as

$$
\begin{aligned}
V_{o,dc(C)} &= \frac{1}{(\pi/3)} \int_0^{\pi/3} V_{LL,p} \sin\left(\omega t + \frac{\pi}{3}\right) d\omega t \\
&= \frac{3}{\pi} V_{LL,p} \left[\cos\left(\omega t + \frac{\pi}{3}\right)\right]_{\pi/3}^0 = \frac{3}{\pi} V_{LL,p} \\
&\approx 0.955 V_{LL,p}.
\end{aligned}
\tag{4.4}
$$

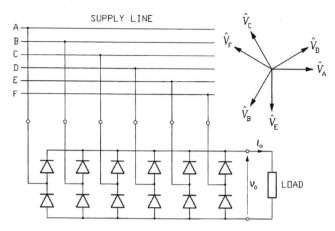

Figure 4.9   Twelve-pulse diode rectifier.

The output voltage is not only of higher quality than that of the three-pulse rectifier, but also has a much bigger dc component. Since the line-to-line voltages are rectified here, a dc output voltage is twice as high as that of a three-pulse rectifier supplied from the same ac line with line-to-neutral voltages.

Comparing the relations between the average value of output voltage and the peak value of input voltage for the two-, three-, and six-pulse diode rectifiers, a general formula for a p-pulse diode rectifier in the continuous conduction mode can be obtained as

$$V_{o,\text{dc(unc)}} = \frac{p}{\pi} V_{i,p} \sin \frac{\pi}{p} \qquad p = 2, 3, \ldots \qquad (4.5)$$

The integer value of p is not limited to 2, 3, or 6. Indeed, both the half-wave and full-wave rectifier topologies described can be expanded by increasing the number of phases of the ac supply source and the corresponding number of diode pairs. As an example, a full-wave 12-pulse rectifier is shown in Figure 4.9. The six-phase supply is obtained from a three-phase line using a transformer with six secondary windings. Clearly, if p increases, $V_{o,\text{dc(unc)}} / V_{i,p}$ approaches unity and the voltage ripple factor approaches zero. Such "superrectifiers" can be found in electrochemical plants, but otherwise, the field of ac-to-dc power conversion is dominated by six-pulse bridge rectifiers because of the common availability of the three-phase electrical power.

An equation for the output current waveform, $i_{o(C)}(\omega t)$, in the continuous conduction mode for the first cycle of this current, that is, for the zero-to-$\pi/3$ interval of ωt, can be derived using the method explained in Chapter 1. For generality, an RLE load is considered. The circuit equation is

$$V_{\text{AB}} = L \frac{di_o}{dt} + Ri_o + E \qquad (4.6)$$

where L, R, and E denote the load inductance, resistance, and EMF, respectively.

The forced component, $i_{o(F)}(\omega t)$, generated by voltage v_{AB} given by Eq. (4.3), is

$$i_{o(F)}(\omega t) = \frac{V_{LL,p}}{Z} \sin\left(\omega t + \frac{\pi}{3} - \varphi\right) - \frac{E}{R} \tag{4.7}$$

where Z is the load impedance, equal to

$$Z = \sqrt{R^2 + (\omega L)^2} \tag{4.8}$$

and φ denotes the load angle, given by

$$\varphi = \tan^{-1}\frac{\omega L}{R} = \cos^{-1}\frac{R}{Z}. \tag{4.9}$$

Since

$$\frac{E}{R} = \frac{V_{LL,p}}{Z}\frac{E}{V_{LL,p}}\frac{Z}{R} = \frac{V_{LL,p}}{Z}\frac{\varepsilon}{\cos\varphi} \tag{4.10}$$

where $\varepsilon \equiv E/V_{LL,p}$ is a load EMF coefficient, Eq. (4.7) can be rewritten as

$$i_{o(F)}(\omega t) = \frac{V_{LL,p}}{Z}\left[\sin\left(\omega t + \frac{\pi}{3} - \varphi\right) - \frac{\varepsilon}{\cos\varphi}\right]. \tag{4.11}$$

The natural component, $i_{o(N)}(\omega t)$, characteristic for a resistive–inductive load, is

$$i_{o(N)}(\omega t) = A_{(C)}e^{-(R/L)t} = A_{(C)}e^{-(R/\omega L)\omega t} = A_{(C)}e^{-(\omega t/\tan\varphi)} \tag{4.12}$$

where $A_{(C)}$ is a constant to be determined from the initial and final conditions. Consequently,

$$i_{o(C)}(\omega t) = i_{o(F)}(\omega t) + i_{o(N)}(\omega t)$$

$$= \frac{V_{LL,p}}{Z}\left[\sin\left(\omega t + \frac{\pi}{3} - \varphi\right) - \frac{\varepsilon}{\cos\varphi}\right] + A_{(C)}e^{\omega t/\tan\varphi}. \tag{4.13}$$

To find $A_{(C)}$, advantage can be taken from the fact that the initial value of the current, at $\omega t = 0$, equals the final value, at $\pi/3$. Thus,

$$i_{o(C)}(0) = \frac{V_{LL,p}}{Z}\left[\sin\left(\frac{\pi}{3} - \varphi\right) - \frac{\varepsilon}{\cos\varphi}\right] + A_{(C)} \tag{4.14}$$

equals

$$i_{o(C)}\left(\frac{\pi}{3}\right) = \frac{V_{LL,p}}{Z}\left[\sin\left(\frac{2\pi}{3} - \varphi\right) - \frac{\varepsilon}{\cos\varphi}\right] + A_{(C)}e^{-\pi/3\,\tan\varphi}. \tag{4.15}$$

Comparing right-hand sides of Eqs. (4.14) and (4.15) yields

$$A_{(C)} = \frac{V_{LL,p}}{Z} \frac{\sin \varphi}{1 - e^{-\pi/3 \tan \varphi}} \tag{4.16}$$

which, when substituted in Eq. (4.13), gives

$$i_{o(C)}(\omega t) = \frac{V_{LL,p}}{Z} \left[\sin \left(\omega t + \frac{\pi}{3} - \varphi \right) - \frac{\varepsilon}{\cos \varphi} + \frac{\sin \varphi}{1 - e^{-\pi/3 \tan \varphi}} e^{-\omega t / \tan \varphi} \right]. \tag{4.17}$$

The discontinuous conduction is possible only when the load EMF, E, exceeds the lowest instantaneous value of output voltage, v_o. Diodes of the rectifier become reverse biased and the load current cannot pass through them. Consequently, until the input voltage increases above E and the currents starts flowing, $v_o = E$. The lowest output voltage occurs at $\omega t = 0$ (see Figure 4.8) and, according to Eq. (4.3), it equals $\sqrt{3}/2 \, V_{LL,p}$. Therefore, the discontinuous conduction mode of operation of the rectifier may happen only when $\varepsilon > \sqrt{3}/2$.

This conclusion can be confirmed analytically. At the borderline between the continuous and discontinuous conduction modes, the minimal value of the output current, occurring at the ends of the zero-to-$\pi/3$ interval of ωt considered reaches zero. Consequently, if the current is to be continuous, it must be greater than zero at $\omega t = 0$, which, according to Eq. (4.17), is tantamount to the condition

$$\sin \left(\frac{\pi}{3} - \varphi \right) - \frac{\varepsilon}{\cos \varphi} + \frac{\sin \varphi}{1 - e^{-\pi/3 \tan \varphi}} > 0 \tag{4.18}$$

that is,

$$\varepsilon < \left[\sin \left(\frac{\pi}{3} - \varphi \right) + \frac{\sin \varphi}{1 - e^{-\pi/3 \tan \varphi}} \right] \cos \varphi. \tag{4.19}$$

The last relation is illustrated in Figure 4.10. Indeed, the minimum value of the load EMF coefficient, ε, for the discontinuous conduction is seen to be $\sqrt{3}/2 \approx 0.866$. Above $\varepsilon = 0.955$, no conduction is possible, because of the permanent reverse bias of the diodes by the load EMF.

Derivation of the expression for the discontinuous output current waveform, $i_{o(D)}(\omega t)$, proceeds similarly to that for the continuous current. The current starts flowing when the waveform of input voltage crosses over the level of load EMF, that is, at

$$\omega t = \sin^{-1} \varepsilon - \frac{\pi}{3} = \alpha_C. \tag{4.20}$$

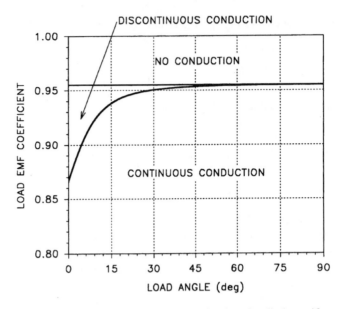

Figure 4.10 Conduction mode areas of a six-pulse diode rectifier.

Angle α_C will subsequently be referred to as a *crossover angle*. The general equation for the current waveform, similar to Eq. (4.13), is

$$i_{o(D)}(\omega t) = \frac{V_{LL,p}}{Z}\left[\sin\left(\omega t + \frac{\pi}{3} - \varphi\right) - \frac{\varepsilon}{\cos\varphi}\right] + A_{(D)}e^{-\omega t/\tan\varphi}. \tag{4.21}$$

Constant $A_{(D)}$ can be found from the initial condition

$$i_{o(D)}(\alpha_C) = 0 \tag{4.22}$$

which yields

$$A_D = -\frac{V_{LL,p}}{Z}\left[\sin\left(\omega t + \frac{\pi}{3} - \varphi\right) - \frac{\varepsilon}{\cos\varphi}\right]e^{-\omega t/\tan\varphi}. \tag{4.23}$$

This, when substituted in Eq. (4.21), gives

$$i_{o(D)}(\omega t) = \frac{V_{LL,p}}{Z}\left\{\sin\left(\omega t + \frac{\pi}{3} - \varphi\right) - \frac{\varepsilon}{\cos\varphi}\right.$$
$$\left. - \left[\sin\left(\alpha_C + \frac{\pi}{3} - \varphi\right) - \frac{\varepsilon}{\cos\varphi}\right]e^{-(\omega t - \alpha_C)/\tan\varphi}\right\}. \tag{4.24}$$

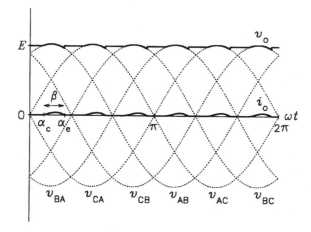

Figure 4.11 Waveforms of output voltage and current in a six-pulse diode rectifier in the discontinuous conduction mode (RLE load).

Output voltage and current waveforms in a case of discontinuous conduction are shown in Figure 4.11. A current pulse begins at the crossover angle, α_C, and ends at the *extinction angle*, α_e. The length, β, of the pulse in the angle domain, called a *conduction angle*, is given by

$$\beta = \alpha_e - \alpha_C. \tag{4.25}$$

The extinction angle, which depends on the load-EMF coefficient, ε, and load angle, φ, can best be calculated using a "brute force" approach, that is, computing the output current according to Eq. (4.24) for sequential values of ωt, starting at α_C and proceeding until the current crosses zero in the negative direction. Clearly, this is valid only when the conditions for discontinuous conduction are satisfied (see Figure 4.10).

All the equations derived for the RLE load can easily be adapted for simpler loads by substituting $\varphi = 0$ for an RE load, $\varepsilon = 0$ for an RL load, and $\varphi = 0$ and $\varepsilon = 0$ for an R load. An RE load, for example, may represent a battery, RL load an electromagnet, and R load an electrochemical process. The RLE load is usually employed to model a dc motor.

The dc component, $V_{o,\mathrm{dc(D)}}$, of output voltage in the discontinuous conduction mode of operation of the rectifier can be found by averaging the output voltage waveform, $v_o(\omega t)$. As

$$v_o(\omega t) = \begin{cases} v_{\mathrm{AB}}(\omega t) & \text{for } \alpha_C < \omega t < \alpha_e \\ E & \text{otherwise} \end{cases} \tag{4.26}$$

then

$$V_{o,\text{dc(D)}} = \frac{1}{\pi 3} \left[\int_0^{\alpha_C} E\, d\omega t + \int_{\alpha_C}^{\alpha_e} V_{\text{LL},p} \sin\left(\omega t + \frac{\pi}{3}\right) d\omega t + \int_{\alpha_e}^{\pi/3} E\, d\omega t \right]$$

$$= \frac{3}{\pi} V_{\text{LL},p} \left[2\sin\left(\alpha_C + \frac{\beta}{2} + \frac{\pi}{3}\right) \sin\frac{\beta}{2} + \varepsilon\left(\frac{\pi}{3} - \beta\right) \right]. \tag{4.27}$$

As seen in Figure 4.11, the complex expression for $V_{o,\text{dc(D)}}$ notwithstanding, the dc output voltage of the rectifier is approximately equal to the peak value of the supply line-to-line voltage. In both conduction modes, the average output current, $I_{o,\text{dc}}$, is given by the simple expression

$$I_{o,\text{dc}} = \frac{V_{o,\text{dc}} - E}{R} \tag{4.28}$$

because no dc voltage can appear across the load inductance.

The superiority of the continuous conduction mode over the discontinuous conduction mode is evident. The average output voltage does not depend on the load, and the ripple factors of output voltage and current are low. Also, the current drawn by the rectifier from the ac supply system is of higher quality with respect to both the harmonic content and the input power factor than that under discontinuous conduction conditions.

Input Current and Power Factor   Limiting the subsequent considerations to the continuous conduction mode and assuming ideal output current, $i_o = I_{o,\text{dc}}$, the phase A line current, i_A, is given by

$$i_A = \begin{cases} I_{o,\text{dc}} & \text{for } 0 < \omega t < \frac{2}{3}\pi \\ -I_{o,\text{dc}} & \text{for } \pi < \omega t < \frac{5}{3}\pi \\ 0 & \text{otherwise} \end{cases} \tag{4.29}$$

as depicted in Figure 4.12, which also shows the fundamental line current, $i_{A,1}$. The rms value, I_A, of i_A is

$$I_A = \sqrt{\frac{1}{2\pi} \left[\int_0^{(2/3)\pi} I_{o,\text{dc}}^2\, d\omega t + \int_\pi^{(5/3)\pi} I_{o,\text{dc}}^2\, d\omega t \right]} = \sqrt{\frac{2}{3}} I_{o,\text{dc}} = 0.82 I_{o,\text{dc}} \tag{4.30}$$

while the rms fundamental current, $I_{A,1}$, can be calculated as

$$I_{A,1} = \frac{1}{\sqrt{2}} \sqrt{I_{A,1c}^2 + I_{A,1s}^2} \tag{4.31}$$

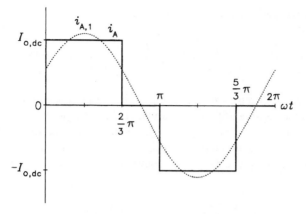

Figure 4.12   Waveform of input current in a six-pulse diode rectifier (assuming an ideal dc output current).

where

$$I_{A,1c} = \frac{1}{\pi}\left[\int_0^{(2/3)\pi} I_{o,dc}\cos \omega t \, d\omega t + \int_\pi^{(5/3)\pi} -I_{o,dc}\cos \omega t \, d\omega t\right] = \frac{3}{\pi}I_{o,dc} \qquad (4.32)$$

and

$$I_{A,1s} = \frac{1}{\pi}\left[\int_0^{(2/3)\pi} I_{o,dc}\sin \omega t \, d\omega t + \int_\pi^{(5/3)\pi} -I_{o,dc}\sin \omega t \, d\omega t\right] = \frac{\sqrt{3}}{\pi}I_{o,dc}.$$

$$(4.33)$$

Thus,

$$I_{A,1} = \frac{1}{\sqrt{2}}\sqrt{\left(\frac{3}{\pi}I_{o,dc}\right)^2 + \left(\frac{\sqrt{3}}{\pi}I_{o,dc}\right)^2} = \frac{\sqrt{6}}{\pi}I_{o,dc} \approx 0.78 I_{o,dc}. \qquad (4.34)$$

The harmonic content, $I_{A,h}$, of the line current is

$$I_{A,h} = \sqrt{I_A^2 - I_{A,1}^2} = \sqrt{\left(\sqrt{\frac{2}{3}}I_{o,dc}\right)^2 \left(\frac{\sqrt{6}}{\pi}I_{o,dc}\right)^2} = \sqrt{\frac{2}{3} - \frac{6}{\pi^2}}I_{o,dc} \approx 0.24 I_{o,dc}$$

$$(4.35)$$

and the total harmonic distortion, THD, is

$$\text{THD} = \frac{I_{A,h}}{I_{A,1}} = \frac{\sqrt{(2/3) - (6/\pi^2)I_{o,dc}}}{(\sqrt{6}/\pi)I_{o,dc}} = \sqrt{\frac{\pi^2}{9} - 1} \approx 0.31 \qquad (4.36)$$

which is quite high, even though the rectifier operates in the continuous conduction mode. Recall the less-than-5% desired value of THD mentioned in Chapter 1.

Both the conversion efficiency, η_c, and input power factor, PF, of a six-pulse diode rectifier in the continuous conduction mode are high. Here, because of the assumed ideal dc output current, the conversion efficiency is 100%. The power factor, determined as

$$\text{PF} = \frac{P_i}{S_i} = \frac{P_o}{S_i} = \frac{V_{o,dc}I_{o,dc}}{\sqrt{3}\,V_{LL}I_L} = \frac{(3/\pi)V_{LL,p}I_{o,dc}}{\sqrt{3}(V_{LL,p}/\sqrt{2})\sqrt{2/3}I_{o,dc}} = \frac{3}{\pi} \approx 0.955$$
$$(4.37)$$

is also close to unity. Comparing Figures 4.11 and 4.12, it can be seen that the fundamental input current, $i_{A,1}$, lags line-to-line voltage v_{AB} by 30°; that is, according to Figure 4.7, it is in phase with line-to-neutral voltage v_{AN}. If current i_A were sinusoidal, this zero phase shift would result in a unity input power factor. All the figures of merit deteriorate dramatically in the discontinuous conduction mode, due to the significant ac component of the output current.

For better understanding of the power relationships in the rectifier, the rms value, $I_{A,1}$, of the fundamental line current will now be determined from the power balance instead of the Fourier series expressions leading to Eq. (4.34). Assuming a balanced set of the input voltages, all three phases of the ac supply source contribute the same amount of real input power, equal to a third of the output power. Hence,

$$V_{AN}I_{A,1} = \frac{1}{3}V_{o,dc}I_{o,dc} \qquad (4.38)$$

where v_{AN} denotes the rms value of line-to-neutral voltage v_{AN}. As $V_{AN} = V_{LL,p}/\sqrt{6}$ and, according to Eq. (4.4), $V_{o,dc} = 3V_{LL,p}/\pi$, then

$$I_{A,1} = \frac{V_{o,dc}I_{o,dc}}{3V_{AN}} = \frac{(3/\pi)V_{LL,p}I_{o,dc}}{3(V_{LL,p}/\sqrt{6})} = \frac{\sqrt{6}}{\pi}I_{o,dc} \qquad (4.39)$$

which confirms the result of Eq. (4.34).

The nonsinusoidal, quasi-square-wave line currents drawn from the power system are rich in low-order harmonics, as illustrated in Figure 4.13, which shows the spectrum of current in Figure 4.12. Harmonic amplitudes are expressed in the per-unit format, the dc output current, $I_{o,dc}$, taken as the base current. The harmonic currents have many adverse effects on the supply system. In particular, they produce extra losses in rotating machines, transformers, and capacitors, interfere with communication systems, and may cause metering errors, malfunctions of control systems,

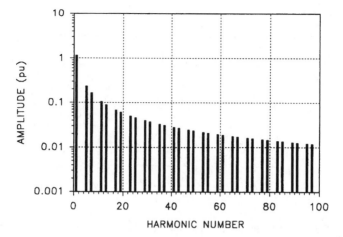

Figure 4.13 Harmonic spectrum of input current in a six-pulse diode rectifier (assuming ideal dc output current).

and excitation of resonances with capacitor banks in the system. The increasingly stringent standards of power quality in power systems enforce usage of input filters (line filters) to reduce the harmonic currents.

It must be pointed out that currents drawn by three-phase rectifiers are free from triple harmonics, that is, those with harmonic numbers being multiples of three. This property can, for example, be discerned in the spectrum in Figure 4.13. The sum of line currents i_A, i_B, and i_C is zero at all instants of time because of the lack of the fourth (neutral) wire. Since the fundamentals of these currents differ from each other by the 120° phase shift, the corresponding phase shift for any triple harmonic would be a multiple of 360°. It means that if these harmonic currents were present, they would all be in phase and their sum would not be zero, which would contradict Kirchhoff's current law. The currents in question are also void of even harmonics because of the half-wave symmetry of the waveforms. Therefore, the most prominent harmonics present are the fifth, seventh, eleventh, and thirteenth.

A filter for preventing currents of frequency $k\omega$ from propagating in the system consists of three inductors and capacitors connected between the three wires of the supply line. The line-to-line inductance and capacitance are denoted by L_f and C_f, respectively. Assuming lossless filter components, if

$$L_f C_f = \frac{1}{(k\omega)^2} \qquad (4.40)$$

the impedance of the filter for the kth harmonic current is zero and the current is shunted from the power system. A typical practical input filter, also called a *harmonic trap*, is shown in Figure 4.14. It is composed of two resonant LC filters, designated filter 1 and filter 2, for the fifth and seventh harmonics, respectively, and a damped resonant filter 3 that presents a low impedance over a wide frequency band,

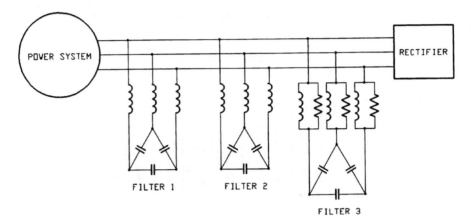

Figure 4.14   Input filter (harmonic trap) for a three-phase rectifier.

with the minimum around 12ω. The reactors and capacitors of each filter are connected in a way that maximizes their utilization. Specifically, in each filter the circuit between every two wires of the three-phase line constitutes a series connection of two inductors and one capacitor shunted by two capacitors in series. Consequently, each inductor needs to have the inductance of only half of the required resonant inductance, L_f, and each capacitor must have the capacitance of two-thirds of the resonant capacitance C_f.

4.2   PHASE-CONTROLLED RECTIFIERS

Phase-controlled rectifiers, which allow adjustment of the average output voltage, $V_{o,\mathrm{dc}}$, have the same topologies as those of diode rectifiers, with the diodes replaced by SCRs or fully controlled semiconductor power switches. The dc component of the output voltage is adjustable, with the maximum available value equal to that of the corresponding diode rectifier and given by Eq. (4.5).

Most practical controlled rectifiers are of the three-phase six-pulse SCR-based type. As semicontrolled switches, the SCRs are impractical for pulse width modulation, but they are perfectly suited for phase control. If a forward-biased SCR is fired, it starts conducting a current. Depending on the conduction mode, the SCR ceases to conduct when either the current drops to zero or it is taken over by another SCR. The more the firing instant is delayed with respect to the instant when the SCR became forward biased, the lower average output voltage is obtained.

Replacing SCRs with fully controlled fast switches such as IGBTs or BJTs allows control of the output voltage by pulse width modulation. The PWM rectifiers are characterized by higher quality input and output currents than those in phase-controlled rectifiers.

With the load EMF, E, of negative polarity, a negative dc output voltage can be produced. The output current cannot be negative because of the unidirectionality of semiconductor switches. Thus, such a situation represents a negative power flow, from the load EMF to the ac supply source. The voltage reversal property, characteristic of controlled rectifiers only, can be augmented by a current reversal feature using two rectifiers in antiparallel configuration. Such dual converters, capable of producing positive and negative dc output voltage and current, are particularly useful in the control of dc motors.

4.2.1   Phase-Controlled Six-Pulse Rectifier

The power circuit of a phase-controlled six-pulse rectifier based on SCRs is shown in Figure 4.15. SCRs are often called thyristors; hence the SCRs of the rectifier in question are designated TA through TC′. The switching sequence of the SCR pairs is the same as that in the diode rectifier described in Section 4.1.2. However, the turn-on of an SCR can be delayed with respect to the natural commutation instant, provided that the SCR is forward biased. As already explained in Section 1.4, that delay in the angle domain is called a *firing angle* and denoted by α_f.

In the considerations presented here, these are SCRs TA and TB′, which are assumed to be turned on in the first cycle of the output voltage, that is, when $0 \leq \omega t < \pi/3$. Therefore, the firing angle for these SCRs is measured from $\omega t = 0$, with the SCRs in question being fired at α_f. Other SCR pairs are fired with appropriate delays that are multiples of 60°. Hence, for example, in the second cycle of the output voltage, SCRs TA and TC′ are fired at $\omega t = \alpha_f + \pi/3$.

Output Voltage and Current   The full sequence of gate pulses firing the individual SCRs is shown in Figure 4.16 for $\alpha_f = 45°$. The corresponding output voltage and current waveforms of the rectifier in the continuous conduction mode are illustrated in Figure 4.17. It can be seen that delaying the firing of SCRs has resulted in reduction of the average output voltage compared with that of a diode rectifier. The dc output

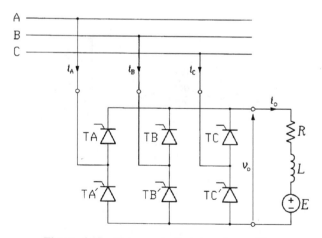

Figure 4.15   Phase-controlled six-pulse rectifier.

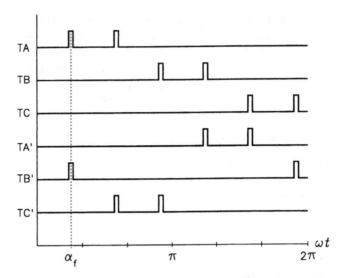

Figure 4.16   Firing pulses in a phase-controlled six-pulse rectifier.

voltage is given by

$$V_{o,\text{dc(C)}} = \frac{1}{\pi/3} \int_{\alpha_f}^{\alpha_f + \pi/3} V_{\text{LL},p} \sin\left(\omega t + \frac{\pi}{3}\right) d\omega t = \frac{3}{\pi} V_{\text{LL},p} \cos \alpha_f. \qquad (4.41)$$

Equation (4.41) constitutes a special case of a general formula for the dc output voltage, $V_{o,\text{dc(cntr)}}$, of all multipulse ($p > 1$) controlled rectifiers in the continuous

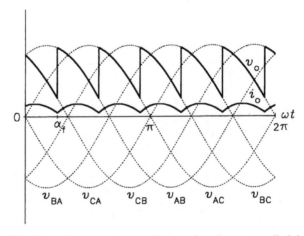

Figure 4.17   Waveforms of output voltage and current in a phase-controlled six-pulse rectifier in the continuous conduction mode ($\alpha_f = 45°$, RLE load).

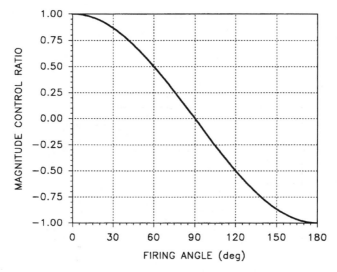

Figure 4.18   Control characteristic of a phase-controlled six-pulse rectifier in the continuous conduction mode.

conduction mode, which is

$$V_{o,\text{dc(cntr)}} = V_{o,\text{dc(unc)}} \cos \alpha_f \tag{4.42}$$

where $V_{o,\text{dc(unc)}}$ is the dc output voltage of the corresponding diode rectifier, given by Eq. (4.5).

The voltage control characteristic expressed by Eq. (4.41) is shown in Figure 4.18. As mentioned before, the dc output voltage can be negative; that is, the power can flow from the load EMF to the ac supply source. For a rectifier to operate in this *inverter mode*, two conditions must be met: (1) there must be a negative load EMF and (2) the firing angle must be greater than 90°. Then, as illustrated in Figure 4.19, it is the load EMF that delivers the power, with the load current having the same polarity as the EMF. Waveforms of the output voltage and current in the inverter mode, with $\alpha_f = 105°$, are shown in Figure 4.20.

It must be stressed that because of the load EMF, which affects the bias of the SCRs, not all values of the firing angle are feasible. Considering, for instance, SCRs TA and TB′, which apply voltage v_{AB} to the output terminals of the rectifier, it is obvious that they cannot be fired when the load EMF, E, is equal to or greater than v_{AB}, that is, when $\alpha_f \leq \alpha_c$ or $\alpha_f \geq \pi/3 - \alpha_c$. Therefore, based on Eq. (4.20), feasible values of the firing angle are those that satisfy the condition

$$\sin^{-1} \varepsilon - \frac{1}{3}\pi < \alpha_f < \frac{2}{3}\pi - \sin^{-1} \varepsilon \tag{4.43}$$

which is illustrated in Figure 4.21.

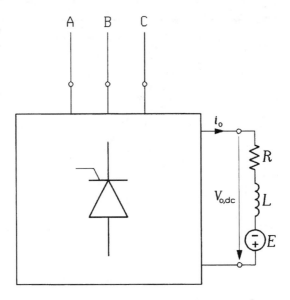

Figure 4.19 Rectifier in the inverter mode ($\alpha_f > 90°$).

An equation for the continuous output current, $i_{o(C)}(\omega t$, within the zero-to-$\pi/3$ interval can be derived similarly to that for the diode rectifier, the only difference being that $i_{o(C)}(\alpha_f) = i_{o(C)}(\alpha_f + \pi/3)$ instead of $i_{o(C)}(0) = i_{o(C)}(\pi/3)$. The current waveform is given by

$$i_{o(C)}(\omega t) = \frac{V_{LL,p}}{Z}\left[\sin\left(\omega t + \frac{\pi}{3} - \varphi\right) - \frac{\varepsilon}{\cos\varphi} + \frac{\sin(\varphi - \alpha_f)}{1 - e^{-\pi/3\tan\varphi}}e^{-(\omega t - \alpha_f/\tan\varphi)}\right]$$

(4.44)

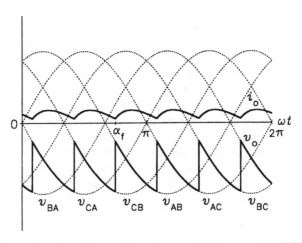

Figure 4.20 Waveforms of output voltage and current in a phase-controlled six-pulse rectifier in the continuous conduction mode ($\alpha_f = 105°$).

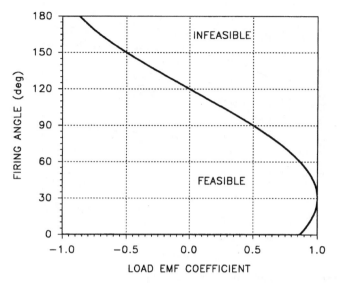

Figure 4.21   Area of feasible firing angles.

and the condition for the continuous conduction is $i_{o(C)}(\alpha_f) > 0$. By virtue of Eq. (4.44), this condition can be expressed as

$$\varepsilon < \left[\sin\left(\alpha_f + \frac{\pi}{3} - \varphi\right) + \frac{\sin(\varphi - \alpha_f)}{1 - e^{\pi/3 \tan \varphi}}\right] \cos \varphi. \tag{4.45}$$

Relation (4.45) is illustrated in Figure 4.22. Clearly, Figure 4.10 for the diode rectifier represents the $\alpha_f = 0$ case. Analogously, if a zero is substituted for α_f in Eqs. (4.44) and (4.45), the corresponding formulas (4.17) and (4.19) for the uncontrolled rectifier are obtained.

In the discontinuous conduction mode, a pulse of the output current begins at $\omega t = \alpha_f$, similarly to the start of current flow at $\omega t = \alpha_c$ in the diode rectifier. Therefore, expressions for the current waveform, $i_{o(D)}(\omega t)$, and average output voltage, $V_{o,dc(D)}$, can be found directly from Eqs. (4.24) and (4.27) by substituting α_f for α_c. This yields

$$i_{o(D)}(\omega t) = \frac{V_{LL,p}}{Z} \left\{ \sin\left(\omega t + \frac{\pi}{3} - \varphi\right) - \frac{\varepsilon}{\cos \varphi} \right.$$
$$\left. - \left[\sin\left(\alpha_f + \frac{\pi}{3} - \varphi\right) - \frac{\varepsilon}{\cos \varphi}\right] e^{-(\omega t - \alpha_f)/\tan \varphi} \right\} \tag{4.46}$$

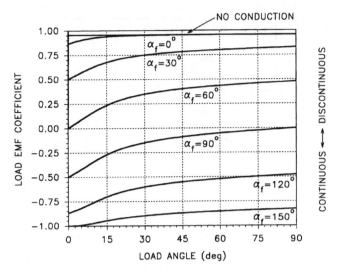

Figure 4.22 Conduction mode areas of a phase-controlled six-pulse rectifier.

and

$$V_{o,\text{dc(D)}} = \frac{3}{\pi} V_{\text{LL},p} \left[2 \sin\left(\alpha_f + \frac{\beta}{2} + \frac{\pi}{3}\right) \sin\frac{\beta}{2} + \varepsilon\left(\frac{\pi}{3} - \beta\right) \right] \qquad (4.47)$$

where the conduction angle, β, is

$$\beta = \alpha_c - \alpha_f. \qquad (4.48)$$

Note that in the continuous conduction mode, $\beta = \pi/3$, which when substituted in Eq. (4.47) yields Eq. (4.41). The discontinuous conduction mode with a positive and negative output voltage is illustrated in Figure 4.23.

Input Current and Power Factor The waveform of the input, line current, i_A, and its fundamental, $i_{A,1}$, in the continuous conduction mode of the rectifier are shown in Figure 4.24. As before, an ideal dc output current is assumed. The waveforms are similar to those in a diode rectifier (see Figure 4.12), but with one important difference, which is a phase shift equal to the firing angle. This shift results in a reduced input power factor. Indeed, replacing $V_{o,\text{dc}}$ in Eq. (4.37) with $V_{o,\text{dc}} \cos\alpha_f$, the power factor of the controlled rectifier is found to be

$$\text{PF} = \frac{V_{o,\text{dc}}}{V_{\text{LL},p}} = \frac{3}{\pi} \cos\alpha_f \approx 0.95 \cos\alpha_f. \qquad (4.49)$$

To improve the power factor and shunt the harmonic currents from the power system, input filters such as those shown in Figure 4.14 are recommended.

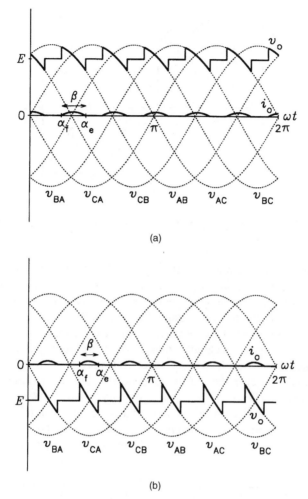

(a)

(b)

Figure 4.23   Waveforms of output voltage and current in a phase-controlled six-pulse rectifier in the discontinuous conduction mode: (a) rectifier operation ($\alpha_f = 45°$); (b) inverter operation ($\alpha_f = 135°$).

Impact of Source Inductance   For simplicity, an ideal ac source was assumed in the considerations presented. In reality, the power system feeding a rectifier introduces a certain amount of resistance and inductance on the supply side. These are mainly resistances and inductances of the system transformers and power lines. In fact, from the utilities' point of view, the source inductance is desirable, as it reduces the high-frequency current harmonics and increases the short-circuit impedance.

For the analysis of a six-pulse phase-controlled rectifier supplied through an inductance, it is convenient to employ the network shown in Figure 4.25. The actual full-wave bridge topology of the rectifier has been replaced here with an equivalent

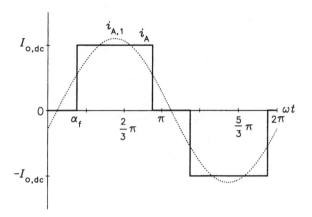

Figure 4.24 Waveform of input current in a phase-controlled six-pulse rectifier (ideal dc output current).

six-pulse half-wave configuration, with the actual line-to-line voltages appearing as phase voltages. The total source resistance and inductance in each path of the actual supply current are represented by lumped parameters R_s and L_s, respectively.

The input current produces a voltage drop across the source resistance, which reduces the output voltage of the rectifier. In practice, this effect is insignificant and comparable with the usually negligible impact of voltage drops across the conducting power switches. Therefore, in subsequent considerations, a zero source resistance will be assumed. However, the source inductance, which prevents rapid changes of the supply currents when one SCR ceases to conduct and another SCR takes over conduction of the output current, strongly affects the operation of a rectifier. This process is called *line-supported commutation*, which is initiated by firing an SCR and results in the turn-off of another SCR. Clearly, this type of commutation happens in

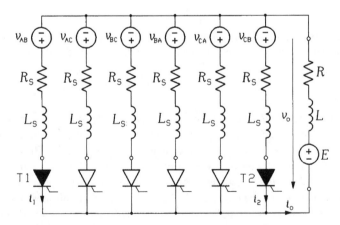

Figure 4.25 Equivalent circuit of a phase-controlled six-pulse rectifier supplied from a practical dc voltage source.

the continuous conduction mode only, while in the discontinuous conduction mode, SCRs cease to conduct, without any external action, when their currents drop to zero.

As illustrated in Figure 4.24, when no source inductance exists, conduction of the input current by an SCR begins and ends rapidly. However, rapid current changes are impossible in inductors, and the commutation in rectifiers fed from practical ac sources is not instantaneous. As a result, during the commutation, the incoming and outgoing SCRs share the current. Figure 4.25 pertains to a situation when SCR T1 is gradually taking over the output current, i_o, from SCR T2. Since T1 is connected to the source of the line-to-line voltage v_{AB} and T2 to the source of v_{CB}, this case represents the commutation between SCRs TA and TC in the actual bridge rectifier. It can be seen in Figures 4.17 and 4.20 that this happens at $\omega t = \alpha_f$, when TA is fired, causing extinction of TC.

Assuming an ideal dc output current, $i_o = I_{o,dc}$, the circuit in Figure 4.25 is described by

$$v_o = v_{AB} - L_s \frac{di_1}{dt} \tag{4.50}$$

$$v_o = v_{CB} - L_s \frac{di_2}{dt} \tag{4.51}$$

and

$$i_o = i_1 + i_2 = I_{o,dc} \tag{4.52}$$

where i_1 and i_2 denote currents conducted by SCRs T1 and T2, respectively, voltage v_{AB} is given by Eq. (4.2), and v_{CB} by

$$v_{CB} = V_{LL,p} \sin\left(\omega t + \frac{2}{3}\pi\right). \tag{4.53}$$

Adding Eqs. (4.50) and (4.51) side by side gives

$$2v_o = v_{AB} + v_{CB} - L_s \left(\frac{di_1}{dt} + \frac{di_2}{dt}\right) = v_{AB} + v_{CB} \tag{4.54}$$

because

$$\frac{di_1}{dt} + \frac{di_2}{dt} = \frac{d}{dt}(i_1 + i_2) = \frac{d}{dt} I_{o,dc} = 0. \tag{4.55}$$

Consequently,

$$v_o = \frac{v_{AB} + v_{CB}}{2} = \frac{1}{2}\left[V_{LL,p} \sin\left(\omega t + \frac{1}{3}\pi\right) + V_{LL,p} \sin\left(\omega t + \frac{2}{3}\pi\right)\right]$$

$$= \frac{\sqrt{3}}{2} V_{LL,p} \cos \omega t. \tag{4.56}$$

To derive equations for currents $i_1(\omega t)$ and $i_2(\omega t)$ during the commutation interval, Eqs. (4.50) and (4.51) are subtracted side by side, yielding

$$
\frac{di_1}{dt} = -\frac{di_2}{dt} = \frac{1}{2L_s}(v_{AB} - v_{CB})
$$

$$
= \frac{1}{2L_s}\left[V_{LL,p}\sin\left(\omega t + \frac{1}{3}\pi\right) - V_{LL,p}\sin\left(\omega t + \frac{2}{3}\pi\right)\right]
$$

$$
= \frac{V_{LL,p}}{2L_s}\sin \omega t. \tag{4.57}
$$

Now, $i_1(\omega t)$ can be calculated as

$$
i_1(\omega t) = \int \frac{V_{LL,p}}{2L_s}\sin \omega t\, dt + A_1 = \frac{V_{LL,p}}{2X_s}\cos \omega t + A_1 \tag{4.58}
$$

where X_s is the source reactance, equal to ωL_s, and A_1 is an integration constant. Since $i_1(\alpha_f) = 0$, then

$$
A_1 = \frac{V_{LL,p}}{2X_s}\cos \alpha_f \tag{4.59}
$$

and

$$
i_1(\omega t) = \frac{V_{LL,p}}{2X_s}(\cos \alpha_f - \cos \omega t) \tag{4.60}
$$

while

$$
i_2(\omega t) = I_{o,dc} - i_1(\omega t) = I_{o,dc} - \frac{V_{LL,p}}{2X_s}(\cos \alpha_f - \cos \omega t). \tag{4.61}
$$

The commutation process is illustrated in Figure 4.26. Beginning at the firing angle, α_f, current i_1 gradually increases and current i_2 decreases, reaching $I_{o,dc}$ and zero, respectively, at $\omega t = \alpha_f + \mu$. Angle μ, called an *overlap angle* or *commutation angle*, represents the length of the commutation interval in the angle domain. Its value can be determined using Eq. (4.60) and taking into account that $i_1(\alpha_f + \mu) = I_{o,dc}$. Thus,

$$
\frac{V_{LL,p}}{2X_s}\left[\cos \alpha_f - \cos(\alpha_f + \mu)\right] = I_{o,dc} \tag{4.62}
$$

and

$$
\mu = \left|\cos^{-1}\left(\cos \alpha_f - 2\frac{X_s I_{o,dc}}{V_{LL,p}}\right) - \alpha_f\right|. \tag{4.63}
$$

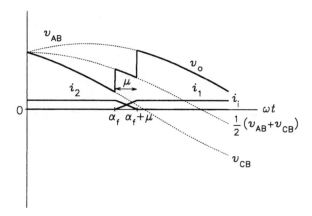

Figure 4.26 Waveforms of voltage and current in a phase-controlled six-pulse rectifier during commutation.

Waveforms of the output voltage and actual, nonideal current of a six-pulse phase-controlled rectifier with a nonzero load inductance are shown in Figure 4.27 for $V_{o,\text{dc}} > 0$ (rectifier mode) and for $V_{o,\text{dc}} < 0$ (inverter mode). The operating conditions of the rectifier are the same as in Figures 4.17 and 4.20. It can be seen that the noninstantaneous commutation has significantly affected both the output voltage and current.

The output voltage, which during commutation is the arithmetical mean of the line-to-line voltages involved, has its dc component reduced. The difference, $\Delta V_{o,\text{dc}}$, between the dc output voltage with $X_s = 0$ and with $X_s > 0$ can be calculated by spreading the area between the respective waveforms over the $\pi/3$ interval of ωt. Using partial results of Eqs. (4.57) and (4.62), an expression for $\Delta V_{o,\text{dc}}$ is obtained as

$$\Delta V_{o,\text{dc}} = \frac{3}{\pi} \int_{\alpha_f}^{\alpha_f+\mu} v_{\text{AB}} - \frac{v_{\text{AB}} + v_{\text{CB}}}{2}\, d\omega t = \frac{3}{\pi} \int_{\alpha_f}^{\alpha_f+\mu} \frac{v_{\text{AB}} - v_{\text{CB}}}{2}\, d\omega t$$

$$= \frac{3}{2\pi} \int_{\alpha_f}^{\alpha_f+\mu} V_{\text{LL},p} \sin \omega t\, d\omega t$$

$$= \frac{3}{2\pi} V_{\text{LL},p}[\cos \alpha_f - \cos(\alpha_f + \mu)] = \frac{3}{\pi} X_s I_{o,\text{dc}}. \tag{4.64}$$

The output current has also been reduced and its lowest instantaneous value has become closer to zero. This implies a certain reduction in the areas of continuous conduction mode in Figure 4.20.

The output voltage of a practical rectifier is also affected, albeit to a lesser degree, by voltage drops across the source resistance, conducting power switches, and wiring

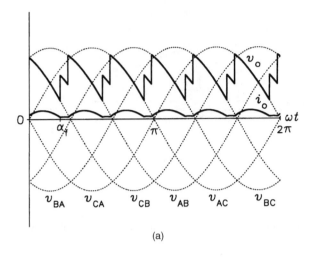

(a)

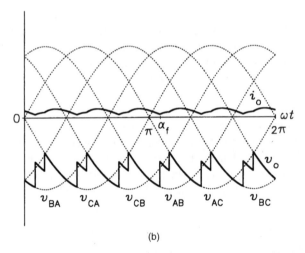

(b)

Figure 4.27 Waveforms of output voltage and current in a phase-controlled six-pulse rectifier supplied from a source with inductance: (a) rectifier mode ($\alpha_f = 45°$); (b) inverter mode ($\alpha_f = 135°$)

of the rectifier, including the cable connecting it to the load. Thus, the apparent internal resistance, R_r, of a bridge rectifier is given by

$$R_r = \frac{3}{\pi} X_S + R_S + 2R_{ON} + R_W \tag{4.65}$$

where R_{ON} is the equivalent on-state resistance of power switches employed in the rectifier (in bridge converters, two switches in series conduct the output current) and R_W is the wiring resistance. Based on Eqs. (4.41) and (4.64), the dc output voltage of a practical rectifier in the continuous conduction mode can be expressed as

$$V_{o,\text{dc}} = \frac{3}{\pi} V_{\text{LL},p} \cos \alpha_f - R_r I_{o,\text{dc}} \qquad (4.66)$$

which, unsurprisingly, indicates that the rectifier constitutes a practical dc voltage source whose terminal voltage decreases with an increase in the drawn current.

Another adverse effect of the source inductance is the *line voltage notching*. Consider, for example, the commutation interval in Figure 4.26. The sinusoidal waveforms of v_{AB} and v_{CB} shown there are those of the source EMFs. However, during commutation, the corresponding line-to-line voltages, v_{ab} and v_{cb}, at the very input to the rectifier substantially differ from the source EMFs. They both acquire the waveform of the output voltage, that is,

$$v_{ab} = v_{bc} = v_o = \frac{1}{2}(v_{AB} + v_{CB}) \qquad (4.67)$$

since SCR TB′ connects supply line B to the negative output terminal of the rectifier, while the simultaneously conducting SCRs TA and TC connect lines A and C to the positive terminal. Consequently, the v_{ab} waveform, shown in Figure 4.28, becomes seriously distorted, the notches appearing at all the commutation intervals involving SCRs TA, TA′, TB, and TB′. Similar distortions affect the other input voltages as well, of course. Although not a problem for the rectifier itself, the notched voltages may disturb the operation of other equipment supplied in parallel with the rectifier. The source inductance affects the operation of all multipulse rectifiers, both controlled and uncontrolled.

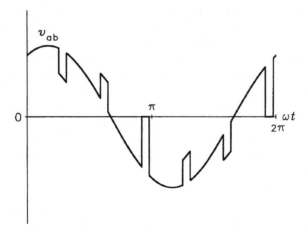

Figure 4.28 Notched waveform of input voltage in a phase-controlled six-pulse rectifier supplied from a source with inductance.

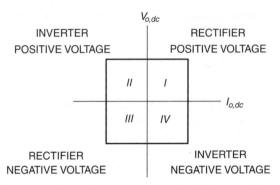

Figure 4.29   Operation plane, operating area, and operating quadrants of a rectifier.

4.2.2   Dual Converters

It was shown in the preceding section that controlled rectifiers can generate a negative dc output voltage if the load includes an EMF of such polarity and magnitude that the electrical power can be transferred to the ac supply source. However, under no circumstances can a negative output current be produced, as it would have to flow from the cathode to the anode in the SCRs. It is said that a controlled rectifier can operate in only two quadrants of an operation plane, which is shown in Figure 4.29. The values $I_{o,\text{dc}}$ and $V_{o,\text{dc}}$ of the dc output current and voltage of a rectifier represent coordinates of an *operating point* in the plane, and all the allowable operating points range out an *operating area*. Clearly, a controlled rectifier can operate only in the first and fourth quadrants, and diode rectifiers can operate in the first quadrant only.

Extension on four quadrants of operation can easily be accomplished by installing a mechanical cross-switch at the output of a controlled rectifier, as shown in Figure 4.30. With the cross-connection between the rectifier and load terminals, the load current, i'_o, equals $-i_o$ and the load voltage, v'_o, equals $-v_o$. In this way, the load can operate in the second or third quadrant, depending on the polarity of the dc output voltage, $V_{o,\text{dc}}$. However, this solution is practical only when the mechanical switching required is infrequent, such as in dc-motor-driven vehicles, because of the limited life span and low operating frequency of mechanical switches. Therefore, in most practical applications, so-called *dual converters* are used.

A dual converter, shown in Figure 4.31, is an antiparallel arrangement of two controlled rectifiers. For generality, different ac supply sources are indicated for each rectifier. There are two basic types of dual converters: *circulating current-free* and *circulating current-conducting* converters. The meaning of *circulating current* is explained later.

Circulating Current-Free Dual Converter   The power circuit of a circulating current-free dual converter is shown in Figure 4.32. The single SCRs of the regular rectifier are replaced here with antiparallel pairs of SCRs. If a positive output current is needed, SCRs TA1 through TC1′ are in operation, while SCRs TA2 through

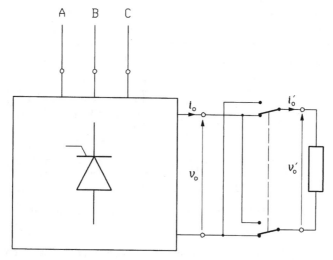

Figure 4.30   Controlled rectifier with a mechanical cross-switch.

TC2′ are all off (not fired). Vice versa, to produce a negative output current, SCRs TA2 through TC2′ are fired in an appropriate sequence, and SCRs TA1 through TC1′ remain unused.

The circulating current-free dual converter is simple and compact, but it has two serious disadvantages. Clearly, only one of the antiparallel connected SCRs can conduct the current, as the other SCR is reverse biased by the voltage drop across the conducting SCR. However, if both SCRs are off, one of them is always forward biased, and improper firing may cause a short circuit. For example, when SCRs TB1 and TC1′are conducting, the voltage, v_{BC}, across SCRs TC1 and TC2′ produces forward bias of the latter SCR. If fired, TC2′ would short lines B and C. This can be prevented by appropriate control of the firing signals, but when a change of polarity of the output current is required, the incoming rectifier must wait until the current in the outgoing rectifier dies out and the conducting SCRs turn off.

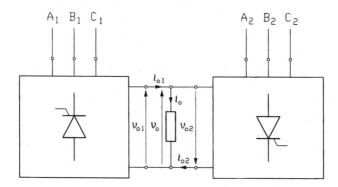

Figure 4.31   Antiparallel connection of two controlled rectifiers.

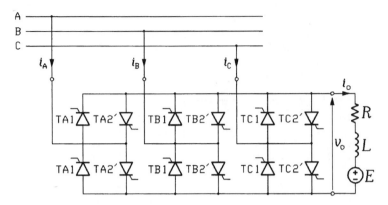

Figure 4.32   Six-pulse circulating current-free dual converter.

This mandatory delay slows the response of the converter to current control commands, which in certain applications is unacceptable.

The other disadvantage, common for all phase-controlled rectifiers, consists in the transition to the discontinuous conduction mode at large firing angles (see Figure 4.22). The continuous conduction area of operation can be expanded by adding inductance to the load. Such inductance would also reduce the current ripple. Generally, the discontinuous conduction is avoided if the currents conducted by the SCRs are sufficiently high for continuous conduction to be maintained no matter how large the firing angle is. This idea is employed ingeniously in the circulating current-conducting dual converters.

Circulating Current-Conducting Dual Converter   In a circulating current-conducting dual converter, both constituent rectifiers operate simultaneously, one in the rectifier mode, with the firing angle, α_{f1}, less than 90°, and the other in the inverter mode, with the firing angle, α_{f2}, equal to $180° - \alpha_{f1}$. According to Eq. (4.41), if both rectifiers operate in the continuous conduction mode, their output voltages have the same dc component, since $\cos\alpha_{f1} = -\cos\alpha_{f2}$. However, the instantaneous values of these voltages differ from each other, and the difference, Δv_o produces a current circulating between the rectifiers. If the rectifiers were directly connected as in Figure 4.31, the circulating current, i_{cr}, limited only by the resistance of wires and conducting SCRs, would be excessive. Therefore, inductors are placed between the rectifiers and the load to attenuate the ac component of the circulating current.

One version of a circulating current-conducting dual converter is shown in Figure 4.33. To avoid short circuits between the ac supply lines, the converter is fed from two secondary windings of a three-phase transformer, with the input voltages for both rectifiers equal with respect to amplitude and phase. Two inductors, L_1 and L_2, separate the rectifiers from the load. Rectifier RCT1 operates in the rectifier mode with a firing angle, α_{f1}, of 45°, and rectifier RCT2 operates in the inverter mode with a firing angle, α_{f2}, of 135°. Current paths at $\omega t = \pi/3$ are indicated by thick lines. Waveforms of the output voltages of both rectifiers and the load current waveform

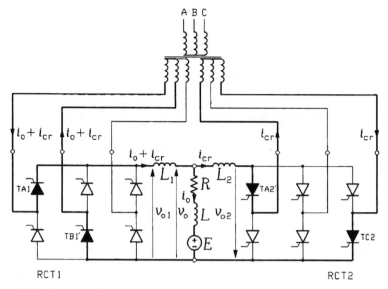

Figure 4.33   Six-pulse circulating current-conducting dual converter supplied from two separate ac sources.

are shown in Figure 4.34. The vertical dashed line indicates the instant of operation considered.

Waveforms of the differential output voltage, $\Delta v_o = v_{o,1} + v_{o,2}$, and circulating current, i_{cr}, are illustrated in Figure 4.35. The average differential voltage is zero, and the waveform of the circulating current indicates a borderline continuous conduction mode. In practice, the sum of the firing angles of the constituent rectifiers is made slightly less than 180°, which results in a nonzero dc component of Δv_o. Resistance

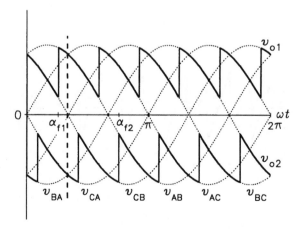

Figure 4.34   Waveforms of output voltages of constituent rectifiers of a circulating current-conducting dual converter ($\alpha_{f1} = 45°$, $\alpha_{f2} = 135°$).

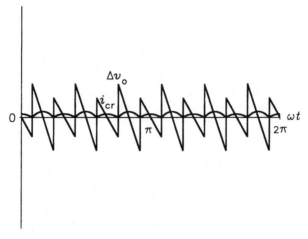

Figure 4.35 Waveforms of differential output voltage and circulating current in a circulating current-conducting dual converter: $\alpha_{f1} + \alpha_{f2} = 180°$.

of the circuit carrying the circulating current limits the circulating current to an allowable level.

In practical dual converters, a closed-loop control circuit maintains the average circulating current at the value desired, typically at 10 to 15% of the rated load current. The control circuit adjusts the firing angle, here α_{f2}, of the rectifier that conducts the circulating current only, while control of the load current is accomplished by adjusting the firing angle, here α_{f1}, of the other rectifier. Waveforms of the differential output voltage and circulating current when $\alpha_{f1} = 45°$ and $\alpha_{f2} = 134°$ are shown in Figure 4.36.

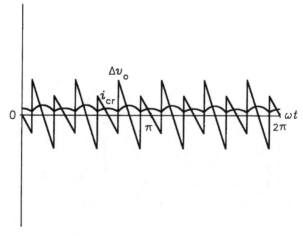

Figure 4.36 Waveforms of differential output voltage and circulating current in a circulating current-conducting dual converter: $\alpha_{f1} + \alpha_{f2} = 179°$.

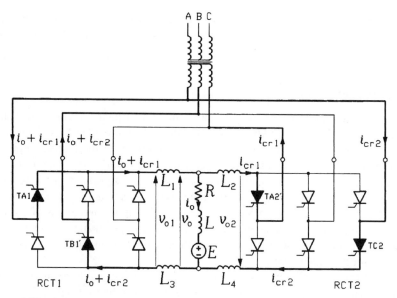

Figure 4.37   Six-pulse circulating current-conducting dual converter supplied from a single ac source.

Circulating current-conducting dual converters have excellent dynamic characteristics, as transitions between the operating quadrants can be made almost instantaneously. The absence of the discontinuous conduction mode allows wide-range control of the output voltage and current. Another version of such a converter is shown in Figure 4.37. Here, both rectifiers are supplied from the same ac line, but the four separating inductors, L_1 through L_4, prevent an interphase short circuit. Two circulating currents, i_{cr1} and i_{cr2}, flow in this converter.

PWM dual converters are theoretically feasible but impractical. Instead, high-quality four-quadrant dc power is obtained from a four-quadrant dc-to-dc converter (chopper), covered in Chapter 6. The input dc source for the chopper can be obtained from a controlled (SCR-based or PWM) rectifier.

4.3   PWM RECTIFIERS

Uncontrolled and phase-controlled rectifiers are classified as *line-commutated converters*, since the turn-off conditions for their switches are determined by the voltages of the ac supply line. As explained in Section 4.2.1, the nonsinusoidal supply currents and dependence of the input power factor on the firing angle constitute major disadvantages of those converters. The availability of fully controlled power switches, such as IGBTs or power MOSFETs, makes it possible to dispose of phase control of the output voltage and use pulse width modulation (PWM) instead. In contrast to line-commutated power electronic converters, those based on fully controlled switches are called *force-commutated.* It will be shown that the adverse effects on the ac supply system can be greatly mitigated employing force-commutated PWM rectifiers with small series–parallel LC input filters.

4.3.1   Impact of Input Filter

The enhancing impact of a series–parallel LC input filter on the quality of the supply current can best be explained by considering the single-phase PWM rectifier shown in Figure 4.38. The distribution of currents at the input to the rectifier represents an ideal situation not fully attainable in practice. The input current, i_i, consists of a fundamental, $i_{i,1}$, and a harmonic current, $i_{i,h}$. Radian frequencies of the harmonic components of $i_{i,h}$ are multiples of the fundamental frequency, ω; that is, they are many times higher than this frequency. As a result, higher harmonics of the input current are much more attenuated by the filter inductance, L_f, than is the fundamental, the inductive reactance of the filter for the kth current harmonic being k times *higher* than that for the fundamental.

In contrast to L_f, the filter capacitance, C_f, offers the current harmonics an easier passage than that for the fundamental, since the capacitive reactance for the kth harmonic is k times *lower* than that for the fundamental. Therefore, the rectifier draws most of the harmonic current from the capacitor. If all significant harmonics of the input current had frequencies approaching infinity, or more realistically, if the low-order harmonics were negligible, the actual distribution of currents would be close to the ideal case illustrated in Figure 4.38.

Clearly, the filter described is impractical for phase-controlled rectifiers, in which the largest harmonics of the input current are the low-frequency harmonics. In contrast, as shown later, frequencies of large harmonic currents in PWM rectifiers are

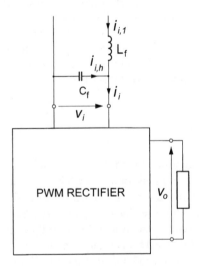

Figure 4.38   Single-phase PWM rectifier with a series–parallel LC input filter.

high enough to be attenuated effectively using small inductances and capacitances of the input filter. In practice, the same source inductance that causes unwanted commutation effects in phase-controlled rectifiers is often sufficient as the inductive part of the input filter for PWM rectifiers. Then, only input capacitors need to be installed as discrete elements of the filter, to provide the capacitive part.

4.3.2   Principles of Pulse Width Modulation

The general idea of pulse width modulation was explained in Chapter 1. In most considerations thereof, the duty ratios of switches of the generic converter were assumed constant, while the variable duty ratios of switches in the generic PWM inverter (see Figure 1.26) were mentioned only briefly. PWM rectifiers are the first practical PWM power electronic converters described in this book, and as with the inverter mentioned, the duty ratios of their switches vary throughout the cycle of input voltage. To cover the issue of pulse width modulation in depth, certain important concepts, which have not yet been covered, will be introduced and explained.

Switching variables, whose values determine the state of a converter, are usually defined as the minimum set of binary variables that allows representation of each state allowed. Considering as an example the generic converter in Figure 1.2, three switching variables, x_{12}, x_{34}, and x_5, suffice to describe all three states allowed. Specifically, the switching variables can be defined as

$$x_{12} = \begin{cases} 0 & \text{if S1 \& S2 = OFF} \\ 1 & \text{if S1 \& S2 = ON} \end{cases} \tag{4.68}$$

$$x_{34} = \begin{cases} 0 & \text{if S3 \& S4 = OFF} \\ 1 & \text{if S3 \& S4 = ON} \end{cases} \tag{4.69}$$

$$x_5 = \begin{cases} 0 & \text{if S5 = OFF} \\ 1 & \text{if S5 = ON} \end{cases} \tag{4.70}$$

Then $x_{12}x_{34}x_5 = 001$, 100, and 010 imply states 0, 1, and 2, respectively. Three binary variables can, in general, describe $2^3 = 8$ states, but in this case the remaining five states are forbidden as resulting in either a short circuit or an open circuit ("floating" output). Note also that the number of switching variables is less than that of switches in the generic converter (which could actually produce as many as $2^5 = 32$ possible states). Such minimal representation of converter states is even more economical in many practical power electronic systems, such as in the popular voltage-source inverters presented in Chapter 7. However, certain control algorithms may require that each switch of a converter be assigned a switching variable, equal to 0 if the switch is OFF (open) and to 1 if the switch is ON (closed).

Clearly, waveforms of switching variables are trains of rectangular pulses with unity magnitude, and PWM indicates the modulation of widths (durations) of individual pulses. Operation of ac-input and ac-output power electronic converters is cyclic, the length of a cycle being determined by the respective input or output frequency. Typically, in PWM converters, the cycle of the input or output ac voltage consists of N subcycles, called *switching intervals* or *switching cycles* (in general, N does not have to be an integer). The reciprocal, f_{sw}, of a switching period (length of a switching cycle), T_{sw}, is called a *switching frequency*. Switching frequencies in PWM power electronic converters are usually in the range of 4 to 20 kHz. The average switching frequency is N times higher than the input or output frequency of the converter.

Although there is only one method of phase control, the number of PWM techniques is countless, and new algorithms, usually modifications of known algorithms, are still being published. In most cases, a single pulse of each switching variable occupies one switching interval. Since the duty ratio, d_x, of switching variable x can be adjusted from zero to unity, it is possible that no pulse is present in the switching interval or that a pulse of x fills the entire interval.

In modern PWM techniques, pulses of switching variables, which are closely associated with actual switching signals of converter switches, are seldom generated directly and independent of each other. Instead, the most common PWM algorithms produce a timed sequence of converter states, each state defined by the values of switching variables.

As an example, consider a simple hypothetical PWM converter described by two switching variables, x_1 and x_2. All possible states are allowed, and each state is designated by the decimal equivalent of the binary number $(x_1x_2)_2$; that is, the converter in question can assume any of the following states: state 0, when $x_1x_2 = 00$; state 1, when $x_1x_2 = 01$; state 2, when $x_1x_2 = 10$; or state 3, when $x_1x_2 = 11$. Let us say that the PWM algorithm dictates the following sequence and timing of states within a single switching interval: $0\,(50\,\mu s)$–$2\,(40\,\mu s)$–$3\,(20\,\mu s)$–$2\,(40\,\mu s)$–$0\,(50\,\mu s)$. State 1 is not used. Clearly, the switching period is 200 μs, which corresponds to the switching frequency of 5 kHz. The switching pattern in question is shown in Figure 4.39. It can be seen that the pulse width of x_1 is 100 μs and that of x_2 is 20 μs, which means that the duty ratio of x_1 is 0.5 and that of x_2 is 0.1. Both pulses are centered about the middle of the switching interval. All that information can be derived from the timed state sequence without specifying each switching variable separately.

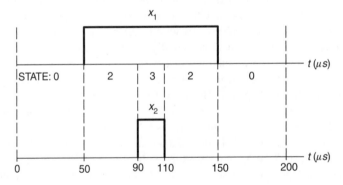

Figure 4.39   Switching pattern of a hypothetical four-state PWM converter.

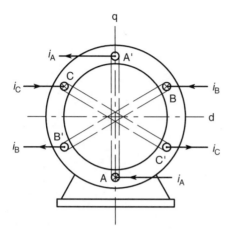

Figure 4.40   Stator of a three-phase electric ac machine.

Another important concept in control of PWM converters is that of voltage and current *space vectors*. Proposed in the 1920s for dynamic analysis of electric machines, in the 1980s they have been adapted successfully for use in PWM methods. *Space vector PWM* (SVPWM) techniques are now most popular. Superficially, space vectors resemble *phasors*, commonly used in analysis of ac circuits. However, space vectors differ from phasors in several aspects, and these two types of electric quantities should not be confused.

To explain the origin of voltage and current space vectors, a simple representation of a stator of a three-phase electric ac machine (generator or motor) is shown in Figure 4.40. Each of the three phase windings is comprised of two conductors perpendicular to the page plane and connected at the back of the stator (connection not shown). The six conductors, marked A, B, C, A', B', and C', form three coils displaced by 120° from each other. Stator currents i_A, i_B, and i_C are considered positive when they enter the stator winding at the front ends of conductors A, B, and C, respectively.

Currents in the coils generate magnetomotive forces (MMFs) $\vec{\mathcal{F}}_A$, $\vec{\mathcal{F}}_B$, and $\vec{\mathcal{F}}_C$, which add up to the stator MMF, $\vec{\mathcal{F}}_S$. A case of balanced currents (same magnitude, 120° mutual displacement) is illustrated in Figure 4.41. As the phasor diagram indicates, currents i_A and i_B are positive, with their instantaneous values equal to half of the peak value and the current i_C at its negative peak. The MMFs, each perpendicular to the coil that produces it, are true vectors, as they have specific magnitude and direction in the physical space of the stator. It can easily be shown that as time progresses and the currents follow the positive phase sequence, $\vec{\mathcal{F}}_S$ rotates counterclockwise with angular velocity, ω, equal to the radian frequency of the stator currents. Thus, stationary vectors $\vec{\mathcal{F}}_A$, $\vec{\mathcal{F}}_B$, and $\vec{\mathcal{F}}_C$ yield a revolving vector $\vec{\mathcal{F}}_S$.

Each of the space vectors considered can be described by its horizontal (direct-axis) and vertical (quadrature-axis) components. If the direct, d, axis of the stator is assumed to be real and the quadrature, q, axis to be imaginary, a MMF vector can be

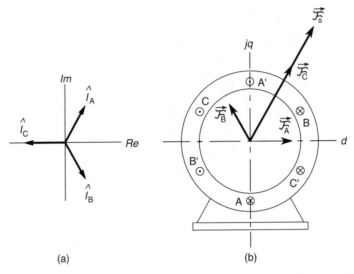

Figure 4.41   Generation of a space vector of the stator magnetomotive forces in a three-phase electric ac machine: (a) a phasor diagram of stator currents; (b) vectors of magnetomotive forces in the stator.

represented by a complex number. For example, as illustrated in Figure 4.42,

$$\vec{\mathcal{F}}_S = \mathcal{F}_{ds} + j\mathcal{F}_{qs} = \mathcal{F}_s e^{i\Theta_s} \tag{4.71}$$

where $\mathcal{F}_{ds}$ and $\mathcal{F}_{qs}$ denote the direct and quadrature components of the vector of stator MMF, respectively, and $\mathcal{F}_s$ and Θ_S are the magnitude and phase angle of that vector.

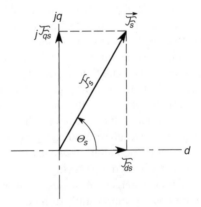

Figure 4.42   Space vector of stator MMFs and its components.

The space vector $\vec{\mathcal{F}}_S$ is given by

$$\vec{\mathcal{F}}_S = \mathcal{F}_{as} + \mathcal{F}_{bs}e^{j120°} + \mathcal{F}_{cs}e^{j240°} \tag{4.72}$$

where $\mathcal{F}_{as}$, $\mathcal{F}_{bs}$, and $\mathcal{F}_{cs}$ are values of the individual phase MMFs. The MMFs in Figure 4.41 and in the per-unit system (assuming the largest possible value as the reference) are $\mathcal{F}_{as} = \mathcal{F}_{bs} = 0.5$ p.u. and $\mathcal{F}_{cs} = -1$ p.u. Substitution of these values in Eq. (4.72) yields $\vec{\mathcal{F}}_s = 0.75 + j1.299 = 1.5e^{j60°}$ p.u.; that is, $\mathcal{F}_{ds} = 0.75$ p.u., $\mathcal{F}_{qs} = 1.299$ p.u., $\mathcal{F}_S = 1.5$ p.u, and $\Theta_S = 60°$. These results clearly agree with those seen in Figures 4.41 and 4.42.

An MMF is a product of a number of turns in a coil and a current in the coil. Therefore, dividing an MMF space vector by the turns number (which carries no physical units) gives a current space vector, $\vec{i}$. Equation (4.72) can be used to express components of that vector, i_d and i_q, in terms the phase currents, i_A, i_B, and i_C:

$$\vec{i} = \begin{bmatrix} i_d \\ i_q \end{bmatrix} = \begin{bmatrix} 1 & -\dfrac{1}{2} & -\dfrac{1}{2} \\ 0 & \dfrac{\sqrt{3}}{2} & -\dfrac{\sqrt{3}}{2} \end{bmatrix} \begin{bmatrix} i_A \\ i_B \\ i_C \end{bmatrix}. \tag{4.73}$$

This $abc \to dq$ conversion, called the *Park transformation*, is valid only for three-wire three-phase systems, in which the phase currents add up to zero; that is, only two currents are independent. Consequently, the three currents, i_A, i_B, and i_C, can be converted into two currents, i_d and i_q, without any loss of information.

The concept of current space vectors is somewhat abstract, as it applies to any three-phase currents, not necessarily those in the winding of an ac machine. The extension of that concept on the space vector of line-to-neutral voltages, given by

$$\vec{v} = \begin{bmatrix} v_d \\ v_q \end{bmatrix} = \begin{bmatrix} 1 & -\dfrac{1}{2} & -\dfrac{1}{2} \\ 0 & \dfrac{\sqrt{3}}{2} & -\dfrac{\sqrt{3}}{2} \end{bmatrix} \begin{bmatrix} v_{AN} \\ v_{BN} \\ v_{CN} \end{bmatrix} \tag{4.74}$$

is even more abstract, as it is difficult to think about voltage "pointing out" in a specific direction. However, the current and voltage space vectors are very convenient tools in the analysis and control of three-phase electric ac machines, and of three-phase power electronic converters. A space vector of line-to-line voltages can, of course, be defined similarly to that of line-to-neutral voltages.

Considerations presented have illustrated differences between current and voltage space vectors and phasors of these quantities. A *phasor*, used to simplify arithmetic operations on *sinusoidal* ac quantities, is a complex number describing a single such quantity, usually current or voltage. Thus, the current phasor shares the magnitude and phase (but not the frequency) with the corresponding single current, and its real part represents the actual value of the current at a given instant of time. In contrast, a current space vector captures information about all three currents in a three-phase three-wire system. Importantly, these currents do not have to be sinusoidal, as the current vector is determined by their instantaneous values.

As illustrated in Figure 4.42, the definition of a space vector expressed by Eq. (4.72) yields a true vector whose magnitude is 1.5 times larger than that of the corresponding phasor. Therefore, an alternative definition of space vectors includes multiplication by two-thirds of the right-hand side of the relevant equation. In such a case, the MMF space vector $\vec{\mathcal{F}}_s$ given by a modified Eq. (4.72) would have a magnitude, $\mathcal{F}_s$, of 1 p.u., as does the phasor of that MMF. The difference between those two definitions of space vectors is insignificant as long as a given definition is used consistently in all considerations.

Depending on the type of three-phase PWM converter, given that three input or output currents or voltages are controllable, which means that their space vector can be made to follow a certain reference vector, $\vec{i}^*$. The control is realized by imposing a specific timed sequence of states of the converter. For examples, consider an unspecified converter whose state X produces current vector $\vec{I}_X$, state Y produces vector $\vec{I}_Y$, and state Z produces vector $\vec{I}_Z$, whose magnitude is zero. These three vectors are to be employed to generate the reference vector $\vec{i}^*$ located between $\vec{I}_X$ and $\vec{I}_Y$, as shown in Figure 4.43.

As time progresses, the reference vectors rotate clockwise, not necessarily at constant speed. Also, their magnitude can vary. However, switching intervals are practically so short that within each of them the reference vector can be assumed stationary and unchanging. As shown in Figure 4.43, the reference vector can be represented as

$$\vec{i}^* = d_X\vec{I}_X + d_Y\vec{I}_Y \tag{4.75}$$

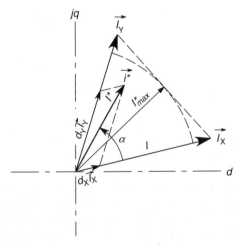

Figure 4.43   Synthesis of a rotating space vector $\vec{i}^*$ from stationary vectors $\vec{I}_X$ and $\vec{I}_Y$.

where d_X and d_Y are duty ratios of states X and Y, respectively, that is, relative durations of these states with respect to the length, T_{sw}, of the switching interval. The maximum realizable magnitude, I^*_{max}, of the reference vector equals the radius of the arc inscribed in the triangle formed by the "framing" vectors $\vec{I}_X$ and $\vec{I}_Y$. In such case the sum of d_X and d_Y equals 1. Otherwise, to fill up the switching interval, a zero vector, I_Z, is enforced with such a duty ratio d_Z that

$$d_X + d_Y + d_Z = 1. \tag{4.76}$$

In this way, a revolving vector $\vec{i}^*$ is generated as a *time average* of fractions of stationary vectors, $\vec{I}_X$, $\vec{I}_Y$, and $\vec{I}_Z$, that is, by maintaining state X for a total of $d_X T_{sw}$ seconds, state Y for a total of $d_Y T_{sw}$ seconds, and state Z for a total of $d_Z T_{sw}$ seconds. In practice, the switching cycle is usually divided into more than three subcycles. Most often, six subcycles are used, each one holding a half of the total time allocated for a given state.

Formulas for calculation of d_X, d_Y, and d_Z are derived from trigonometric equations describing the vector diagram in Figure 4.43. In particular, if current vectors $\vec{I}_X$ and $\vec{I}_Y$ have the same magnitude I and differ in phase by 60°, which is a common situation in practical three-phase converters, then

$$d_X = m \sin(60° - \alpha) \tag{4.77}$$

$$d_Y = m \sin \alpha \tag{4.78}$$

where m denotes the *modulation index*, and, according to Eq. (4.76),

$$d_Z = 1 - d_X - d_Y. \tag{4.79}$$

In certain PWM converters (but not in all of them), the modulation index has the same meaning as the magnitude control ratio, M, introduced previously [see Eq. (1.40)]. The term *modulation index* has been inherited from terminology describing modulation in communication systems (e.g., amplitude modulation (AM)) such as the radio, and is commonly used with respect to PWM power electronic converters. Considering current vectors in Figure 4.43, the modulation index is defined as

$$m = \frac{I^*}{I^*_{max}} \tag{4.80}$$

where I^* denotes the magnitude of the reference vector $\vec{i}^*$ and I^*_{max} is the maximum available value of this magnitude. The sequence of states is so determined as to obtain the best quality or efficiency of operation of the converter. The former condition is satisfied when individual switches of the converter are switched in a possibly regular manner, that is, when time intervals between switchings are possibly uniform.

The highest efficiency is achieved when the number of switchings per cycle of the ac voltage is minimized.

Certain control schemes for PWM converters utilize the concept of *rotating reference frame* for space vectors of currents and voltages. Note that as these vectors rotate in the stationary *dq* set of coordinates, their *d* and *q* components are sinusoidal functions of time, that is, ac variables. They are inconvenient to use in the control schemes, which usually employ dc quantities. If a given revolving vector is described by *D* and *Q* coordinates of a frame rotating with an angular velocity ω identical or close to that of the vector, these coordinates become dc variables. Traditionally, a superscript *e* is used to denote space vectors defined in a rotating reference frame, which in the theory of electric machinery is often referred to as the *excitation frame*.

Consider a voltage space vector $\vec{v}$, which in the stationary *dq* reference frame is expressed as

$$\vec{v} = v_d + jv_q. \tag{4.81}$$

The same vector in the rotating *DQ* reference frame is given by

$$\vec{v}^e = \vec{v}e^{-j\omega t} = v_D + jv_Q \tag{4.82}$$

and the relation between *dq* and *DQ* components is described by matrix equation

$$\begin{bmatrix} v_D \\ v_Q \end{bmatrix} = \begin{bmatrix} \cos\omega t & \sin\omega t \\ -\sin\omega t & \cos\omega t \end{bmatrix} \begin{bmatrix} v_d \\ v_q \end{bmatrix}. \tag{4.83}$$

Clearly,

$$\vec{v} = \vec{v}^e e^{j\omega t} \tag{4.84}$$

and

$$\begin{bmatrix} v_d \\ v_q \end{bmatrix} = \begin{bmatrix} \cos\omega t & -\sin\omega t \\ \sin\omega t & \cos\omega t \end{bmatrix} \begin{bmatrix} v_D \\ v_Q \end{bmatrix}. \tag{4.85}$$

The concept of rotating reference frame described is illustrated in Figure 4.44.

4.3.3 Current-Type PWM Rectifier

A circuit diagram of a current-type PWM rectifier is shown in Figure 4.45. It is also referred to as a current-source rectifier, which can be misleading because the input capacitors imply a voltage-source supply of the converter (see Figure 1.6). Another name is *buck rectifier*, which in this case means that the average output dc voltage, V_o, cannot be higher than that of a corresponding diode rectifier. The reference to current in the name of the rectifier comes from the fact that similarly to a current source, the

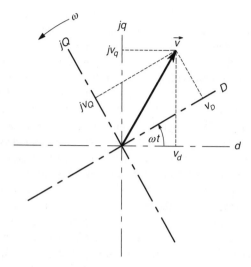

Figure 4.44   Voltage space vector in the stationary and rotating reference frames.

output current, i_o, cannot be reversed. Thus, as in the phase-controlled rectifiers, a negative output voltage is required for a negative (load to source) power flow. The inductance shown in Figure 4.45 as connected in series with the load smoothes the output current. It does not have to be a distinct device if the internal load inductance is deemed sufficiently large.

As indicated by Eq. (4.4), the maximum value of V_o in a three-phase diode rectifier, or an SCR rectifier with a firing angle of zero, equals about 95% of the peak value of the supply line-to-line ac voltage. As illustrated in Figure 1.20a, PWM rectifiers display notches in the output voltage waveform, which make the maximum available average output voltage, $V_{o,\max}$, about 3% lower.

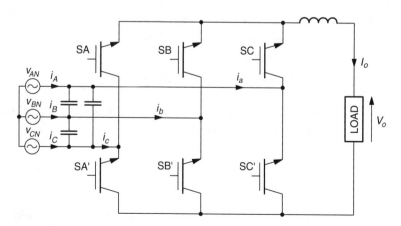

Figure 4.45   Current-type PWM rectifier.

Two major advantages of a PWM rectifier are (1) the feasibility of sinusoidal input currents with the unity input factor, and (2) a continuous output current, thanks to the narrow pulses and notches of the waveform of output voltage. With a sufficiently high switching frequency, the input currents and output current are only slightly rippled. The magnitude of the output voltage is controlled by the value of the modulation index, equal here to the magnitude control ratio, $m = V_o/V_{o,\max}$. Using the SVPWM principles (see Section 4.3.2), the space vector of input currents is made to follow the reference vector $\vec{i}^*$ revolving in synchronism with the space vector, $\vec{v}_i$, of input line-to-neutral voltages. Installing sensors of those voltages allows easy determination of $\vec{v}_i$. Note that it is only the phase angle, β^*, of $\vec{i}^*$ that is of interest, as the magnitude, I^*, depends on the load-dependent output current. This current is assumed constant and equal to its average value, I_o, thanks to the ripple-attenuating load inductance.

Two and only two switches of the rectifier are allowed to conduct at any time, one in the upper row and the other in the lower row. If, for example, switches SA and SB were ON, the potential of the upper bus of the rectifier would be undetermined, as the bus would be connected simultaneously to supply lines A and B. Also, the split of the output current, I_o, between those switches would be undefined. Thus, switching variables a, b, c, a', b', and c' of switches SA through SC' must satisfy the condition

$$a + b + c = a' + b' + c' = 1. \tag{4.86}$$

It can be seen that the condition above limits the number of allowable states of the rectifier to nine, namely:

State 1: $a = b' = 1$ (conducting switches: SA & SB')
State 2: $a = c' = 1$ (conducting switches: SA & SC')
State 3: $b = c' = 1$ (conducting switches: SB & SC')
State 4: $b = a' = 1$ (conducting switches: SB & SA')
State 5: $c = a' = 1$ (conducting switches: SC & SA')
State 6: $c = b' = 1$ (conducting switches: SC & SB')
State 7: $a = a' = 1$ (conducting switches: SA & SA')
State 8: $b = b' = 1$ (conducting switches: SB & SB')
State 9: $c = c' = 1$ (conducting switches: SC & SC')

In state 1, currents i_A, i_B, and i_C equal I_o, $-I_o$, and 0, respectively. Thus, according to Eq. (4.73), the space vector of input currents in this state is

$$\vec{I}_1 = \frac{3}{2}I_o - j\frac{\sqrt{3}}{2}I_o. \tag{4.87}$$

Similarly, vectors associated with the remaining states can be determined. The nonzero (active) vectors for states 1 through 6 of the rectifier are shown in

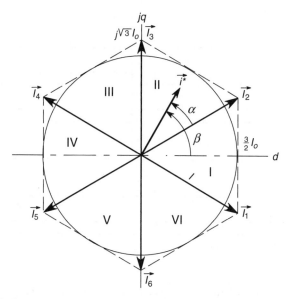

Figure 4.46 Reference current vector in the vector space of input currents of a current-type PWM rectifier.

Figure 4.46. States 7, 8, and 9 produce zero vectors of input currents:

$$\vec{I_7} = \vec{I_8} = \vec{I_9} = 0. \tag{4.88}$$

It can be seen that the active vectors have a magnitude, I, of $\sqrt{3}\,I_o$ and that the circle limiting the magnitude, I^*, of the reference current vector, $\vec{i^*}$, has a radius of $1.5I_o$. The vectors divide the dq plane into six sectors, designated I through VI. The reference current vector is given by

$$\vec{i^*} = I^*e^{j\beta} \tag{4.89}$$

and its in-sector angle is denoted by α. For an input power factor of unity, the rectifier must be so controlled that $\vec{i^*}$ is in phase with the space vector, $\vec{v_i}$, of the input line-to-neutral voltages, or it lags that vector by 180°. In the former case, the rectifier operates in the rectifier mode, and in the latter, in the second quadrant (see Figure 4.29), in the inverter mode. The voltage vector is easily determined using voltage sensors at the input of the rectifier.

Based on formulas (4.77) through (4.79), durations of states X and Y framing a sector in which the reference current vector is currently located are given by

$$T_X = mT_{sw}\sin(60° - \alpha) \tag{4.90}$$

$$T_Y = mT_{sw}\sin\alpha \tag{4.91}$$

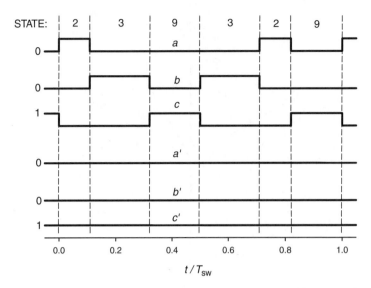

Figure 4.47 Example waveforms of switching variables in one switching cycle of a current-type PWM rectifier.

and the duration of a zero-vector state Z by

$$T_Z = T_{sw} - T_X - T_Y. \tag{4.92}$$

To minimize the number of commutations (switchings), the following state sequences are used in individual sectors of the dq plane:

Sector I: States 1–2–7–2–1–7 ···
Sector II: States 2–3–9–3–2–9 ···
Sector III: States 3–4–8–4–3–8 ···
Sector IV: States 4–5–7–5–4–7 ···
Sector V: States 5–6–9–6–5–9 ···
Sector VI: States 6–1–8–1–6–8 ···

It can be seen that within a switching cycle each state appears twice, and each appearance lasts half of the allotted time. A switching cycle is illustrated in Figure 4.47. It represents a situation when $m = 0.65$ and $\beta = 70°$, which implies location of the reference current vector in sector II. Thus, $\alpha = 40°$, X = 2, Y = 3, Z = 9, and, according to Eqs. (4.90) through (4.92), $T_2 = 0.22T_{sw}$, $T_3 = 0.42T_{sw}$, and $T_9 = 0.36T_{sw}$.

Analysis of waveforms of the switching variables in Figure 4.47 indicates that three switches, here SA, SB, and SC, turn on and off twice per switching cycle, while the remaining three switches, SA', SB', and SC', do not change their state. Generally,

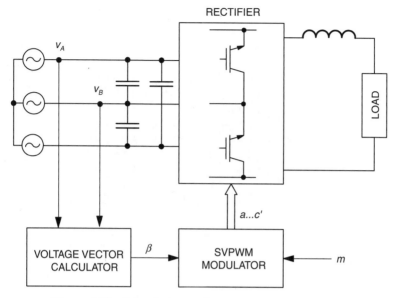

Figure 4.48   Control scheme of a current-type PWM rectifier.

in even sectors the commutating switches are those in the upper row of the rectifier bridge, while in odd sectors switches in the lower row are commutating. On average, there is one turn-on and one turn-off per switch per switching cycle. The control scheme described is illustrated in Figure 4.48. If the inverter mode of operation is required, it can be signalized by a negative value of the desired modulation index, m, with $|m|$ substituted for m in Eqs. (4.90)–(4.92) for computation of intervals T_X, T_Y, and T_Z.

To illustrate the impact of the modulation index, waveforms of the output voltage, v_o, and current, i_o, in a current-type PWM rectifier operating in the rectifier mode are shown in Figure 4.49, and waveforms of the input current, i_a, and its fundamental, $i_{a,1}$, in Figure 4.50. The corresponding line current, i_A (not shown), is similar to $i_{a,1}$, but with a certain amount of ripple depending on the size of the filter capacitors. Inverter operation of the rectifier is illustrated in Figure 4.51. Harmonic spectra of the input current are shown in Figure 4.52 to demonstrate the impact of switching frequency, f_{sw}, on current harmonics.

Sensors of input voltages provide information about the angle of the reference current vector. Two sensors are sufficient, as in a three-wire system the phase voltages add up to zero, so only two of them are independent variables. In practical rectifiers, additional sensors can be employed if required by the control algorithm. A sensor of the output dc voltage is very common, this voltage being the main quantity controlled. On the other hand, sensors are believed to increase the cost and decrease the reliability of a converter. Various efforts have thus been made to estimate, rather than measure, certain variables using *sensorless control schemes*.

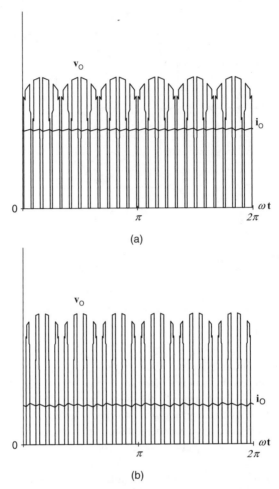

Figure 4.49 Waveforms of the output voltage and current in a current-type PWM rectifier: (a) $m = 0.75$; (b) $m = 0.35$ ($f_{sw}/f_0 = 24$, RLE load).

4.3.4   Voltage-Type PWM Rectifier

In the voltage-type, or *boost*, PWM rectifier shown in Figure 4.53, the output current is reversible but the output voltage is always positive. The rectifier is supplied through input inductors and the capacitor across the load smoothes the output current. Note that the fully controlled switches of the rectifier are upside-down with respect to those in the current-type PWM rectifier in Figure 4.45, and that each switch is equipped with a freewheeling diode. The name *boost* indicates that the output dc voltage of the rectifier is always higher than that of a corresponding diode rectifier.

In practice, voltage-type PWM rectifiers enjoy higher popularity than the current-type rectifiers described previously, thanks to their "true" dc output voltage waveform,

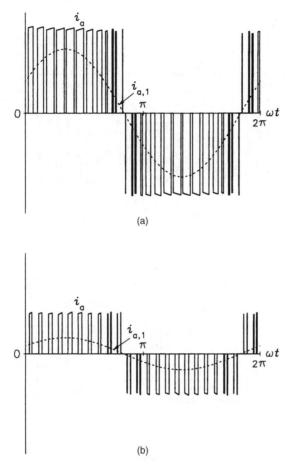

Figure 4.50 Waveforms of the input the current and its fundamental in a current-type PWM rectifier: (a) $m = 0.75$; (b) $m = 0.35$ ($f_{sw}/f_0 = 24$, RLE load).

whose ripple is maintained at a low level by a properly sized capacitor. Also, the boost property allows high values of the output voltage without the necessity of a step-up transformer at the rectifier's ac side. Two basic methods of control of the converter will be presented subsequently. Importantly, except for excessively low values of the modulation index, m, the magnitude control ratio, M, in the voltage-type PWM rectifier is a *reciprocal* of that index. As a rule of thumb, when $M = m = 1$, the dc output voltage of the rectifier equals the peak value of the line-to-line input voltage.

Voltage-Oriented Control with Space Vector PWM   The phase A branch of the rectifier is shown in Figure 4.54. Note that thanks to the existence of two switches, SA and SA′, the branch can theoretically assume a total of four states. However, if both switches were on, the output capacitor would be dangerously short-circuited. On the other hand, if both switches were off, the voltage of point A would depend on the

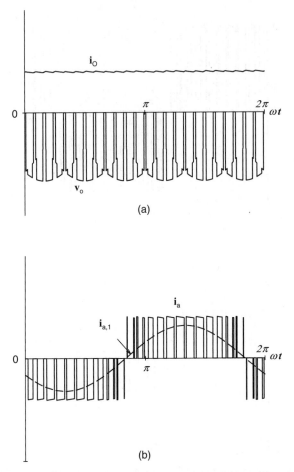

Figure 4.51 Inverter mode of operation of a current-type PWM rectifier: (a) waveforms of the output voltage and current (b) waveforms of the input current and its fundamental ($m = 0.75, f_{sw}/f_0 = 24$, RLE load).

polarity of the input current i_A. It would flow through either DA or DA′, making the terminal voltage equal to zero or V_o, respectively. Thus, if the relation between the voltage in question and the output voltage is to be determined by the state of switches, only two states of the rectifier branch can be allowed: SA = ON and SB = OFF, or, vice versa, SA = OFF and SB = ON. Consequently, a single switching variable, a, defined as

$$a = \begin{cases} 0 & \text{if SA = OFF \& SA′ = ON} \\ 1 & \text{if SA = ON \& SA′ = OFF} \end{cases} \qquad (4.93)$$

is sufficient to describe the state of the branch. Similarly defined switching variables b and c apply to the other two branches of the rectifier.

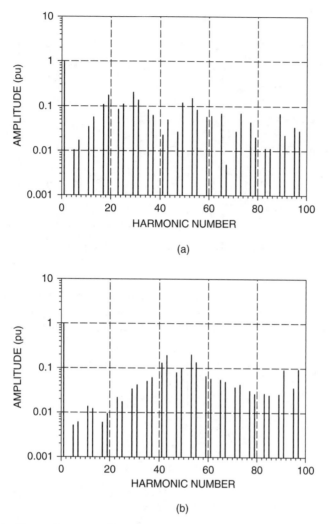

Figure 4.52   Harmonic spectra of input current in a current-type PWM rectifier: (a) $f_{sw}/f_0 =$ 24; (b) $f_{sw}/f_0 = 48$ ($m = 1$, ideal dc output current).

When $a = 0$ and $i_A > 0$, then $v_a = 0$, as switch SA' connects terminal A' to ground. When $a = 0$ and $i_A < 0$, the current must flow through diode DA', as SA is off. Again, terminal A' is connected to ground and $v_a = 0$. Vice versa, when $a = 1$, then independent of the polarity of i_A, $v_a = V_o$. Extension of these observations on the other two branches gives

$$\begin{bmatrix} v_a \\ v_b \\ v_c \end{bmatrix} = V_o \begin{bmatrix} a \\ b \\ c \end{bmatrix}. \tag{4.94}$$

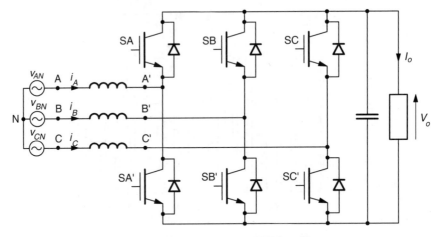

Figure 4.53 Voltage-type PWM rectifier.

As $v_{ab} = v_a - v_b$, $v_{bc} = v_b - v_c$, $v_{ca} = v_c - v_a$, $v_{an} = (v_{ab} + v_{ac})/3$, $v_{bn} = (v_{ba} + v_{bc})/3$, and $v_{cn} = (v_{ca} + v_{cb})/3$, where v_{ab}, v_{bc}, and v_{ca} denote the line-to-line input voltages and v_{an}, v_{bn}, and v_{cn} are the line-to-neutral input voltages, then

$$\begin{bmatrix} v_{ab} \\ v_{bc} \\ v_{ca} \end{bmatrix} = V_o \begin{bmatrix} 1 & -1 & 0 \\ 0 & 1 & -1 \\ -1 & 0 & 1 \end{bmatrix} \begin{bmatrix} a \\ b \\ c \end{bmatrix} \tag{4.95}$$

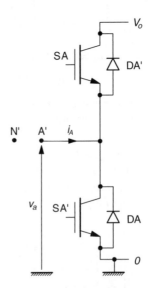

Figure 4.54 Phase A branch of a voltage-type PWM rectifier.

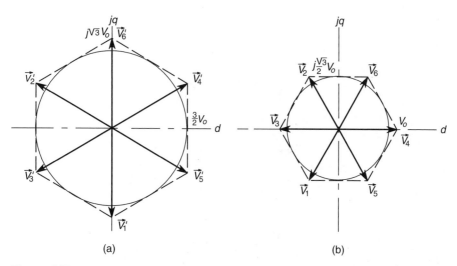

Figure 4.55 Input-voltage space vectors of a voltage-type PWM rectifier: (a) line-to-line voltages; (b) line-to-neutral voltages.

and

$$
\begin{bmatrix} v_{an} \\ v_{bn} \\ v_{cn} \end{bmatrix} = \frac{V_o}{3} \begin{bmatrix} 2 & -1 & -1 \\ -1 & 2 & -1 \\ -1 & -1 & 2 \end{bmatrix} \begin{bmatrix} a \\ b \\ c \end{bmatrix}. \tag{4.96}
$$

The three switching variables imply eight states of the rectifier. Naming a state with a decimal number $(abc)_{10}$, the states can be listed as 0 $(abc = 000)$ through 7 $(abc = 111)$. States 0 and 7 will subsequently be referred to as *zero states*, because they make all voltages equal zero. All three input terminals of the rectifier are then connected simultaneously to either the top or the bottom dc bus. Based on the $ABC \rightarrow dq$ transformation [Eq. (4.74)], voltage space vectors corresponding to the active states, 1 through 6, can be determined. The vectors of line-to-line voltages (marked by a prime) are shown in Figure 4.55a and those of line-to-neutral voltages in Figure 4.55b. Notice the similarity of the diagram of the line-to-line voltage vectors to that of the line current vectors in Figure 4.46.

The SVPWM principle with respect to control of voltages at the input to the rectifier is illustrated in Figure 4.56, which again shows space vectors of line-to-neutral input voltages. The reference voltage vector, $\vec{v}^* = V^* \angle \beta$ (here in sector III), is synthesized from vectors $\vec{V}_X$ (here $\vec{V}_2$), $\vec{V}_Y$ (here $\vec{V}_3$), and $\vec{V}_Z$ ($\vec{V}_0$ or $\vec{V}_7$) using formulas (4.91) to (4.93).

The *voltage-oriented control principle*, an ingenious concept that employs the rotating reference frame DQ, is employed in the rectifier. Figure 4.57 shows space vectors $\vec{v}$ and $\vec{i}$ of the input voltage and current, respectively, in both the stationary and revolving reference frames. It can be seen that for the unity power factor to be

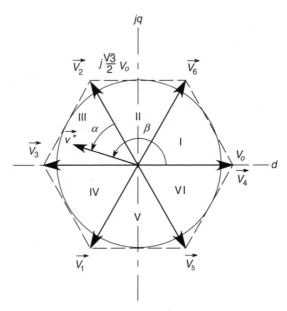

Figure 4.56 Reference voltage vector in the vector space of line-to-neutral input voltages of a voltage-type PWM rectifier.

maintained, those two vectors must be aligned. Then the angle φ between $\vec{v}$ and $\vec{i}$ equals zero, and the input power factor, $\text{PF} = \cos\varphi$, equals 1. Thus, with the D-axis aligned with the vector $\vec{v}$, that is, v_Q equal to zero, the Q-component, i_Q, of vector $\vec{i}$ should be forced to zero, too.

Figure 4.58 is a block diagram of the control system of the rectifier. Input voltages and currents are sensed and converted into space vectors. Components i_d and i_q of

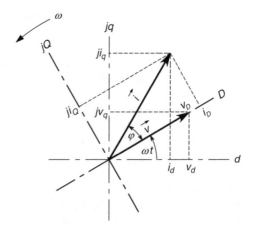

Figure 4.57 Principle of voltage-oriented control in a voltage-type PWM rectifier.

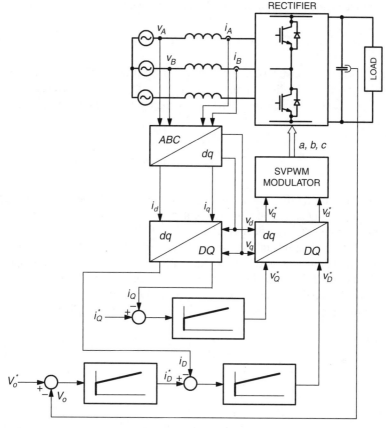

Figure 4.58 Control system of a voltage-type PWM rectifier using a rotating reference frame and SVPWM.

the current vector $\vec{i}$ are then transformed to components i_D and i_Q in the rotating reference frame. The reference value, i_Q^*, of the Q-component of current vector is set to zero, while that of the D-component, i_D^*, is obtained at the output of a proportional-plus-integral (PI) controller of the output voltage, V_o, of the rectifier. The reference value, V_o^*, of that voltage is compared with the feedback signal of V_o at the input to the controller. Similar PI controllers produce signals v_D^* and v_Q^*, which, finally, are transformed into corresponding components v_d^* and v_q^* of the reference voltage vector to be realized by the rectifier using the SVPWM. Components v_d and v_q of the input voltage vector $\vec{v}$ allow determination of the running angle, ωt, for the $dq \rightarrow DQ$ and $DQ \rightarrow dq$ transformations using Eqs. (4.83) and (4.85).

Direct Power Control   The unity power factor condition is tantamount to the zero amount of reactive power at the input to the rectifier. As known from the theory of three-phase ac circuits, the phasor, $\bar{S}$, of complex power can be calculated as

$$\bar{S} = 3\bar{V}_{AN}\bar{I}_A^* = P + jQ \tag{4.97}$$

where $\bar{V}_{AN}$ denotes an rms phasor of the line-to-neutral voltage v_{AN}, $\bar{I}_A^*$ is the rms conjugate phasor of the line current i_A, and P and Q denote the average values of the real and reactive power, respectively. An analogous equation for a vector, $\vec{s}$, of complex power is

$$\vec{s} = \frac{2}{3}\vec{v}\vec{i}^* = p + jq \tag{4.98}$$

where p and q are the instantaneous values of the real and reactive power (do not confuse this q with that denoting the quadrature axis). The coefficient $\frac{2}{3}$ in Eq. (4.98) is 4.5 times smaller than its counterpart of 3 in Eq. (4.97). It is so because, as shown in Figure 4.41, the magnitude of a space vector is 1.5 greater than that of its three phase components, and this magnitude represents a peak value. Each of the rms phasors in Eq. (4.97) is thus $1.5\sqrt{2}$ smaller than the corresponding vector, and their product is $(1.5\sqrt{2})^2 = 4.5$ smaller than that of the vectors in Eq. (4.98).

Substituting in Eq. (4.98) $v_d + jv_q$ for $\vec{v}$ and $i_d - ji_q$ for $\vec{i}^*$ yields

$$p = \frac{2}{3}(v_d i_d + v_q i_q) \tag{4.99}$$

and

$$q = \frac{2}{3}(v_q i_d - v_d i_q). \tag{4.100}$$

Applying the ABC $\rightarrow$ dq transformation given by Eqs. (4.73) and (4.74), the instantaneous real and reactive powers can now be expressed as

$$p = v_{AN}i_A + v_{BN}i_B + v_{CN}i_C \tag{4.101}$$

and

$$q = \frac{1}{\sqrt{3}}(v_{BC}i_A + v_{CA}i_B + v_{AB}i_C). \tag{4.102}$$

Thus, sensing the input voltages and currents allows calculation of the real and reactive powers drawn by the rectifier from the supply line.

A block diagram of the direct power control (DPC) of voltage-type PWM rectifier is shown in Figure 4.59. The next state of the rectifier is determined by the values of three control variables, x, y, and z. Variable x, an integer in the range 1 to 12, indicates the 30°-wide sector of the dq plane in which the space vector, $\vec{v}$, of the input voltage

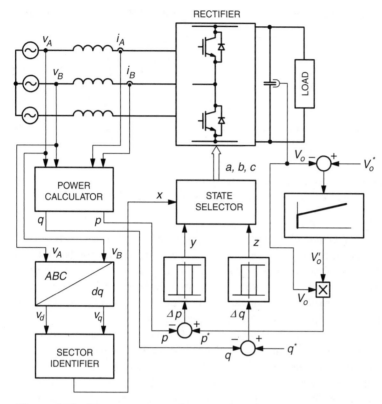

Figure 4.59 Direct power control system of a voltage-type PWM rectifier.

is currently located. If the phase of that vector is denoted by β, then

$$x = \text{int}\left(\frac{\beta}{30°}\right) + 1. \tag{4.103}$$

For example, if $\beta = 107°$, then $x = 4$, which means that $\vec{v}$ is in sector 4. Angle β is determined from the v_d and v_q components of the voltage vector. Logic variables y and z are obtained at the outputs of bang-bang controllers of the real and reactive powers. The width of the hysteresis loop of the controllers constitutes a tolerance band for the control errors, Δp and Δq, of these powers. Thus, it is so adjusted as to result in the average switching frequency desired.

The real power, p, is compared with the reference value, p^*, of this power, which is obtained from the control circuit of the output voltage, V_o, of the rectifier. As the real power at the input to the rectifier is closely related to the output power, which is proportional to the square of the output voltage, the reference signal p^* is taken as a product of V_o and the output signal, V_o', of a PI controller of that voltage. The reference value, q^*, of the instantaneous reactive power is set to zero and compared

TABLE 4.1   State Selection in the Voltage-Type PWM Rectifier with Direct Power Control

x:		1	2	3	4	5	6	7	8	9	10	11	12
y = 0	z = 0	6	4	4	5	5	1	1	3	3	2	2	6
	z = 1	2	6	6	4	4	5	5	1	1	3	3	2
y = 1	z = 0	0	4	7	5	0	1	7	3	0	2	7	6
	z = 1	0	0	7	7	0	0	7	7	0	0	7	7

with the actual reactive power, q. Based on the values of x, y, and z, the rectifier state that best counteracts the control errors is selected according to Table 4.1.

Voltage and Current Waveforms   As already mentioned, a voltage-type PWM rectifier is a "true" rectifier, in the sense that the input currents and, obviously, the voltages, are sinusoidal, while both the output voltage and current are true dc waveforms, minor ripple notwithstanding. Waveforms of the input voltages and currents satisfying the unity power factor condition are shown in Figure 4.60, and Figure 4.61 depicts the corresponding output voltage and current waveforms.

As an aside it is worth mentioning that both the rotating reference frame and direct power control employed in voltage-type PWM rectifiers are concepts that were first used to control three-phase ac motors. The rotating reference frame is a fundamental tool of the field orientation technique, which allows shaping the supply currents such that the magnetic flux and mechanical torque developed in the motor can be controlled independent of each other. In direct torque control, similar to the direct power control described in Section 4.3.4, sequential states of the inverter feeding the motor are selected on the basis of flux and torque control errors.

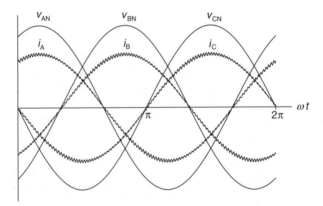

Figure 4.60   Waveforms of input voltage and current in a voltage-type PWM rectifier at unity power factor.

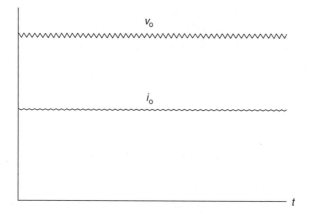

Figure 4.61   Waveforms of output voltage and current in a voltage-type PWM rectifier.

4.4   DEVICE SELECTION FOR RECTIFIERS

The voltage rating, V_{rat}, of semiconductor power switches in rectifiers (and, for that matter, in all power electronic converters) should exceed the highest instantaneous voltage possible to appear between any two points of a power circuit. In rectifiers, it is the peak value, $V_{i,p}$, of input voltage. Since semiconductor devices are vulnerable to overvoltages, even those of very short duration, substantial safety margins should be assumed in the design. The safety margins depend on the voltage level, the higher ones used in low-voltage converters. Thus, denoting a voltage safety margin by s_V, the condition to be satisfied is

$$V_{rat} \geq (1 + s_V)V_{i,p}. \tag{4.104}$$

In six-pulse rectifiers, $V_{i,p} = V_{LL,p}$.

The current rating, I_{rat}, which is the maximum allowable average current in a power switch, must be greater than the actual maximum average current, $I_{ave(max)}$. In six-pulse rectifiers, each power switch conducts the output current within one-third of the cycle of input voltage. Therefore, $I_{ave(max)} = I_{o,dc(rat)}/3$, where $I_{o,dc(rat)}$ denotes the rated dc output current, usually determined as the ratio of the rated power of a rectifier to the rated output voltage. Consequently, denoting the current safety margin by s_I, the condition to meet is

$$I_{rat} \geq \frac{1}{3}(1 + s_I)I_{o,dc(rat)}. \tag{4.105}$$

Typically, lower safety margins are used for the current than for the voltage. Temporary current overloads are less destructive than overvoltages, thanks to the thermal inertia of devices always equipped with heat sinks. In dual converters, an additional allowance should be made for the circulating current.

Fully controlled semiconductor power switches for PWM rectifiers are switched many times within a single cycle of input voltage. Therefore, dynamic characteristics of these devices must be taken into account. Specifically, for crisp voltage and current pulses to be produced, the shortest on-time of the rectifier should be significantly longer than the turn-on time of the switch. An analogous rule applies to the off- and turn-off times.

4.5 COMMON APPLICATIONS OF RECTIFIERS

Diode rectifiers are used to provide a fixed voltage to dc-powered equipment such as electromagnets or electrochemical plants. They are also employed as supply sources for dc-input power electronic converters, that is, inverters and choppers. Controlled rectifiers are used primarily for dc motor control, in high-voltage dc transmission lines, as battery chargers, as exciters for synchronous ac machines, and in certain technological processes requiring dc current control, such as electric arc welding. Also, if an inverter or a chopper is to provide a bidirectional flow of energy, it must be supplied from a controlled rectifier, which when operated in the inverter mode, allows transmission of power back to the ac supply system.

PWM rectifiers entered the mainstream power electronics market only recently. Current-type rectifiers are a worthy alternative to phase-controlled ac-to-dc converters, whose input currents are far from ideal (see Figure 4.24). Voltage-type rectifiers cannot be employed for control of dc motors, because the output voltage can only be adjusted up from the minimum level, roughly equal to the peak value of the supply line-to-line voltage. Therefore, they are used primarily in frequency changers for control of ac motors. A fixed three-phase supply ac voltage is first rectified, under the unity power factor condition, and then, in an inverter, converted into an adjustable three-phase voltage applied to the stator of the motor being controlled.

Besides controlled rectifiers, other power electronic converters, such as inverters, choppers, and cycloconverters, can also operate electrical machinery. The converters and machines share a common purpose: conversion of one form of energy into another. Therefore, operation of a controlled rectifier supplying a dc motor will be described to illustrate the basic relations between the operating modes of these two subsystems.

A dc machine, here assumed to be of the separately excited type, can be represented as in Figure 4.62. The armature circuit is comprised of the armature resistance, R_a, armature inductance, L_a, and armature EMF, E_a. The latter quantity is proportional to the speed, n, of the machine (in r/min), and the torque developed, T, is proportional to the armature current, i_a. For reference, the speed and torque are assumed positive when clockwise. Using T and n as operating variables of the machine, an operation plane can be set up as shown in Figure 4.63. The four quadrants of the plane correspond to the four possible modes of operation, from motoring clockwise (first quadrant) to generating counterclockwise (fourth quadrant). The idea of operating quadrants extends to all revolving machines, not necessarily electrical machines.

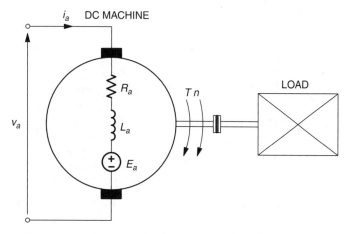

Figure 4.62 Electromechanical representation of a dc machine.

A dc machine supplied from a controlled rectifier or dual converter form a dc drive system, usually equipped with automatic control of speed and torque. For generality, an ac-to-dc converter in the form of a rectifier with a mechanical switch, or a dual converter, are assumed in subsequent considerations. In the steady state of the drive, the dc component, $V_{a,dc}$, of the armature voltage has the same polarity as the armature EMF, E_a, the two quantities differing only by the small voltage drop across the armature resistance. As the armature voltage and current are supplied by the converter, the operating quadrants of the machine correspond to their counterparts in the converter plane of operation in Figure 4.29. In particular, the motoring mode of the machine is associated with the rectifier mode of the converter, with the power flow from the ac source, through the converter and machine, to the load. Conversely, the generating mode of the machine results from the inverter mode of the converter, with the power flow reversed.

As an example, an electric locomotive supplied from a three-phase overhead line and driven by dc motors is considered. Today, dc motors in traction have largely been

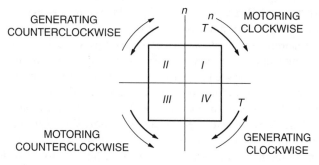

Figure 4.63 Operation plane, operating area, and operating quadrants of a rotating machine.

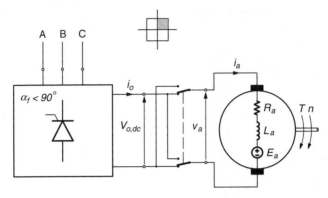

Figure 4.64 Dc motor supplied from a rectifier with a mechanical switch: first-quadrant operation.

superseded by ac motors, but an electric dc locomotive was one of the first high-power applications of SCRs. Several motors operate the wheels of a practical locomotive. For simplicity, only one of these motors, assumed to be fed from a rectifier with a mechanical switch, is considered.

Figure 4.64 shows the rectifier–switch–motor cascade when the locomotive hauls a train with the motor rotating clockwise, that is, with positive speed and torque. The switch provides a direct connection between the rectifier and the motor. Thus, the average armature voltage, V_a, and instantaneous armature current, i_a, of the motor are equal to the dc output, $V_{o,\text{dc}}$, and output current, i_o, of the rectifier, respectively. Both the converter and the machine operate in the first quadrant, which means a rectifier mode with positive dc output voltage for the converter and a motoring clockwise mode for the machine. The firing angle, α_f, is kept below 90°.

The kinetic energy of a train running at full speed is enormous, and using only friction brakes to slow down or stop the train is ineffective and wasteful. Therefore, an electric braking scheme is implemented by transition to the second quadrant, in which the dc machine operates as a generator and the converter works in the inverter mode. This situation is illustrated in Figure 4.65. The firing angle has been increased to over 90° and the mechanical switch has been switched to the cross-connecting position, making $V_{a,\text{dc}} = -V_{o,\text{dc}}$ and $i_a = -i_o$. Although the direction of speed and the polarity of motor EMF have not changed, the EMF is now seen as negative from the rectifier terminals. The conditions for the inverter operation are thus satisfied, and the negative armature current produces a braking torque opposing the motion of the motor. The momentum of the train provides the driving torque for the dc machine, and the electric power generated is transferred via the rectifier to the ac supply line.

The third and fourth quadrants of operation represent the motoring and braking modes, respectively, when the locomotive runs backward and the motor rotates counterclockwise. The development of connection diagrams similar to those in Figures 4.64 and 4.65 is left to the reader.

It is worth noting that certain drives operate in the first and fourth quadrants only, so that the rectifier can be connected directly to the motor. Lift-type drives, such as

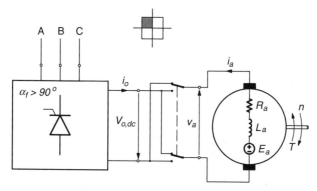

Figure 4.65 Dc motor supplied from a rectifier with a mechanical switch: second-quadrant operation.

those of elevators, are the typical example. No matter whether the load moves up or down, the motor torque must counter the torque imparted by the gravity. Therefore, the torque never changes its polarity, and the output current of the rectifier does not need to be reversed. When stopping from upward motion, the gravity force, possibly supplemented with friction brakes, is sufficient to provide the braking torque.

High-voltage dc (HVDC) transmission lines link distant power systems, using dc current as a medium for energy exchange between the systems. Therefore, synchronous operation of the systems is not required (they may even operate with different frequencies), and the easy control of transferred power improves the stability of the systems. HVDC lines allow efficient transmission of electric power because the inductance and capacitance of the line do not affect the voltage and current.

A typical HVDC transmission system is illustrated in Figure 4.66. The link between the interconnected power systems consists of 12-pulse rectifiers RCT1 and RCT2, transformers TR1 and TR2, smoothing inductors L1 through L4, and a transmission line. Each rectifier comprises two six-pulse SCR-based rectifiers connected in series, one supplied from a wye-connected secondary winding of the transformer and the other from a delta-connected secondary winding. Such 12-pulse rectifiers are characterized by a very smooth output voltage, with a minimal ripple factor.

The dc line is usually of the two-wire type, although single-wire lines, with the ground as a return path, are also feasible. If the power is transmitted from system 1 to system 2, rectifier RCT1 operates in the rectifier mode and rectifier RCT2 in the inverter mode. As a result, the three-phase ac power drawn from system 1 is converted into dc power, transmitted over the line to RCT2, and converted back into three-phase ac power in system 2. The rectifiers interchange modes of operation if the power is transmitted in the opposite direction. Notice that the dc current can flow in only one direction, and the reversal of power flow is accomplished by reversing the polarity of the dc voltage.

A single switch of an HVDC rectifier is composed of tens of high-rating SCRs connected in series and in parallel to withstand the high voltages and currents of the line. Individual SCRs are fired by photothyristors, to which the firing signals

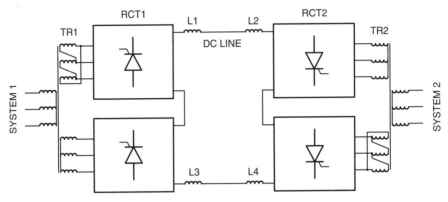

Figure 4.66 High-voltage dc transmission system.

are supplied via fiber-optic cables to ensure simultaneous turn-on. The apparatus required affects the cost of HVDC lines, which are economical only when voltages of several hundred kilovolts, powers of several hundred to several thousand megawatts, and distances of several hundred miles are involved.

4.6 SUMMARY

Ac-to-dc power conversion is performed by rectifiers, which in practice are mostly of the six-pulse three-phase bridge topology. Diode rectifiers produce a fixed dc voltage whose value depends on the amplitude of the ac input voltage. The output current in diode rectifiers is continuous unless the load EMF has a value approaching the amplitude of the input voltage and the load inductance is low. The power flow is unidirectional, from the ac source, via the rectifier, to the load.

Controlled rectifiers are more flexible operationally, allowing control of the dc output voltage and being capable of reversing the power flow. In phase-controlled SCR-based rectifiers, the voltage control is realized by adjusting the firing angle of the SCRs. In the continuous conduction mode, the dc output voltage is proportional to the cosine of the firing angle. The inverter mode is possible only when the load EMF is negative and the firing angle is greater than $90°$.

The load and source inductances affect the operation of rectifiers. Resonant input filters are recommended to reduce the harmonic content of the input currents drawn from the supply system by uncontrolled and phase-controlled rectifiers.

An extension to four-quadrant operation of controlled rectifiers is possible by installing a mechanical switch at the output, which allows for a negative load current. Dual converters represent a more convenient solution. A dual converter consists of two rectifiers connected antiparallel. Circulating current-free dual converters are simpler but operationally inferior to circulating current-conducting converters.

The quality of the input currents can be greatly enhanced using PWM rectifiers, based on fully controlled semiconductor power switches. The PWM rectifiers, of either the current (buck) or voltage (boost) type, can be controlled for a unity input power factor. The output currents are also of high quality. The maximum available volt–ampere ratings of PWM rectifiers are, however, lower than those of phase-controlled rectifiers with SCRs.

The selection of semiconductor power switches for rectifiers is based primarily on the amplitude of the ac supply voltage and the maximum average output current. Switches for PWM rectifiers must be sufficiently fast to realize the frequent switching required.

Dc motor control, HVDC transmission lines, battery charging, and supply of dc-input power electronic converters are the most common applications of rectifiers. They are also employed in various technological processes, such as electric arc welding or electrolysis.

EXAMPLES

Example 4.1   A three-pulse and a six-pulse diode rectifier are supplied from a 460-V ac line. Compare the dc output voltages available from these rectifiers.

Solution: According to Eq. (4.2), the dc voltage, $V_{o,dc(3)}$, of the three-pulse rectifier is

$$V_{o,dc(3)} = \frac{3\sqrt{3}}{2\pi} V_{LN,p}$$

where the peak value, $V_{LN,p}$, of the input line-to-neutral voltage is

$$V_{LN,p} = \frac{\sqrt{2} \times 460}{\sqrt{3}} = 375.6 \text{ V}$$

(a rated voltage of any three-phase apparatus is always given as the rms value of line-to-line voltage). Thus,

$$V_{o,dc(3)} = \frac{3\sqrt{3}}{2\pi} = 375.6 = 310.6 \text{ V}.$$

Following Eq. (4.4), the dc voltage of the six-pulse rectifier,

$$V_{o,dc(6)} = \frac{3}{\pi} V_{LL,p} = \frac{3}{\pi} \times \sqrt{2} \times 460 = 621.2 \text{ V}$$

that is, exactly twice as high as that of the three-pulse rectifier.

Example 4.2   A 270-V battery pack is charged from a six-pulse diode rectifier supplied from a 230-V ac line. The internal resistance of the battery pack is 0.72 Ω. Calculate the dc charging current.

Solution: The load EMF coefficient, ε, is

$$\varepsilon = \frac{270}{\sqrt{2} \times 230} = 0.83$$

and the load angle, φ, is zero, since no load inductance is assumed. Therefore, condition (4.19) for continuous conduction is

$$\varepsilon < \sin \frac{\pi}{3} = \frac{\sqrt{3}}{2} \approx 0.866$$

and it is satisfied. Consequently, the dc output voltage, $V_{o,dc}$, of the rectifier can be calculated from Eq. (4.4) as

$$V_{o,dc} = \frac{3}{\pi} \times \sqrt{2} \times 230 = 310.6 \text{ V}$$

and, according to Eq. (4.28), the dc output current, $I_{o,dc}$, is

$$I_{o,dc} = \frac{310.6 - 270}{0.72} = 56.4 \text{ A.}$$

Example 4.3   A dc motor with the armature resistance, R_a, of 0.6 Ω and armature inductance, L_a, of 4 mH is supplied from a phase-controlled six-pulse rectifier fed from a 460-V 60-Hz line. The motor rotates with such speed that the armature EMF, E_a, is 510 V. Find the dc output voltage, $V_{o,dc}$, and current, $I_{o,dc}$, of the rectifier when the firing angle, α_f, is 30° and 60°.

Solution: First, the feasibility of the assumed firing angles must be checked. The load EMF coefficient, ε, is

$$\varepsilon = \frac{510}{\sqrt{2} \times 460} = 0.784$$

and condition (4.43) for the allowable values of firing angle gives

$$-0.146 \text{ rad} < \alpha_f < 1.193 \text{ rad}$$

which is satisfied by both $\alpha_f = 30°$ and $\alpha_f = 60°$. This is also confirmed in Figure 4.21.

Next, the conduction modes must be determined for proper selection of formulas for the dc output voltage. The load angle, φ, is

$$\varphi = \tan^{-1} \frac{\omega L_a}{R_a} = \tan^{-1} \frac{120\pi \times 4 \times 10^{-3}}{0.6} = 1.192 \text{ rad} = 68.3°$$

and when $\alpha_f = 30° = \pi/6$ rad, condition (4.45) for a continuous output current is

$$\varepsilon < \left[\sin\left(\frac{\pi}{6} + \frac{\pi}{3} - 1.192\right) + \frac{\sin(1.192 - \pi/6)}{1 - e^{\pi/3\tan(1.192)}} \right] \cos(1.192) = 0.81$$

which is satisfied. Similar calculations for $\alpha_f = 60°$ yield $\varepsilon < 0.447$, which is not satisfied, that is, the rectifier operates in the discontinuous conduction mode. The same conclusions could be reached by inspecting Figure 4.22.

Using Eqs. (4.41) and (4.28), the dc output voltage and current for $\alpha_f = 30°$ are calculated as

$$V_{o,\text{dc}} = \frac{3}{\pi} \times \sqrt{2} \times 460 \times \cos\frac{\pi}{6} = 538 \text{ V}$$

and

$$I_{o,\text{dc}} = \frac{538 - 510}{0.6} = 46.7 \text{ A.}$$

Calculation of $V_{o,\text{dc}}$ for $\alpha_f = 60°$, when the output current, i_o, is discontinuous, requires knowledge of the conduction angle, β. It can be determined by computing $i_o(\omega t)$ for sequential values of ωt, starting at $\omega t = \alpha_f$ and ending when the current drops to zero at $\omega t = \alpha_e = \alpha_f + \beta$. The impedance, Z, of the load is

$$Z = \sqrt{R_a^2 + (\omega L_a)^2} = \sqrt{0.6^2 + (120\pi \times 4 \times 10^{-3})^2} = 1.623 \ \Omega$$

and Eq. (4.46) gives

$$i_o(\omega t) = 400.8 \left[\sin(\omega t - 0.145) - 2.12 + 1.335e^{\frac{(\omega t - \pi/3)}{2.513}} \right].$$

A computer should be employed to calculate the current. Beginning with $\omega t = 60°$ and proceeding $0.1°$ increments (smaller increments could, of course, be used as well), the following printout is obtained:

Angle	Current
60.0°	0.0000000 A
60.1°	0.0614110 A
60.2°	0.1221164 A
⋮	⋮
75.8°	0.1172305 A
75.9°	0.0513202 A
76.0°	− 0.015876 A

Thus, the extinction angle, α_e, is about 76°, and the conduction angle, β, is 76° − 60° = 16°, that is, 0.279 rad. Using Eq. (4.47), the dc output voltage is obtained as

$$V_{o,\text{dc}} = \frac{3}{\pi}\sqrt{2} \times 460 \left[2\sin\left(\frac{\pi}{3} + \frac{0.279}{2} + \frac{\pi}{3}\right) \sin\left(\frac{0.279}{2}\right) + 0.784\left(\frac{\pi}{3} - 0.279\right) \right]$$
$$= 510.4 \text{ V}. \tag{4.106}$$

The dc output current is

$$I_{o,\text{dc}} = \frac{510.4 - 510}{0.6} = 0.7 \text{ A}$$

that is, two orders of magnitude smaller than that in the continuous conduction mode with $\alpha_f = 30°$.

Note that if the discontinuous conduction mode had not been identified and Eq. (4.41) is used for calculation of $V_{o,\text{dc}}$ instead of Eq. (4.47), the value 310.6 V would be lower than the load EMF, implying an impossible negative output current.

Example 4.4 A phase-controlled six-pulse rectifier is supplied from a 460-V 60-Hz power line whose inductance, L_s, seen from the input terminals of the rectifier is 1 mH/ph. The rectifier operates with a firing angle, α_f, of 30°, producing an average output current, $I_{o,\text{dc}}$, of 140 A. Neglecting resistances of the supply circuit, find the overlap angle, μ, and average output voltage, $V_{o,\text{dc}}$, of the rectifier.

Solution: The source reactance, X_s, is given by

$$X_s = \omega L_s = 120\pi \times 10^{-3} = 0.377 \ \Omega/\text{ph}$$

and the peak value of the line-to-line input voltage, $V_{LL,p}$, is $\sqrt{2} \times 460 = 650.5$ V. Substituting the values of α_f, X_s, $I_{o,\text{dc}}$, and $V_{LL,p}$ in Eq. (4.63) yields

$$\mu = \left| \cos^{-1}\left[\cos\frac{\pi}{6} - 2\frac{0.377 \times 140}{650.5} \right] - \frac{\pi}{6} \right| = 0.267 \text{ rad} = 15.3°.$$

If the source inductance were zero, the average output voltage, $V_{o,\text{dc}(0)}$, would be

$$V_{o,\text{dc}(0)} = \frac{3}{\pi} \times 650.5 \times \cos\frac{\pi}{6} = 538 \text{ V}.$$

However, as seen from Eq. (4.64), the load inductance causes reduction of this voltage by

$$\Delta V_{o,\text{dc}} = \frac{3}{\pi} \times 0.377 \times 140 = 50.4 \text{ V}$$

so that the actual dc output voltage is 538 V − 50.4 V = 487.6 V.

Example 4.5   A current-type PWM rectifier is supplied from a 460-V line and is to produce 500 V of dc voltage at its output. At a certain instant, the angle, β, of the vector of the input voltages is 235°. Find the switching pattern of the rectifier if the switching frequency is 5 kHz.

Solution: The maximum dc voltage of the rectifier equals about 92% of the peak line-to-line supply voltage, that is,

$$V_{o(\text{max})} = 0.92 \times \sqrt{2} \times 460 = 598 \text{ V}.$$

Thus, the modulation index, m, equals

$$m = \frac{V_o}{V_{o,\text{max}}} = \frac{500}{598} = 0.84.$$

For unity power factor, the space vector of input currents is to be aligned with the input voltage vector, which is in sector V, which extends from 210° to 270° (see Figure 4.46). Hence, the in-sector angle, α, of that vector is 235° − 210° = 25°, and states X, Y, and Z are states 5, 6, and 9, respectively. With the switching period, $T_{\text{sw}} = 1/5 \text{ kHz} = 200 \, \mu\text{s}$, durations of these states are:

$$T_S = 0.84 \times 200 \times \sin(60° - 25°) = 96.4 \, \mu\text{s}$$
$$T_6 = 0.84 \times 200 \times \sin(25°) = 71 \, \mu\text{s}$$
$$T_9 = 200 - 96.4 - 71 = 32.6 \, \mu\text{s}.$$

The state sequence in sector V, as listed in Section 4.3.3, is 5–6–9–6–5–9 ⋯, and the corresponding time intervals, in microseconds, are 48.2–35.5–16.3–35.5–48.2–16.3 ⋯. Switches that turn on and off are those in the lower row, that is, SA′, SB′, and SC′, while SA and SB are off and SC is on all the time. The switching pattern described is shown in Figure 4.67.

Example 4.6   A voltage-type PWM rectifier is controlled such that its input current vector lags the input voltage vector by 30°. At a certain instant, the input line-to-neutral voltages v_{AN} and v_{BN} are 261.5 and −90.8 V, while the input line currents i_A and i_B are 94.0 and −76.6 A. After retuning the control system so that the rectifier operates with unity power factor, i_A and i_B at an instant when the voltages are at the

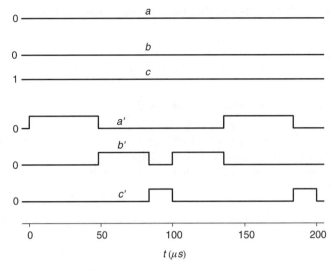

Figure 4.67   Switching pattern in Example 4.5.

same levels as before are 98.5 and −34.2 A, respectively. Calculate the instantaneous real, reactive, and apparent input power in both cases.

Solution: The line-to-neutral and line-to-line voltages are

$$v_{AN} = 261.5 \text{ V}, \ v_{BN} = -90.8 \text{ V}, \quad v_{CN} = -v_{AN} - v_{BN} = -170.7 \text{ V}, v_{AB} = v_{AN} -$$

$$v_{BN} = 352.3 \text{ V}, \ v_{BC} = v_{BN} - v_{CN} = 79.9 \text{ V}, \ v_{CA} = -v_{AB} - v_{BC} = -432.2 \text{ V}$$

In the first case, the line currents are

$$i_A = 94.0 \text{ A}, \ i_B = -76.6 \text{ A}, \ i_C = -i_A - i_B = -17.4 \text{ A}$$

and according to Eqs. (4.102) and (4.103), $p = 34,506$ W and $q = 19,911$ VAr. The instantaneous apparent power, s, which is a geometric sum (square root of sum of squares) of p and q, equals 39,839 VA.

In the second case,

$$i_A = 98.5 \text{ A}, \ i_B = -34.2 \text{ A}, \ i_C = -i_A - i_B = -64.3 \text{ A}$$

and the real, reactive, and apparent power values are

$$p = 39,839 \text{ W}, \quad q = 0, \quad s = 39,839 \text{ VA}.$$

It can be seen that the reactive power from the preceding case has been converted into real power and added to the initial real power, so that now $p = s$.

PROBLEMS

P4.1 A three-pulse diode rectifier fed from a 230-V ac line supplies a 10-Ω resistive load. Calculate the average output voltage and current of the rectifier.

P4.2 The rectifier in Problem 4.1 charges a 180-V battery pack. Is the charging current continuous or discontinuous? Sketch the output voltage and current waveforms.

P4.3 A six-pulse diode rectifier fed from a 460-V ac line charges a 480-V battery pack. Is the charging current continuous or discontinuous? Sketch the output voltage and current waveforms.

P4.4 A smoothing inductor of 50 mH has been connected between the rectifier in Problem 4.3 and the battery pack. The internal resistance of the battery is 2.1 Ω. Determine the conduction mode and find the dc output voltage and current of the rectifier.

P4.5 The rectifier in Problem 4.3 charges a 600-V battery pack. Find the crossover, extinction, and conduction angles.

P4.6 Sketch the output voltage waveform of a phase-controlled three-pulse rectifier operating in the continuous conduction mode with a firing angle of 30°.

P4.7 Assuming the continuous conduction mode, sketch the output voltage waveforms of a phase-controlled six-pulse rectifier corresponding to firing angles of 0°, 60°, 90°, 150°, and 180°.

P4.8 What firing angle is always feasible in an SCR-based rectifier if the load EMF does not exceed the peak value of line-to-line input voltage?

P4.9 Sketch the output voltage and current waveforms of a phase-controlled six-pulse rectifier supplying an RLE load with the load EMF equal to 80% of the peak line-to-line input voltage, and operating in the discontinuous conduction mode with a firing angle of 60°.

P4.10 A phase-controlled six-pulse rectifier fed from a 230-V 60-Hz line supplies a dc motor whose armature resistance and inductance are 0.15 Ω and 0.7 mH, respectively. The speed of the motor is such that the EMF induced in the armature is 260 V. Determine the conduction mode of the rectifier and calculate the dc armature voltage and current if the rectifier operates with a firing angle of 25°.

P4.11 Determine the conduction mode for the rectifier in Problem 4.10 if it operates with a firing angle of 45°. Find the dc output voltage and current.

P4.12 The rectifier in Problem 4.10 operates in the inverter mode with a firing angle of 120° and the motor working as a dc generator. What armature EMF is needed to maintain continuous armature current?

P4.13   A phase-controlled six-pulse rectifier operates in the continuous conduction mode with a firing angle of 60°. What is the input power factor of the rectifier?

P4.14   A phase-controlled six-pulse rectifier fed from a 460-V 60-Hz line operates with a firing angle of 40°. The source inductance is such that the dc output voltage of the rectifier is reduced by 10% compared to that with a zero-inductance source. Assuming the continuous conduction mode, find the overlap angle and dc output voltage of the rectifier.

P4.15   A circulating current-conducting dual converter fed from a 460-V line operates in the continuous conduction mode with a 60° firing angle in one of the constituent rectifiers. What is the firing angle of the other rectifier, and what is the average output voltage of the converter? Sketch the output voltage waveforms of both rectifiers of the converter and find the value of the differential voltage at $\omega t = \pi/12$ rad.

P4.16   At a certain instant, voltages v_{AB} and v_{AC} in a three-wire power line supplying a three-phase rectifier are 249.2 and 305.7 V, respectively, and currents i_A and i_C in the same line are 144.9 and -106.1 A, respectively. Determine and sketch space vectors, $\vec{V}_{LL}$ and $\vec{I}_L$, of the line-to-line voltage and line current.

P4.17   A current-type PWM rectifier operates with the switching frequency of 5 kHz and a modulation index of 0.65. Determine and sketch the waveforms of switching variables when the reference current vector is in sector IV of the dq plane.

P4.18   Find the instantaneous real and reactive powers drawn by the rectifier in Problem 4.16 from the supply line.

P4.19   The SCRs in a phase-controlled six-pulse rectifier are rated at 1200 V and 100 A. Assuming voltage and current safety margins of 40% and 20%, respectively, find the maximum allowable voltage of the supply ac line and the minimum allowable resistance of the load. Remember that the voltage of a three-phase line is always meant as the rms value of line-to-line voltage.

P4.20   SCRs rated at 5 kV are used in a multipulse converter at one end of a 400-kV dc transmission line. The converter can be assumed to produce ideal dc voltage equal to the peak line-to-line supply voltage times the cosine of the firing angle (see Eqs. (4.5) and (4.42)). When the interconnected systems operate with the rated ac voltage, the firing angle of the converter is 30° to provide a margin for control of the dc voltage when the ac voltage strains from its rated level. The ac voltage may vary from 90 to 110% of the rated value. Find the rated ac voltage at the input to the converter and, assuming a voltage safety margin of 20%, determine how many SCRs are connected in series in a single power switch of the rectifier.

COMPUTER ASSIGNMENTS

*CA4.1 Run PSpice program *Contr_Rect_1P.cir* for a single-pulse phase-controlled rectifier. Observe the voltage and current waveforms with various firing angles.

CA4.2 Use computer graphics to produce a background sheet for sketching voltage waveforms of six-pulse rectifiers. The sheet should contain waveforms of all six input line-to-line voltages drawn using dashed lines, as, for example, in Figure 4.8.

*CA4.3 Run PSpice program *Diode_Rect_3P.cir* program for a three-pulse diode rectifier. Set appropriate values of the coefficient of load EMF and perform simulations for the continuous and discontinuous operation modes of the rectifier. In both cases, find for the output voltage and current:

(a) The dc component

(b) The rms value

(c) The rms value of the ac component

(d) The ripple factor

Observe oscillograms of the input voltages and currents.

*CA4.4 Run PSpice program *Diode_Rect_6P.cir* for a six-pulse diode rectifier. Set appropriate values of the coefficient of load EMF and perform simulations for the continuous and discontinuous operation modes of the rectifier. In both cases, find for the output voltage and current:

(a) The dc component

(b) The rms value

(c) The rms value of the ac component

(d) The ripple factor

Observe oscillograms of the input voltages and currents.

*CA4.5 Run PSpice program *Diode_Rect_6P_F.cir* for a six-pulse diode rectifier with an input filter. To evaluate the impact of the input filter, find for the current drawn from the supply source (before the filter) and current at the input terminals of the rectifier (after the filter):

(a) The total harmonic distortion

(b) The amplitudes of the first, fifth, seventh, eleventh, and thirteenth harmonics

*CA4.6 Run PSpice program *Contr_Rect_6P.cir* for a phase-controlled six-pulse rectifier. Find for both the output voltage and current:

(a) The dc component

(b) The rms value

(c) The rms value of the ac component

(d) The ripple factor

Observe oscillograms of the input voltages and currents.

*CA4.7 Run PSpice program *Rect_Source_Induct.cir* for a six-pulse controlled rectifier supplied from a source with inductance. For reference, run the *Contr_Rect_6P.cir* program for a similar rectifier fed from an ideal source. Compare the output voltage waveforms and their average values.

CA4.8 Develop a computer program for analysis of the phase-controlled six-pulse rectifier. For specified values of the supply voltage, parameters of the RLE load, and firnig angle, the program should:

(a) Determine the feasibility of the firing angle

(b) Determine the conduction mode

(c) Calculate the output voltage waveform

(d) Calculate the output current waveform

It is advised that the waveform calculations are done for one subcycle of the input voltage and then extended on the remaining subcycles. Storing the waveform data will make it possible to create plots similar to Figures 4.8, 4.11, 4.17, 4.20, and 4.23.

*CA4.9 Run PSpice program *Dual_Conv.cir* for a six-pulse circulating current-conducting dual converter. Observe oscillograms of the output voltages of the constituent rectifiers, output voltage of the converter, differential output voltage, and circulating current.

*CA4.10 Run PSpice program *PWM_Rect_CT_NF.cir* for an ideal current-type PWM rectifier (dc current source as a load, no input filter). Find for the output voltage:

(a) The dc component

(b) The rms value

(c) The rms value of the ac component

(d) The ripple factor

Observe oscillograms of the input current.

*CA4.11 Run PSpice program *PWM_Rect_CT_F.cir* for a current-type PWM rectifier with an input filter. Determine:

(a) The total harmonic distortion of the input current

(b) The dc output power

(c) The real input power

(d) The apparent input power

(e) The input power factor

Observe oscillograms of the output voltage and current.

*CA4.12 Run PSpice program *PWM_Rect_VT.cir* for a voltage-type PWM rectifier with an input filter. Determine:

    (a) The total harmonic distortion of the input current

    (b) The dc output power

    (c) The real input power

    (d) The apparent input power

    (e) The input power factor

       Observe oscillograms of the output voltage and current.

CA4.13 Develop a computer program emulating the SVPWM modulator for a six-pulse current-type PWM rectifier (see Figure 4.48). Based on information about the modulation index, switching frequency, and angle of the input voltage vector, the program should determine the switching pattern for the rectifier, that is, the waveforms of switching signals for the all six switches within a switching cycle.

CA4.14 Develop a computer program that will determine the minimum required voltage and current ratings of power switches in a six-pulse rectifier. The program should ask for the volt–ampere and voltage ratings of the rectifier, and produce the voltage and current ratings of the switches, taking into account specified safety margins.

LITERATURE

[1] Hingorani, N. G., High-voltage dc transmission: a power electronics workhorse, *IEEE Spectrum*, Apr. 1996, pp. 63–72.

[2] Malesani, L., and Tenti, P., Three-phase ac/dc PWM converter with sinusoidal ac currents and minimum filter requirements, *IEEE Transactions on Industry Applications*, vol. 23, no. 1, pp. 71–77, 1987.

[3] Malinowski, M., Kazmierkowski, M., and Trzynadlowski, A. M., A comparative study of control techniques for PWM rectifiers in AC adjustable speed drives, *IEEE Transactions on Power Electronics*, vol. 18, no. 6, pp. 1390–1396, 2003.

[4] Rashid, M. H., *Power Electronics Handbook*, 2nd ed., Academic Press, San Diego, CA, 2007, Chaps. 10–12.

5 AC-to-AC Converters

Power electronic converters supplied from an ac source and producing voltage of adjustable magnitude and frequency are presented in this chapter. Next, power circuits, control methods, and characteristics of ac voltage controllers, cycloconverters, and matrix converters are described. Selection of power switches for ac-to-ac converters is explained and typical applications of the converters are reviewed.

5.1 AC VOLTAGE CONTROLLERS

If a fixed ac voltage needs to be adjusted, an ac voltage controller is used. No frequency control is possible, and the fundamental output frequency equals the input frequency. Single-phase controllers are supplied from a single-phase ac voltage source and have a single-phase output. In three-phase controllers, both the input and output are of three-phase type, usually in the three-wire version.

Ac voltage controllers are based on pairs of antiparallel-connected power switches. When SCRs or triacs are used, phase control is employed, and the controller belongs in the class of line-commutated power electronic converters. Fully-controlled switches are used in force-commutated, PWM ac voltage controllers, whose advantages are similar to those of PWM rectifiers.

5.1.1 Phase-Controlled Single-Phase AC Voltage Controller

Figure 5.1 is a circuit diagram of a phase-controlled single-phase ac voltage controller. In low-power converters, a triac is used instead of the two antiparallel-connected SCRs shown. An RL load is assumed in subsequent considerations. When one of the SCRs is conducting, the other SCR is reverse biased by the voltage drop across the conducting SCR. Only one path of the current exists, so the input current, i_i, equals the output current, i_o. With either SCR conducting, the input and output terminals of the converter are connected directly, and the output voltage, v_o, equals the input voltage, v_i.

Output Voltage and Current Selected voltage and current waveforms are shown in Figure 5.2. When SCR T1 is forward biased and fired at a certain firing angle, α_f, the current conducted initially increases thanks to the positive input voltage, but eventually drops to zero, somewhat later than the voltage does. Now the input voltage

Introduction to Modern Power Electronics, Second Edition, by Andrzej M. Trzynadlowski
Copyright © 2010 John Wiley & Sons, Inc.

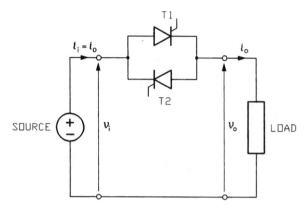

Figure 5.1   Single-phase ac voltage controller.

is negative, and it is SCR T2 that is forward biased and ready for firing. The firing occurs at $\omega t = \alpha_f + \pi$, where ω denotes the radian frequency, common for the input and output. The waveform of current through T2 is a mirror image of that through T1, so both the output current and voltage waveforms have the half-wave symmetry and no dc component. When a triac is used, firing pulses are applied to its gate with a delay of α_f after each zero crossing of the input voltage, that is, every half-cycle of this voltage.

In Figure 5.2a, the firing angle is 45° and in Figure 5.2b it is 135°. An RL load with a load angle, φ, of 30° is assumed. This is relevant information since, as shown later, the minimum feasible firing angle equals the load angle. It can be seen that when the firing angle increases, the conduction angle, β, decreases and the current pulses become smaller and shorter. Consequently, the rms value of the current decreases, as does the rms value of output voltage. Note that the current is always discontinuous, and waveforms of both the current and output voltage are strongly distorted compared with pure sinusoids, especially with high firing angles. Therefore, ac voltage controllers are used sparingly in applications requiring high-quality ac currents such as ac motors.

Accounting for the half-wave symmetry of the output voltage waveform, $v_o(\omega t)$, the rms value of this voltage can be found as

$$
V_o = \sqrt{\frac{1}{\pi} \int_0^\pi v_o^2(\omega t)\, d\omega t} = \sqrt{\frac{1}{\pi} \int_0^\pi (V_{i,p} \sin \omega t)^2\, d\omega t}
$$

$$
= V_{i,p} \sqrt{\frac{1}{\pi} \left[\alpha_e - \alpha_f - \frac{1}{2}(\sin 2\alpha_e - \sin 2\alpha_f) \right]}
$$

$$(5.1)$$

where $V_{i,p}$ and V_i denote peak and rms values of the input voltage, respectively, and α_e is the extinction angle. The extinction angle depends on the firing angle, α_f, and load angle, φ.

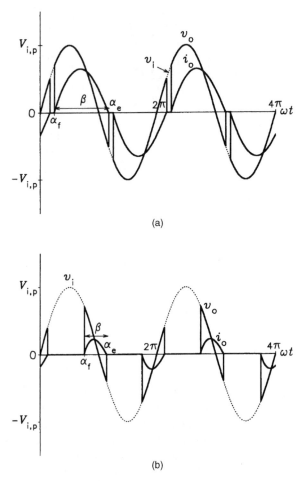

Figure 5.2 Waveforms of output voltage and current of a single-phase ac voltage controller ($\varphi = 30°$): (a) $\alpha_f = 45°$; (b) $\alpha_f = 135°$.

The expression for the load current waveform, $i_o(\omega t)$, within the α_f-to-α_e interval of ωt can be derived in a way similar to that used for a phase-controlled rectifier in the discontinuous conduction mode (see Section 4.2.1). The waveform is given by

$$i_o(\omega t) = \frac{V_{i,p}}{Z} \left[\sin(\omega t - \varphi) - e^{(\omega t - \alpha_f)/\tan \varphi} \sin(\alpha_f - \varphi) \right] \tag{5.2}$$

where Z denotes the load impedance, as in Eq. (4.11). If $\omega t = \alpha_e$, the current has just reached zero, which allows numerical determination of the extinction angle. If $\alpha_f = \varphi$, the right-hand term in brackets disappears, and the current becomes purely sinusoidal, as if the load were connected directly to the supply source. This observation confirms the statement made previously about the minimum feasible firing angle. The expression for the negative pulse of the current differs from Eq. (5.2) by a minus sign only.

Voltage Control   The magnitude control ratio, M, for ac voltage controllers is defined with respect to the rms value, V_o, of the output voltage, whose maximum available value equals the rms value, V_i, of the input voltage. Since there is no closed-form expression for the extinction angle, α_e, which appears in Eq. (5.1), no closed-form expression for the voltage control characteristic can be derived. This is not a serious problem, since in the practical loads the load angle, necessary for computation of the extinction angle, is usually unknown anyway. Therefore, only an envelope of control characteristics for load angles in the realistic (zero to $\pi/2$) range can be determined. If $\varphi = 0$ (purely resistive load), Eq. (5.2) yields

$$i_o(\omega t) = \frac{V_{i,p}}{R} \sin \omega t \tag{5.3}$$

and the extinction angle is π radians. Conversely, if $\varphi = \pi/2$ (purely inductive load), then

$$i_o(\omega t) = \frac{V_{i,p}}{\omega L}(\cos \alpha_f - \cos \omega t) \tag{5.4}$$

and the extinction angle is $2\pi - \alpha_f$, as $\cos \alpha_f - \cos(2\pi - \alpha_f) = 0$.

Substituting the values of α_e obtained in Eq. (5.1), the envelope of control characteristics can be expressed as

$$V_{o(\varphi=0)}(\alpha_f) \leq V_o(\alpha_f) \leq V_{o(\varphi=\pi/2)}(\alpha_f) \tag{5.5}$$

where

$$V_{o(\varphi=0)}(\alpha_f) = V_i\sqrt{\frac{1}{\pi}\left(\pi - \alpha_f + \frac{1}{2}\sin 2\alpha_f\right)} \tag{5.6}$$

and

$$V_{o(\varphi=\pi/2)}(\alpha_f) = \sqrt{2}\, V_{o(\varphi=0)}(\alpha_f). \tag{5.7}$$

Graphic representation of Eq. (5.5) is shown in Figure 5.3 in the $M = f(\alpha_f)$ form.

As already mentioned, the load angle is usually unknown. Interestingly, the behavior of an ac voltage controller when the firing angle is less than the load angle depends on the firing technique. As explained in Section 2.3.1, SCRs and triacs can be fired using a short single pulse of the gate current, i_g, or a long multipulse. Operation of the controller with these two firing schemes and $\alpha_f < \varphi$ is illustrated in Figure 5.4. In the case of single gate pulses (Figure 5.4a), only one SCR is fired, by gate pulse i_{g1}, since firing of the other SCR, by gate pulse i_{g2}, is attempted when it is

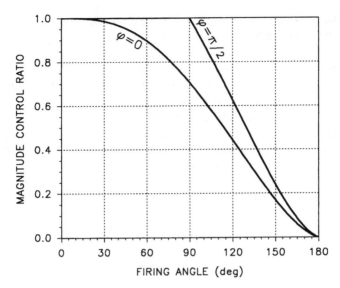

Figure 5.3 Envelope of control characteristics, $V_o = f(\alpha_f)$, of a single-phase ac voltage controller.

reverse biased by the first, still conducting SCR. As a result, the converter operates faultily as a single-pulse rectifier. If multipulses, whose width should not be less than $\pi/2$ radians, are employed (Figure 5.4b), each SCR is fired successfully by one of the constituent pulses of a multipulse immediately after the other SCR has ceased to conduct. Thus, the converter acts as a closed switch between the supply voltage and the load, and the actual firing angle equals the load angle. For this reason, multipulse gate signals are used in all phase-controlled ac voltage controllers, including the three-phase controllers covered in the next section.

An ac voltage controller with the multipulse firing of SCRs operates correctly no matter what firing angle, measured from the zero crossing of the input voltage to the beginning of the multipulse, is used. However, changes in the firing angle in the range zero to φ do not affect the output voltage and current, both of which stay at their maximum rms values. This dead zone, whose width varies with the load angle, is undesirable, especially in feedforward control schemes. Therefore, an alternative approach to the phase control has been devised, in which the firing angle is defined with respect to the instant when the current, not voltage, crosses zero. This technique requires current sensing, which actually may be needed anyway for the control or protection purposes.

As seen in Figure 5.5, the angle interval, α'_f, between an end of a current pulse and the next firing instant is given by

$$\alpha'_f = \pi - \beta = \pi - \alpha_e + \alpha_f. \tag{5.8}$$

Angle α'_f, which represents a firing delay measured from the last zero crossing of current, will subsequently be called a *control angle*. Based on Eq. (5.8), the firing

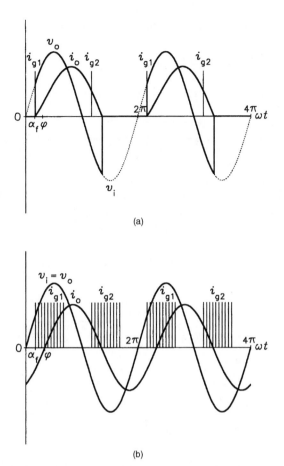

(a)

(b)

Figure 5.4 Operation of a single-phase ac voltage controller with (a) a single-pulse gate signal; (b) a multipulse gate signal.

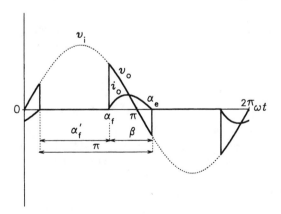

Figure 5.5 Definition of a control angle.

angle, α_f, can be expressed in terms of the control angle as

$$\alpha_f = \alpha'_f + \alpha_e - \pi \tag{5.9}$$

and substituted in Eq. (5.1), to yield

$$V_o = V_i \sqrt{\frac{1}{\pi}[\pi - \alpha'_f + \sin \alpha'_f \, \cos(2\alpha_e + \alpha'_f)]}. \tag{5.10}$$

It can be seen that with $\alpha'_f = 0$, $V_o = V_i$, and with $\alpha'_f = \pi$, $V_o = 0$, independent of the load angle. As before, an envelope of control characteristics corresponding to various load angles can be determined by substituting $\alpha_e = \pi$ in Eq. (5.10) for the $\varphi = 0$ case and $\alpha_e = 2\pi - \alpha_f = 3(\pi - \alpha'_f)/2$ for the $\varphi = \pi/2$ case. Then

$$V_{o(\varphi=\pi/2)}(\alpha'_f) \leq V_o(\alpha'_f) \leq V_{o(\varphi=0)}(\alpha'_f) \tag{5.11}$$

where

$$V_{o(\varphi=\pi/2)}(\alpha'_f) = V_i \sqrt{\frac{1}{\pi}(\pi - \alpha'_f - \sin \alpha'_f)} \tag{5.12}$$

and

$$V_{o(\varphi=0)}(\alpha'_f) = V_i \sqrt{\frac{1}{\pi}\left(\pi - \alpha'_f + \frac{1}{2}\sin 2\alpha'_f\right)}. \tag{5.13}$$

Relation (5.11) is illustrated in Figure 5.6. No dead zone exists in the control characteristics and, particularly with highly inductive loads, the rms output voltage decreases almost linearly when the control angle increases.

Power Factor   The input power factor, PF, of a lossless ac voltage controller can be expressed as

$$\text{PF} = \frac{P_o}{S_i} = \frac{R I_o^2}{V_i I_i} = \frac{Z I_o}{V_i} \cos \varphi. \tag{5.14}$$

When $\alpha_f \leq \varphi$ and, consequently, $I_o = V_i/Z$, the power factor equals that of the load, that is, $\cos \varphi$. When the firing angle increases beyond the load angle, the rms output current, I_o, decreases and so does the power factor. An ac voltage controller with a purely resistive load has the highest power factor, while a purely inductive load results in a power factor of zero for all values of α_f.

The relation between the input power factor, PF, and firing angle, α_f, for an R load ($\varphi = 0$) can easily be derived substituting R for Z and V_o/R for I_o in Eq. (5.14).

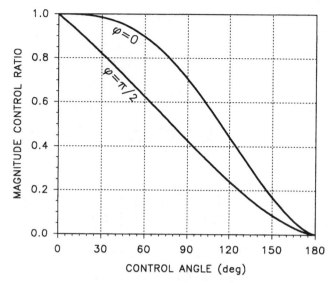

Figure 5.6 Envelope of control characteristics, $V_o = f(\alpha'_f)$, of a single-phase ac voltage controller.

Then $PF = V_o/V_i = M$ (magnitude control ratio). Thus, based on Eq. (5.6),

$$PF_{(\varphi=0)}(\alpha_f) = \sqrt{\frac{1}{\pi}\left(\pi - \alpha_f + \frac{1}{2}\sin 2\alpha_f\right)}.$$

(5.15)

It can be seen that similar to phase-controlled rectifiers, phase-controlled ac voltage controllers are characterized by the decreasing quality of the input current with an increase in the firing angle. The input power factor decreases and the total harmonic distortion of the input current increases.

5.1.2 Phase-Controlled Three-Phase AC Voltage Controllers

Several configurations of three-phase ac voltage controllers are possible. Here, only the most common, *fully controlled* topology with a wye-connected load is analyzed in detail. Other types of three-phase controllers are described briefly, with the stress on only their unique features.

Fully Controlled Three-Phase AC Voltage Controller   Operation of the fully controlled three-phase ac voltage controller, shown in Figure 5.7, is more complicated than that of the single-phase converter. Note, for example, that to get the controller started, two triacs must be fired simultaneously to provide the path for current necessary to maintain the on-state of triacs.

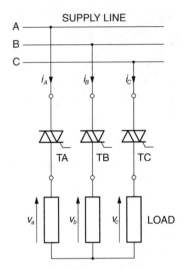

Figure 5.7   Fully controlled three-phase ac voltage controller.

Operation with a single triac conducting a current is impossible. Therefore, the controller can operate with two or three triacs conducting or none at all. If no triac is conducting, the load is cut off from the supply and all currents and output voltages are zero. Two- and three-triac conduction cases are illustrated in Figure 5.8. A balanced load is assumed and, for better visualization, the triacs are represented by generic

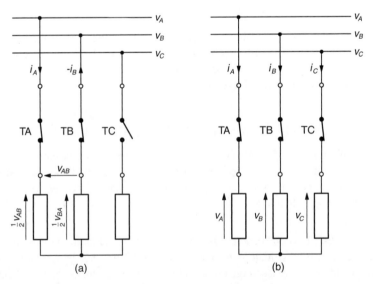

Figure 5.8   Voltage and current distribution in a fully controlled three-phase ac voltage controller: (a) two triacs conducting; (b) three triacs conducting.

switches. It can be seen in Figure 5.8a that with two triacs, TA and TB, conducting, the line-to-line voltage, v_{AB}, is split equally between the load impedances in phases A and B. Consequently, $v_a = v_{AB}/2$, $v_b = -v_{AB}/2 = v_{BA}/2$, and $v_c = 0$. When all three triacs are conducting, as in Figure 5.8b, the output voltages, v_a, v_b, and v_c, equal their input counterparts, v_A, v_B, and v_C.

Switching variables a, b, and c can be introduced for triacs TA, TB, and TC, respectively, and defined as equal to 1 when a given triac is conducting and equal 0 otherwise. It can easily be demonstrated that the output voltages of the controller are given by

$$
\begin{bmatrix} v_a \\ v_b \\ v_c \end{bmatrix} = \frac{1}{2} \begin{bmatrix} a & -b & -c \\ -a & b & -c \\ -a & -b & c \end{bmatrix} \begin{bmatrix} v_A \\ v_B \\ v_C \end{bmatrix}.
\tag{5.16}
$$

Indeed, if no triac is conducting, then $a = b = c = 0$ and $v_a = v_b = v_c = 0$. When, for examples, triacs TA and TB are conducting, then $a = b = 1, c = 0$ and $v_a = (v_A - v_B)/2 = v_{AB}/2$, $v_b = (-v_A + v_B)/2 = v_{BA}/2$, and $v_c = (-v_A - v_B)/2 = v_c/2$. The last equation is only true when $v_c = 0$. Finally, if all triacs are conducting, then $a = b = c = 1$ and $v_a = (v_A - v_B - v_C)/2 = [2v_A - (v_A + v_B + v_C)]/2 = (2v_A - 0)/2 = v_A$, and, similarly, $v_b = v_B$, and $v_c = v_C$.

Operation of the fully controlled ac voltage controller with a purely resistive, wye-connected load (R load) will be considered. Depending on the firing angle, three modes of operation of the controller can be distinguished:

Mode 1 $(0° \leq \alpha_f < 60°)$: two or three triacs conducting (in either direction)
Mode 2 $(60° \leq \alpha_f < 90°)$: two triacs conducting
Mode 3 $(90° \leq \alpha_f < 150°)$: none or two triacs conducting

Mode 1 is illustrated in Figure 5.9. For clarity, only the waveform of phase A output voltage, v_a, is shown (because of the resistive load, the output current waveform has the same shape). In Figure 5.9a, the firing angle, α_f, is zero, and the controller operates in mode 1, with all triacs conducting all the time and connecting the source with the load. All three switching variables have a value of 1. With the firing angle of 30°, as in Figure 5.9b, the controller still operates in mode 1, but the triacs cease to conduct when their currents reach zero. As a result, either two or three triacs are conducting simultaneously and the output voltage, v_a, sequentially equals v_A, $v_{AB}/2$, v_A, $v_{AC}/2$, and zero.

To explain mode 2, the 0-to-90° interval of ωt is considered. The line-to-neutral input voltages are given by

$$
v_A = V_{LN,p} \sin \omega t
$$
$$
v_B = V_{LN,p} \sin \left(\omega t - \tfrac{2}{3}\pi \right)
\tag{5.17}
$$
$$
v_B = v_{LN,p} \sin \left(\omega t - \tfrac{4}{3}\pi \right)
$$

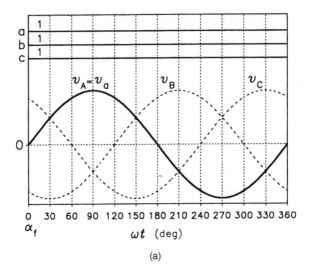

(a)

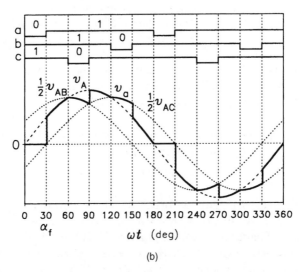

(b)

Figure 5.9 Output voltage waveforms in a fully controlled three-phase ac voltage controller in mode 1: (a) $\alpha_f = 0°$; (b) $\alpha_f = 30°$ (R load).

and the line-to-line voltages by

$$v_{AB} = V_{LL,p} \sin\left(\omega t + \tfrac{1}{6}\pi\right)$$
$$v_{BC} = V_{LL,p} \sin\left(\omega t - \tfrac{1}{2}\pi\right) \qquad (5.18)$$
$$v_{CA} = v_{LL,p} \sin\left(\omega t - \tfrac{7}{3}\pi\right)$$

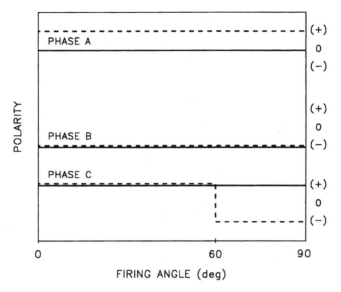

Figure 5.10   Polarities of output voltages and currents in a fully controlled three-phase ac voltage controller in mode 2 before firing triac TA (solid line) and following the firing.

Triacs TB and TC are conducting, and the nonconducting triac TA is fired at $\omega t = \alpha_f$. According to Eq. (5.16), the output voltages just prior to the firing are

$$
\begin{aligned}
v_a(\alpha_f^-) &= 0 \\
v_b(\alpha_f^-) &= \tfrac{1}{2} v_{BC}(\alpha_f) = \tfrac{1}{2} V_{LL,p} \sin\left(\alpha_f - \tfrac{1}{2}\pi\right) \\
v_c(\alpha_f^-) &= -v_b(\alpha_f^-) = -\tfrac{1}{2} V_{LL,p} \sin\left(\alpha_f - \tfrac{1}{2}\pi\right)
\end{aligned}
\tag{5.19}
$$

while just after the firing, they change to

$$
\begin{aligned}
v_a(\alpha_f^+) &= v_A(\alpha_f) = V_{LN,p} \sin \alpha_f \\
v_b(\alpha_f^+) &= v_B(\alpha_f) = V_{LN,p} \sin\left(\alpha_f - \tfrac{2}{3}\pi\right) \\
v_c(\alpha_f^+) &= v_C(\alpha_f) = v_{LN,p} \sin\left(\alpha_f - \tfrac{4}{3}\pi\right).
\end{aligned}
\tag{5.20}
$$

As illustrated in Figure 5.10, if the firing angle is less than 60°, the firing of TA does not change polarities of voltages v_b and v_c and, consequently, polarities of currents i_b and i_c. However, when the firing angle exceeds 60°, the output voltage and current in phase C change their polarities, and triac TC is extinguished by the current reversal caused by firing triac TA. Clearly, this process represents the line-supported commutation, already encountered in phase-controlled rectifiers. Operation of the controller in mode 2 is illustrated in Figure 5.11a, with the firing angle of 75°.

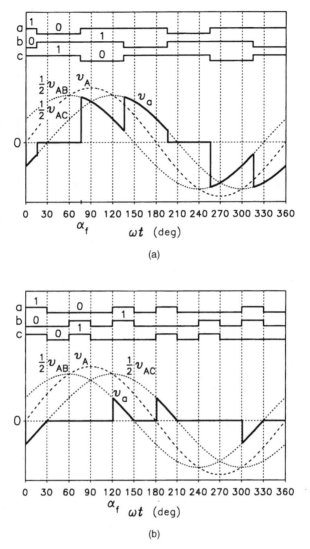

Figure 5.11 Output voltage waveforms in a fully controlled three-phase ac voltage controller: (a) $\alpha_f = 75°$, mode 2; (b) $\alpha_f = 120°$, mode 3.

Another mechanism causes transition from mode 2 to mode 3 when the firing angle is increased to 90° and above. Now, when triac TA is fired, the other two triacs have already been extinguished because their currents had reached zero. Thus, the attempt at firing TA is unsuccessful, and no triac is conducting until a gate pulse is applied to TB, 60° later. It must be kept in mind that as multipulse firing is employed, triac TA is still subjected to gate pulses. Therefore, it turns on simultaneously with TB. Mode 3 of operation is illustrated in Figure 5.11b for the firing angle of 120°.

Inspection of the figure leads to the observation that if the firing angle were increased further, no triac could be fired at $\alpha_f \geq 150°$.

Analysis of operation of the controller with an RL load is difficult, because the extinction angle, α_e, and the *limit angle*, α_{lim}, both not expressible in a closed form, must be known. Mode 2, characterized by rapid changes of the output currents (see Figure 5.10), is impossible due to the load inductance. The ranges of the two remaining operation modes are $\varphi \leq \alpha_f < \alpha_{lim}$ for mode 1 and $\alpha_{lim} \leq \alpha_f < 150°$ for mode 3. The limit angle can be determined numerically from the equation

$$\frac{\sin\left(\alpha_{lim} - \varphi - \frac{4}{3}\pi\right)}{\sin(\alpha_{lim} - \varphi)} = \frac{2e^{-\pi/3\ \tan\varphi} - 1}{2 - e^{-\pi/3\ \tan\varphi}}. \tag{5.21}$$

Readers interested in the detailed analysis of this and other three-phase ac voltage controllers are referred to the book of Rombaut et al. [3]. For completeness, equations for the rms output voltage, V_o, of the fully controlled ac voltage controller with purely resistive and purely inductive loads are provided below without derivation.

Resistive load:

$$V_o = V_i \sqrt{\frac{1}{\pi}\left(\pi - \frac{3}{2}\alpha_f + \frac{3}{4}\sin 2\alpha_f\right)} \tag{5.22}$$

for $0 \leq \alpha_f < 60°$,

$$V_o = V_i \sqrt{\frac{1}{\pi}\left[\frac{\pi}{2} + \frac{3\sqrt{3}}{4}\sin\left(2\alpha_f + \frac{\pi}{6}\right)\right]} \tag{5.23}$$

for $60° \leq \alpha_f < 90°$, and

$$V_o = V_i \sqrt{\frac{1}{\pi}\left[\frac{5}{4}\pi - \frac{3}{2}\alpha_f + \frac{3}{4}\sin\left(2\alpha_f + \frac{\pi}{3}\right)\right]} \tag{5.24}$$

for $90° \leq \alpha_f < 150°$.

Inductive load:

$$V_o = V_i \sqrt{\frac{1}{\pi}\left(\frac{5}{2}\pi - 3\alpha_f + \frac{3}{2}\sin 2\alpha_f\right)} \tag{5.25}$$

for $90° \leq \alpha_f < 120°$, and

$$V_o = V_i \sqrt{\frac{1}{\pi} \left[\frac{5}{2}\pi - 3\alpha_f + \frac{3}{2} \sin\left(2\alpha_f + \frac{\pi}{3}\right) \right]} \tag{5.26}$$

for $120° \leq \alpha_f < 150°$.

The envelope of control characteristics given by Eqs. (5.22) through (5.26) is shown in Figure 5.12. To avoid the dead zone, as in the single-phase controller, the control angle measured from the last zero crossing of current in the given phase of a controller can be employed in place of the firing angle. Relations between the control angle, α'_f, and firing angle, α_f, are given by

$$\alpha'_f = \begin{cases} \alpha_f & \text{for } 0° \leq \alpha_f < 60° \\ 60° & \text{for } 60° \leq \alpha_f < 90° \\ \alpha_f - 30° & \text{for } 90° \leq \alpha_f < 150° \end{cases} \tag{5.27}$$

for a resistive load, and

$$\alpha'_f = 2(\alpha_f - 90°) \tag{5.28}$$

for an inductive load. The envelope of voltage control characteristics when the control angle is used is shown in Figure 5.13. With a purely resistive load, the control characteristic is discontinuous at $\alpha' = 60°$. Therefore, this control method should be employed only when the load contains substantial inductance. The sensitivity of the rms output voltage to changes of the control angle is higher than that in a single-phase ac voltage controller, since the range of the control angle is 120° instead of 180°.

When a three-phase load of a fully controlled ac voltage controller is connected in delta instead of wye, the control characteristics remain unchanged, although different waveforms of the output voltage are generated. The input power factor of all ac voltage controllers is given by the general equation (5.14), where V_i represents the rms line-to-neutral voltage, V_{LN}, for wye-connected loads, and the rms line-to-line voltage, V_{LL}, for delta-connected loads. With a purely resistive load, the power factor, PF, equals the magnitude control ratio, M.

Other Types of Three-Phase AC Voltage Controllers   Some other types of phase-controlled three-phase ac voltage controllers are shown in Figures 5.14 and 5.15. In high-power three-phase ac voltage controllers, actual SCRs must be used, because of the low current ratings of the triacs available. In such controllers, the *half-controlled topology*, depicted in Figure 5.14a, can be employed. Instead of another SCR, each of the three SCRs has a diode connected antiparallel. When a current flows through a diode in one phase, it is controlled indirectly by the conducting SCRs in another phase. Interestingly, the maximum firing angle in the half-controlled ac voltage controller is 210°.

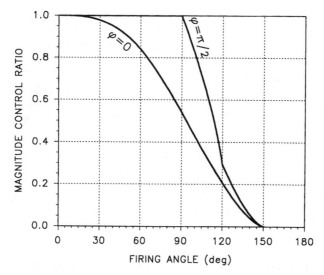

Figure 5.12 Envelope of control characteristics, $V_o = f(\alpha_f)$, of a fully controlled three-phase ac voltage controller.

Three triacs and phase loads can be connected in delta, as shown in Figure 5.14b. Characteristics of each phase controller are the same as those of the single-phase controller, but triple harmonics are absent in the line currents drawn from the ac supply line. The controllers shown in Figure 5.15 are connected *after* the load, which must have all six terminals accessible. On the other hand, the controllers have only

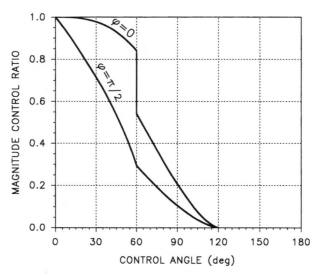

Figure 5.13 Envelope of control characteristics, $V_o = f(\alpha'_f)$, of a fully controlled three-phase ac voltage controller.

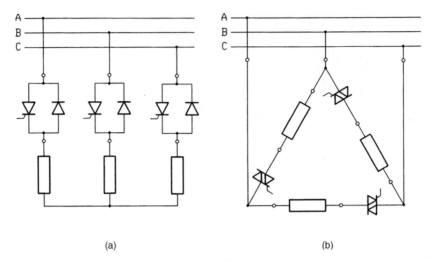

Figure 5.14 Three-phase ac voltage controllers connected before the load: (a) half-controlled; (b) delta-connected.

three terminals. The triacs in the wye-connected controller in Figure 5.15a have a common node, which simplifies the control circuitry. A half-controlled configuration as in Figure 5.14a is also possible. Such configuration is, however, infeasible in the delta-connected controller in Figure 5.15b, since the diodes connecting the load terminals at the controller side would make the control impossible. The advantage

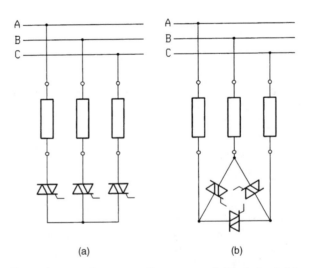

Figure 5.15 Three-phase ac voltage controllers connected after the load: (a) wye-connected; (b) delta-connected.

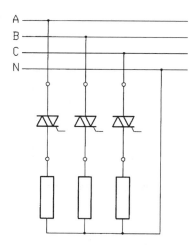

Figure 5.16 Three-phase four-wire ac voltage controller.

of the delta topology consists in low-ampere ratings of the triacs, which carry lower currents than those in the load.

The circuit diagram of a four-wire ac voltage controller is shown in Figure 5.16. The load is connected in wye and the neutrals of the load and supply line are connected. As such, the converter operates as three independent single-phase ac voltage controllers, whose operating properties have been described in Section 5.1.1. In practice, four-wire controllers are used only when there is a need to control individual phase loads independently and when the load has the neutral point accessible. Otherwise, even with unbalanced loads, the arrangement in Figure 5.14b is preferable.

5.1.3 PWM AC Voltage Controllers

The circuit diagram of a single-phase PWM ac voltage controller, also known as an *ac chopper*, is shown in Figure 5.17. An input filter such as that used for PWM rectifiers is required to attenuate the high-frequency harmonic currents drawn from the power system. The inductance provided by the power system is often sufficient, so that only the capacitor is installed. Ac choppers have been developed to improve the input power factor, control characteristics, and quality of the output current.

Fully controlled main switches S1 and S2 connected antiparallel allow controlling rms values of the output voltage and current, while freewheeling switches S3 and S4 are required for providing a path for this current when both S1 and S2 are off. The switches are turned on and off many times within a cycle of the input voltage. Assigning switching variables x_1 through x_4 to switches S1 through S4, respectively, the duty ratios for switches S1 and S2 in the nth switching interval, $d_{1,n}$ and $d_{2,n}$, are

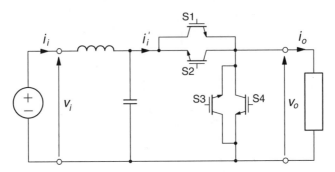

Figure 5.17 Single-phase ac chopper with an input filter.

given by

$$d_{1,n} = \begin{cases} F(m, \alpha_n) & \text{for } 0 < \alpha_n \leq \pi \\ 0 & \text{otherwise} \end{cases}$$

$$d_{2,n} = \begin{cases} F(m, \alpha_n) & \text{for } \pi < \alpha_n \leq \pi \\ 0 & \text{otherwise} \end{cases} \tag{5.29}$$

and for switches S3 and S4

$$x_3 = \bar{x}_1$$
$$x_4 = \bar{x}_2. \tag{5.30}$$

Symbol $F(m,\alpha_n)$ in Eq. (5.29) denotes the value of the *modulating function* $F(m,\omega t)$ at $\omega t = \alpha_n$, where m is the modulation index and α_n is the central angle of the nth switching interval. The modulating function determines the magnitude of output voltage and improves the quality of output current. It can be as simple as $F(\omega t) = m$, or more complex, for example, $F(\omega t) = m|\sin \omega t|$. Duty ratios of switches S1 and S2 in the nth switching interval are denoted by $d_{1,n}$ and $d_{2,n}$, respectively.

In practice, it is convenient to control the switches by applying the on and off switching signals continuously to all switches, so that the *intended* duty ratios and switching variables are

$$d_{1,n}^* = d_{2,n}^* = F(m, \alpha_n)$$
$$x_3^* = x_4^* = \bar{x}_1^* \tag{5.31}$$

Because of the bias of individual switches during periods of positive and negative input voltage, the *actual* duty ratios and switching variables are as described by Eqs. (5.29) and (5.30). Unsuccessful attempts at firing reverse-biased switches do not affect the operation of the controller. In contrast with controlled rectifiers, this control scheme does not require synchronization of firing signals with the input voltage.

Operation of an ac chopper with 12 switching intervals per cycle, a modulation index of 0.8, and the simplest modulating function $F(m,\omega t) = m$ is illustrated in

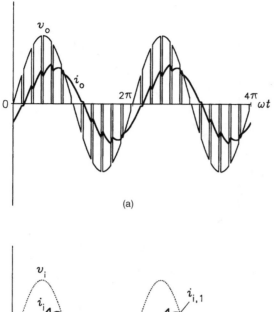

(a)

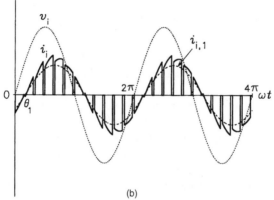

(b)

Figure 5.18 Waveforms of voltages and currents in a single-phase ac chopper: (a) output voltage and current; (b) input voltage and current (after the input filter) and the fundamental output current.

Figure 5.18. Clearly, the output current is of higher quality than that in phase-controlled ac voltage controllers. The output voltage waveform is identical with that in Figure 1.21b for a generic PWM controller (see Figure 1.23 for a typical harmonic spectrum). A pulsed current, i_i, is drawn from the filter. The fundamental, i_{i1}, of this current lags the input voltage, v_i, by angle θ_1, which is practically equal to the load angle φ. Thanks to the input capacitor, the power factor at the terminals of the filter is higher than $\cos\theta_1$. The input current, i_i, delivered by the supply system is sinusoidal, with only minor ripple.

It can be proved that the magnitude control ratio, M, equals $\sqrt{m}$; that is,

$$V_o = \sqrt{m}\, V_i \tag{5.32}$$

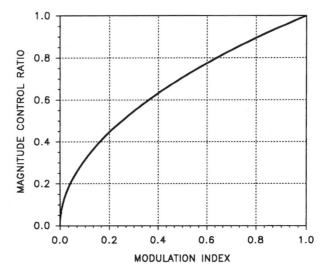

Figure 5.19   Control characteristic of an ac chopper.

which results in the control characteristic shown in Figure 5.19. However, the ratio $V_{o,1}/V_{i,1}$ of fundamentals of the output and input voltage equals the modulation index m.

The simple modulation technique described is most popular, but not exclusive. Several advanced PWM techniques with variable duty ratios have also been developed, for the purpose of reducing the filter size and improving the quality of input and output currents. Three-phase PWM ac voltage controllers in the wye and delta configurations are shown in Figures 5.20 and 5.21, respectively.

The pulse width modulation mode cannot extend over the entire theoretical 0-to-1 range of the modulation index, because the minimum on- and off-times of switches would have to approach zero. Therefore, the actual modulation is performed with the values of the modulation index between $m_{min} > 0$ and $m_{max} < 1$. If the modulation index desired falls below m_{min}, it is rounded up to m_{min}, limiting the minimum available value of the magnitude control ratio to $\sqrt{m_{min}}$. When the desired value of m is greater than m_{max}, the modulation index is set to unity, causing the controller to operate as a closed ac switch.

5.2   CYCLOCONVERTERS

Cycloconverters are ac-to-ac power converters in which the output frequency is a fraction of the input frequency. A single-phase two-pulse generic cycloconverter in a simple trapezoidal mode of operation resulting in an integer ratio of the input frequency to output frequency was shown in Examples 1.1 and 1.2. Practical cyclo-converters have a three-phase output, although a hypothetical single-phase six-pulse cycloconverter will be used to illustrate the operating principles. The number of

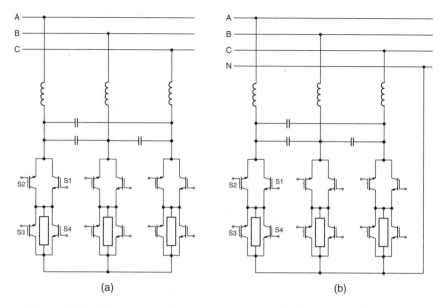

Figure 5.20   Wye-connected three-phase ac choppers: (a) three-wire; (b) four-wire.

phases of a cycloconverter indicated applies to the *output*, the input being typically of the three-phase type.

As the single-phase six-pulse cycloconverter, the circulating current-conducting dual converter (see Figures 4.33 and 4.37) can be used. A situation where different types of power conversion can be realized using the same converter topology, but employing different control laws, is quite common in power electronics.

Any ac-output power converter operates in all four quadrants of the current–voltage plane, because both the output voltage and current change their polarities within each cycle. As the dual converter is capable of such operation, the ac-to-ac conversion required is feasible. For subsequent considerations it is convenient to introduce the *local-averaged* output voltage, $v_{o,av}$. It is the output voltage of a converter averaged over one 60°-long subcycle of the input voltage. In a cycloconverter, the firing angles of the constituent rectifiers are varied such that the local-averaged output voltage changes sinusoidally in time with the radian frequency ω_o. This frequency is the output frequency of the converter and, inherently, it cannot be higher than the input frequency, ω. To maintain a reasonable quality of output current, it is recommended that the ω/ω_o ratio be at least 3.

It follows from Eq. (4.42) that the dc output voltage of a dual converter can be expressed as

$$V_{o,dc} = V_{o,dc(max)} \cos \alpha_f \qquad (5.33)$$

where $V_{o,dc(max)}$ denotes the maximum available value of this voltage, corresponding to the firing angle of zero. If the dual converter operates as a single-phase

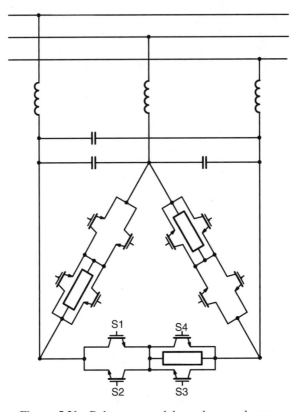

Figure 5.21   Delta-connected three-phase ac chopper.

cycloconverter, the local-averaged output voltage should vary according to the equation

$$v_{o,av}(t) = V_{o,1,p} \sin \omega_o t \qquad (5.34)$$

where $V_{o,1,p}$ is the desired peak value of the fundamental output voltage of the cycloconverter.

Based on Eqs. (5.33) and (5.34), the required variations of the firing angle can be expressed as

$$\cos[\alpha_f(\omega_o t)] = \frac{V_{o,1,p}}{V_{o,dc(max)}}. \qquad (5.35)$$

Given that the maximum available peak value of the local-averaged output voltage equals the maximum available dc voltage, $V_{o,dc(max)}$, the $V_{o,1,p}/V_{o,dc(max)}$ ratio in Eq. (5.35) represents the magnitude control ratio, M. Thus, the firing angle as a function

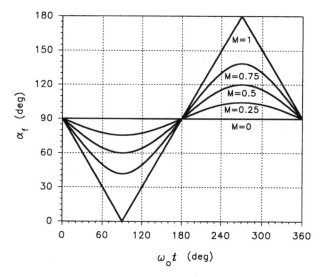

Figure 5.22 Changes in the firing angle in a cycloconverter.

of $\omega_o t$ is given by

$$\alpha_f(\omega_o t) = \cos^{-1}(M \sin \omega_o t) \tag{5.36}$$

and illustrated in Figure 5.22 for various values of M.

Waveforms of the output voltage, $v_{o,1}$, of one of the constituent rectifiers of a six-pulse cycloconverter are shown in Figure 5.23 for a frequency ratio ω/ω_o of 5. The magnitude control ratio is 1 in Figure 5.23a and 0.5 in Figure 5.23b. It is assumed that the cycloconverter is of the circulating current-conducting type. Otherwise, the changeover from one constituent rectifier to another, when the output current crosses zero, would have to be accompanied by a short delay period to allow for a complete turn-off of the SCRs in the outgoing rectifier. Circulating current-conducting cycloconverters require separating inductors, but they make the output voltage waveforms smoother and closer to ideal sinusoids. Consequently, the output currents are also of high quality. On the other hand, voltage drops across the inductors reduce the rms output voltage compared with that of the constituent rectifiers.

Practical cycloconverters are three-phase high-power converters, usually supplied from dedicated transformers. The transformer establishes the maximum available output voltage of the converter, and for certain types of cycloconverters, provides isolation of the supply sources for individual phases.

The circuit diagram of a three-phase three-pulse cycloconverter is shown in Figure 5.24, and two types of three-phase six-pulse cycloconverters are depicted in Figures 5.25 and 5.26. Phase loads of the cycloconverter in Figure 5.25 are isolated from each other, and the converter is supplied from a single three-phase source. When the

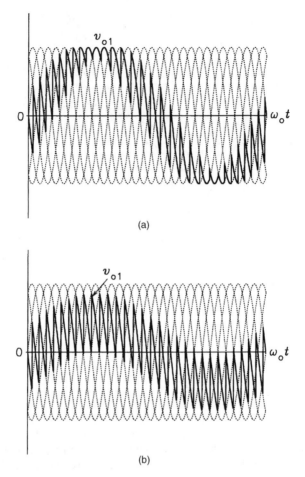

Figure 5.23 Output voltage waveforms in a six-pulse cycloconverter: (a) $M = 1$; (b) $M = 0.5$ ($\omega/\omega_o = 5$).

phase loads are connected, as in Figure 5.26, the interphase isolation is provided by a supply transformer with three secondary windings. All three cycloconverters shown are of the circulating current-conducting type. Otherwise, the separating inductors could be disposed of.

Neglecting the voltage drops across the separating inductors, the rms value, $V_{o,\mathrm{LN},1}$, of the fundamental line-to-neutral output voltage in the three-pulse cycloconverter in Figure 5.24 is given by

$$V_{o,\mathrm{LN},1} = \frac{3\sqrt{3}}{2\pi} M V_{\mathrm{LN}} \approx 0.827 \, M V_{\mathrm{LN}} \tag{5.37}$$

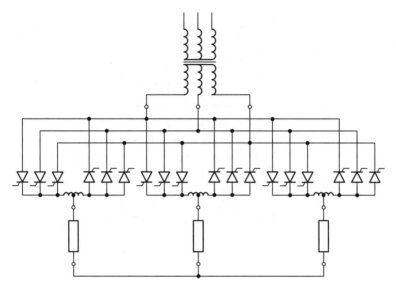

Figure 5.24   Three-phase three-pulse cycloconverter.

while the same value in the six-pulse cycloconverters in Figures 5.25 and 5.26 is

$$V_{o,\text{LN},1} = \frac{3}{\pi} M V_{\text{LL}} \approx 0.955\, M V_{\text{LL}} \tag{5.38}$$

where V_{LN} and V_{LL} denote rms line-to-neutral and line-to-line input voltages, respectively. As in the respective rectifiers, the output voltage of a six-pulse cycloconverter is twice as high as that of a three-pulse cycloconverter.

Cycloconverters are perfectly suited for high-power low-frequency applications, because the quality of the output current increases with a decrease in output frequency. In contrast with inverters, they provide direct ac-to-ac power conversion with an adjustable output frequency. They are inherently capable of four-quadrant oper-

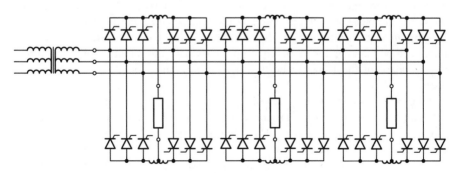

Figure 5.25   Three-phase six-pulse cycloconverter with isolated phase loads.

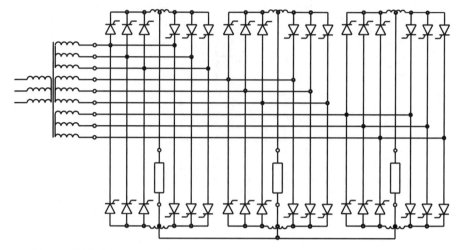

Figure 5.26   Three-phase six-pulse cycloconverter with interconnected phase loads.

ation and are quite reliable, since even with a failure of one SCR a cycloconverter still continues to function, albeit with somewhat increased distortion of the output voltage.

Besides the reduced range of the output frequency, the disadvantages of cycloconverters include the high number of SCRs and the associated complexity of control. Also, as in phase-controlled rectifiers, the input power factor is generally low, particularly when a cycloconverter operates with a low magnitude control ratio. Correction of the power factor is inconvenient and expensive, because a high-power cycloconverter requires a proportionally sized large input filter.

5.3   MATRIX CONVERTERS

Although first proposed in the late 1970s, matrix converters still struggle for a significant share of the power electronic market. The concept of matrix converter represents an extension of the principle of generic converter, introduced in Chapter 1. Each phase of a K-phase ac voltage source is connected with each phase of an L-phase load by *bidirectional* fully controlled switches. Thus, the voltage of any input terminal can be made to appear at any output terminal or terminals, while the current in any phase of the load can be drawn from any phase or phases of the supply source.

Figure 5.27 is a circuit diagram of the most practical three-phase to three-phase $(3\Phi-3\Phi)$ matrix converter. The converter consists of nine switches, denoted S_{Aa} through S_{Cc}. An input filter is employed to screen the supply system from harmonic currents generated in the converter, and the load is assumed to contain inductance that maintains the continuity of the output currents. Clearly, at any time, one and only one switch in each row must be closed. Otherwise, either the supply lines would

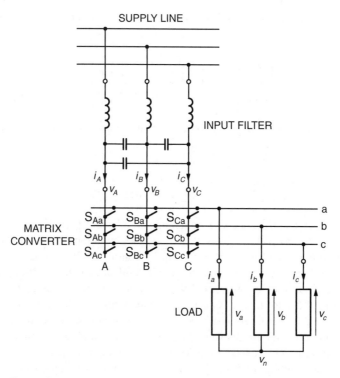

Figure 5.27 Three-phase to three-phase (3Φ–3Φ) matrix converter.

be shorted, or one or more of the output currents would be interrupted. It is easy to determine that of the theoretically possible 512 (2^9) states of the converter, only 27 are permitted.

The voltages v_a, v_b, and v_c at the output terminals are given by

$$
\begin{bmatrix} v_a \\ v_b \\ v_c \end{bmatrix} = \begin{bmatrix} x_{Aa} & x_{Ba} & x_{Ca} \\ x_{Ab} & x_{Bb} & x_{Cb} \\ x_{Ac} & x_{Bc} & x_{Cc} \end{bmatrix} \begin{bmatrix} v_A \\ v_B \\ v_C \end{bmatrix}
\tag{5.39}
$$

where x_{Aa} through x_{Cc} denote switching variables of switches S_{Aa} through S_{Cc}, and v_A, v_B, and v_C are the voltages at the input terminals. It can be proved that with a balanced linear wye-connected load, the voltage v_n at the common point of the load is given by

$$
v_n = \frac{1}{3}(v_a + v_b + v_c).
\tag{5.40}
$$

Consequently, the line-to-neutral output voltages, v_{an}, v_{bn}, and v_{cn}, can be expressed as

$$
\begin{bmatrix} v_{an} \\ v_{bn} \\ v_{cn} \end{bmatrix} = \frac{1}{3} \begin{bmatrix} 2 & -1 & -1 \\ -1 & 2 & -1 \\ -1 & -1 & 2 \end{bmatrix} \begin{bmatrix} v_A \\ v_B \\ v_C \end{bmatrix}. \tag{5.41}
$$

The input currents, i_A, i_B, and i_C, are related to the output currents, i_a, i_b, and i_c, as

$$
\begin{bmatrix} i_A \\ i_B \\ i_C \end{bmatrix} = \begin{bmatrix} x_{Aa} & x_{Ab} & x_{Ac} \\ x_{Ba} & x_{Bb} & x_{Ab} \\ x_{Ca} & x_{Cb} & x_{Cc} \end{bmatrix} \begin{bmatrix} i_a \\ i_b \\ i_c \end{bmatrix}. \tag{5.42}
$$

It can be seen that the matrix of switching variables in Eq. (5.42) is a transpose of the respective matrix in Eq. (5.39). Based on those equations, fundamentals of both the output voltages and input currents can be controlled. It is done by employing appropriately timed sequences of either the individual switching variables or, as in the case of space-vector-controlled PWM rectifiers, entire states of the converter (see Sections 4.2.1 to 4.2.3). As a result of such control, the fundamental output voltages should be balanced and have the desired frequency and amplitude, while the fundamental input currents should also be balanced and have the required phase shift, usually zero, with respect to the corresponding input voltages.

There exist a number of control methods for matrix converters, some quite complex, and a variety of solutions, including closed-loop control of voltages and currents, have been proposed. The feasibility of independent control of output voltages and input currents has been demonstrated by several researchers. It has been found for the $3\Phi-3\Phi$ matrix converter that the maximum obtainable voltage gain, defined here as the ratio of the peak fundamental output voltage to peak input voltage, is $\sqrt{3}/2 \approx 0.866$. It is so because the set of six input line-to-line voltages (see, e.g., Figure 4.8), which constitutes "raw material" for the output voltages, dips every 60° to $\sqrt{3}/2$ of peak value of the input voltages. This is considered a disadvantage, since other means of direct or indirect (with an intermediate dc stage) ac-to-ac power conversion yield voltage gains close to unity. On the other hand, credit must be given to a matrix converter for the inherent capability for bidirectional power flow, lacking in certain popular indirect ac-to-ac conversion schemes.

Variants of the space vector PWM technique, outlined in Section 4.3, are most often used in the control of matrix converters. It is based on representation of the three-phase input currents and output voltages as space vectors. An arrangement of two six-switch matrix converters equivalent to the nine-switch converter under consideration is shown in Figure 5.28. The six-switch converters are placed between the source and the load, and connected via lines P (positive) and N (negative), which constitute a *virtual dc link*. Converter 1 (CONV 1) is of the $3\Phi-1\Phi$ type and converter 2 (CONV 2) is of the $1\Phi-3\Phi$ type. The topology of CONV 1 is identical with that of a six-pulse current-type PWM rectifier; thus, it can be so operated that the voltage between lines P and N has a specified dc component, V_{dc}, while input currents to the

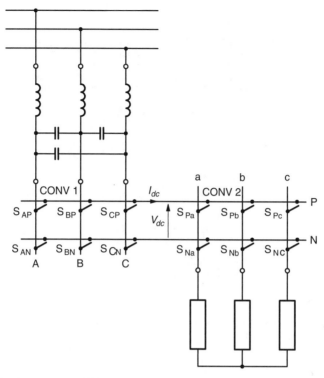

Figure 5.28   Arrangement of 3Φ–1Φ and 1Φ–3Φ matrix converters, equivalent to a 3Φ–3Φ matrix converter.

converter are sinusoidal and satisfy the unity power factor requirement. Conversely, CONV 2 can be operated as a generic PWM three-phase inverter, producing balanced ac currents in the load. Subsequently, CONV 1 and CONV 2 are called a *virtual rectifier* and a *virtual inverter*, respectively.

One and only one switch in each row of the virtual rectifier, and one and only one switch in each column of the virtual inverter, must be conducting at any time. Thus, six different states are permitted for the rectifier and eight for the inverter, and the total number of allowable state of the combined 12-switch converter is 48. However, it can be shown that the connections between the source and load terminals corresponding to each state can be realized using the original nine-switch 27-state 3Φ–3Φ matrix converter in Figure 5.27. Considering, for example, a situation when switches S_{AP}, S_{BN}, S_{Pa}, S_{Pb}, and S_{Nc} are closed, it can be seen that phase A of the source is connected, through line P, to phases a and b of the load, and phase B, through line N, to phase c. The same connections can be made directly in the original matrix converter by closing switches S_{Aa}, S_{Ab}, and S_{Bc}.

The same example can be analyzed in a more general way. In subsequent considerations, each of the allowed states of the matrix converter is assigned a three-letter code, specifying which input terminals, A, B, or C, are connected to output terminals

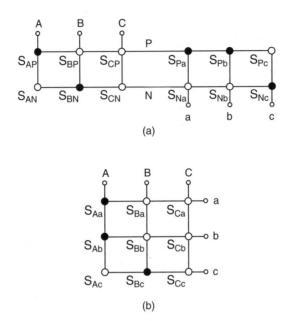

Figure 5.29 State AAB as realized by activation of switches in (a) a virtual rectifier and inverter; (b) a matrix converter.

a, b, and c, respectively. The state described above can thus be designated AAB. Note that the virtual rectifier connects terminal A to line P and terminal B to line N, while terminal C remains unconnected. Therefore, the rectifier state can be called PN0. The virtual inverter connects terminals a and b to line P and terminal c to line N, so its state can be designated PPN. Examining the rectifier state indicates an association of P with A and N with B. Thus, the inverter state PPN can be translated into state AAB of the entire matrix converter, and to connect the input terminals AAB with the output terminals abc, switches S_{Aa}, S_{Ab}, and S_{Bc} must be activated. Realization of the state described is illustrated in Figure 5.29.

Zero states of the virtual rectifier cause $V_{dc} = 0$ that is, the P and N lines are shorted by the switches in either of the input phases, A, B, or C. These states are subsequently called Z00, 0Z0, and 00Z. State 0Z0, for example, represents the situation when only switches S_{BP} and S_{BN} are closed. Zero states of the virtual inverter cause zero output voltages of the inverter, which happens when all three output terminals, a, b, and c, are clamped to either the P or N line. Consequently, these are inverter states PPP and NNN, the latter state, for instance, meaning that switches S_{Na}, S_{Nb}, and S_{Nc} are closed.

When the virtual rectifier is in state PN0, input currents i_A, i_B, and i_C, equal I_{dc}, $-I_{dc}$, and 0, respectively, and the corresponding space vector of input currents is

$$\vec{I}_{PN0} = \frac{3}{2}I_{dc} + j\frac{\sqrt{3}}{2}I_{dc} \qquad (5.43)$$

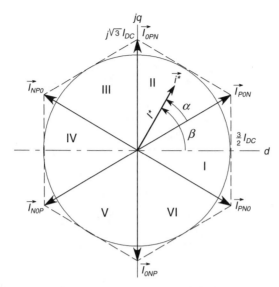

Figure 5.30 Reference current vector in the vector space of input currents of the virtual rectifier.

where I_{dc} denotes the dc current in the virtual dc link (see Figure 5.28). Current vectors for the remaining five states of the virtual rectifier can be determined similarly. All six stationary current vectors and the revolving reference vector, $\vec{i}^*$, are shown in Figure 5.30, which, unsurprisingly, is practically identical with Figure 4.46 for the real current-type PWM rectifier. As in that rectifier, in order to satisfy the unity power factor condition in the matrix converter under consideration, the reference current vector should be aligned with the space vector of input voltages.

Considering state PPN of the virtual inverter, it can be seen that $v_a = v_b = v_P$ and $v_c = v_N$, where v_P and v_N denote the potentials of lines P and N, respectively. According to Eq. (5.41), $v_{an} = v_{bn} = (v_P - v_N)/3 = V_{dc}/3$ and $v_{cn} = -2V_{dc}/3$ Employing Eq. (4.74), the space vector, $\vec{V}_{PPN}$ of the line-to-neutral output voltages is found to be

$$\vec{V}_{PPN} = \frac{1}{2}V_{dc} + j\frac{\sqrt{3}}{2}V_{dc} \tag{5.44}$$

The remaining five active voltage vectors are shown in Figure 5.31 with the reference voltage vector $\vec{v}^*$, which represents the desired line-to-line voltages of the matrix converter. Figure 5.31 is almost identical to Figure 4.56 for the voltage-type rectifier, which, as explained in Section 7.1.3, has the same topology as that of the voltage-source inverter.

Equations (4.77) to (4.79) for duty ratios of states of a converter controlled by space vector PWM can be adapted to the virtual rectifier and inverter. The question is how to specify the modulation indexes for these converters. Let the rectifier modulation

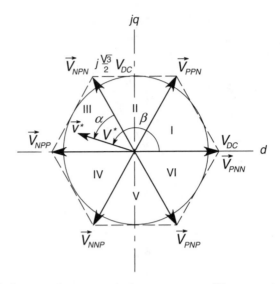

Figure 5.31 Reference voltage vector in the vector space of line-to-neutral output voltages of the virtual inverter.

index, m_{rec}, be defined as

$$m_{\text{rec}} = \frac{I_{i,p}}{I_{\text{dc}}} \tag{5.45}$$

where $I_{i,p}$ denotes the peak value of the input currents (a balanced load is assumed for the matrix converter). Assuming that losses in the virtual rectifier are negligible, the input power, P_i, equals the power, P_{DC}, in the virtual dc link, that is,

$$\sqrt{3}\, V_i I_i \, \cos \varphi_i = V_{\text{dc}} I_{\text{dc}} \tag{5.46}$$

where V_i and I_i denote rms values of the input line-to-line voltage and current, respectively, and $\cos \varphi_i$ is the power factor of the rectifier. Consequently,

$$V_{\text{dc}} = \frac{\sqrt{3}}{2} m_{\text{rec}} V_{i,p} \, \cos \varphi_i \tag{5.47}$$

where $V_{i,p}$ denotes the peak value of the input line-to-line voltages (a balanced supply source is assumed).

Now let the inverter modulation index, m_{inv} be defined as

$$m_{\text{inv}} = \frac{V_{o,p}}{V_{\text{dc}}} \tag{5.48}$$

where $V_{o,p}$ denotes the peak value of fundamental line-to-line output voltage. Substituting Eq. (5.47) in Eq. (5.48) gives

$$m_{\text{inv}} = \frac{2V_{o,p}}{\sqrt{3}V_{i,p}m_{\text{rec}}\cos\varphi_i} = \frac{2m}{\sqrt{3}m_{\text{rec}}\cos\varphi_i} \tag{5.49}$$

where

$$m \equiv \frac{V_o}{V_i} \tag{5.50}$$

is the modulation index (and magnitude control ratio) of the entire matrix converter. Equation (5.49) can be rearranged to

$$m = \frac{\sqrt{3}}{2}m_{\text{rec}}m_{\text{inv}}\cos\varphi_i. \tag{5.51}$$

Equation (5.51) confirms the already mentioned $\sqrt{3}/2$ limit on the voltage gain of the matrix converter. To maximize the control range of the matrix converter, m_{rec} is set to 1. Note that the magnitude of vector of the input currents depends on the load of the converter, and it is only the phase shift of that vector with respect to the vector of input voltages that is to be adjusted for the unity power factor.

Let the states that produce vectors framing the reference current vector in Figure 5.30 be denoted X_I and Y_I, and those that produce vectors framing the reference voltage vector in Figure 5.31 be X_V and Y_V. The corresponding zero states are designated Z_X and Z_V. One of the commonly used switching patterns, in which the switching cycle is divided into nine subcycles, is presented in Table 5.1. The sequential number of a subcycle is denoted by n and its duration, relative to the switching period T_{sw}, by t_n. Duty ratios of individual states are designated by d with an appropriate subscript (e.g., d_{XV} for state X_V).

Typical waveforms of the output voltage and current of a matrix converter supplying an RL load are illustrated in Figure 5.32. For reference, waveforms of the

TABLE 5.1 Switching Pattern for 3Φ–3Φ Matrix Converter with Space Vector PWM

Switching Subcycle	Rectifier State	Inverter State	t_n/T_{sw}
1	X_I	X_V	$d_{XI}d_{XV}/2$
2	X_I	Y_V	$d_{XI}d_{YV}/2$
3	Y_I	Y_V	$d_{YI}d_{YV}/2$
4	Y_I	X_V	$d_{YI}d_{XV}/2$
5	Z_I	Z_V	$1 - (d_{XI} + d_{YI})(d_{XV} + d_{YV})$
6	Y_I	X_V	$d_{YI}d_{XV}/2$
7	Y_I	Y_V	$d_{YI}d_{YV}/2$
8	X_I	Y_V	$d_{XI}d_{YV}/2$
9	X_I	X_V	$d_{XI}d_{XV}/2$

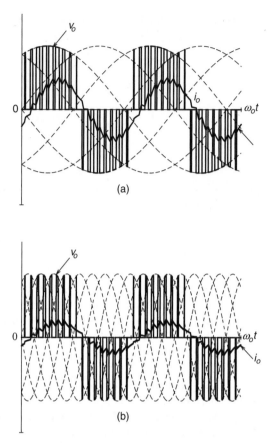

Figure 5.32 Output voltage and current waveforms in a $3\Phi-3\Phi$ matrix converter: (a) $N = 48$, $m = 0.7$, $\omega/\omega_o = 2.8$; (b) $N = 12$, $m = 0.35$, $\omega/\omega_o = 0.7$.

six line-to-line input voltages are also shown. The output frequency, ω_o, in Figure 5.32a is 2.8 times higher than the input frequency, ω, the number N of switching intervals per cycle of the input voltage is 48, and the modulation index, m, is 0.75. The respective parameters in Figure 5.32b are 0.7, 12, and 0.35.

Practical implementation of matrix converters has been somewhat hampered by the unavailability of bidirectional fully controlled semiconductor power switches. Recently, several small companies started advertising such switches, whose scheme is based on two IGBTs, in the common-emitter connection, and two diodes, as shown in Figure 5.33a. Another solution involves a combination of one IGBT and four diodes, depicted in Figure 5.33b. In each case, the circuit diagram of a $3\Phi-3\Phi$ matrix converter contains a relatively large number of semiconductor devices (18 IGBTs and 18 diodes, or 9 IGBTs and 36 diodes) and the associated drivers. Therefore, integration of the devices in a single case per switch represents valuable progress.

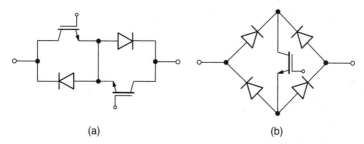

(a)   (b)

Figure 5.33 Bidirectional semiconductor power switches: (a) two IGBTs and two diodes; (b) one IGBT and four diodes.

Because of the absence of large energy storage components (the input filter employs low-value inductors and capacitors), operation of matrix converters is sensitive to grid disturbances. Therefore, in practical applications, the converter must be equipped with fast-acting protection systems.

5.4 DEVICE SELECTION FOR AC-TO-AC CONVERTERS

The highest instantaneous voltage appearing in an ac voltage controller is the peak value of the input voltage. Thus, the switches selected must satisfy condition (4.104) in Section 4.4. For three-phase ac voltage controllers, the peak value, $V_{LL,p}$, of the line-to-line supply voltage is substituted for $V_{i,p}$.

The rules of device selection with respect to the rated current depend on whether triacs or separate, antiparallel-connected devices are used in the controller. For triacs, the rated current is meant as the maximum allowable *rms value* of a sustained sinusoidal ac current, whereas is in the other semiconductor power switches the current ratings involve the allowable *average* current. Therefore, for triac-based ac voltage controllers, the condition to be satisfied is

$$I_{\text{rat}} \geq (1 + s_I)I_{o(\text{rat})} \tag{5.52}$$

where s_I denotes the safety margin for the current and $I_{o(\text{rat})}$ is the rated rms output current of the controller.

The maximum average current in a single power switch of an ac voltage controller occurs at the full-wave conduction, that is, when the output current is sinusoidal and has the maximum rms value representing the rated current, $I_{o(\text{rat})}$, of the controller. The corresponding maximum average value, $I_{\text{ave(max)}}$, of the current conducted by a single switch is given by

$$I_{\text{ave(max)}} = \frac{1}{2\pi} \int_0^\pi \sqrt{2}\, I_{o(\text{rat})} \sin \omega t \, d\omega t = \frac{\sqrt{2}}{\pi} I_{o(\text{rat})} \approx 0.45 I_{o(\text{rat})}. \tag{5.53}$$

Consequently, the condition for the rated current of switches is

$$I_{\text{rat}} \geq \frac{\sqrt{3}}{\pi}(1 + s_I)I_{o(\text{rat})}. \tag{5.54}$$

If the switches are connected in delta after the load, as in Figure 5.15b, the rms current conducted by them is lower from the rms load current by a factor of $\sqrt{3}$. Therefore, the rated current of these switches can be reduced by $\sqrt{3}$ in comparison with that given by condition (5.54).

Usually, the freewheeling switches in PWM ac voltage controllers carry much lower currents than do the main switches. If the application of a given controller is restricted to a passive load, the current rating of the freewheeling switches can be half that of the main switches. However, in general, it is prudent to use identical semiconductor devices for both the main and freewheeling switches. This avoids the danger of overloading the latter switches when the controller supplies an active, power-generating load (e.g., an ac motor that can operate as a generator).

For ac choppers, the minimum on-time and off-time, $t_{\text{ON(min)}}$ and $t_{\text{OFF(min)}}$, of the main switches are

$$t_{\text{ON(min)}} = m_{\text{min}}T_{\text{sw}} \tag{5.55}$$

and

$$t_{\text{OFF(min)}} = (1 - m_{\text{min}})T_{\text{sw}} \tag{5.56}$$

where m_{min} and m_{max} denote the minimum and maximum values of the modulation index. The same rules apply to switches in matrix converters. Respective times for the freewheeling switches in ac choppers are obtained from Eqs. (5.55) and (5.56) by interchanging the ON and OFF subscripts.

Cycloconverters are composed of controlled rectifiers. Therefore, the selection rules for power switches are the same as for rectifiers (see Section 4.4).

5.5 COMMON APPLICATIONS OF AC-TO-AC CONVERTERS

Ac voltage controllers can be used for adjustment of the rms value of ac voltages and currents, and as *static ac switches*. In the latter case, a controller connects the supply and load for a number of cycles, passing an uncontrolled, sinusoidal current. Depending on the power transferred, ac switches are based on SCRs or triacs. In the on-state, an ac switch operates with the minimum firing angle, while the off-state is initiated by removing the gate signals.

Ac voltage controllers are used primarily in lighting and heating control and as *soft starters* for induction motors. A soft starter, connected between a supply line and a motor, reduces the starting current, which otherwise could be excessive. Ac switches are employed as transformer tap changers, allowing voltage control in power systems. Another typical application of ac switches involves speed control of high-inertia induction-motor drives, such as large centrifuges. The driving motor is switched on when the speed of the centrifuge drops below the minimum allowable level, and it is turned off when the speed reaches the maximum allowable value. With the deenergized motor, the centrifuge is driven by its momentum, and the high mechanical inertia and low friction result in a low deceleration rate. In this simple way, the average speed is maintained at a constant level, and the instantaneous speed does not stray from the allowable tolerance band. A similar mode of control can be used for temperature control, with an ac switch intermittently turning an electric heater or air conditioner on and off.

Cycloconverters are used exclusively for control of large ac motors, usually synchronous motors, in low-speed high-power drive systems such as those of rolling mills in metal industries or kilns in cement factories. The speed of ac motors is proportional to the supply frequency, and the low output frequency of a cycloconverter translates into a low speed of the motor, allowing direct gearless drive of the load. Cycloconverter-fed ac drives are capable of rapid acceleration and deceleration, and a regenerative operation mode (with the reversed power flow) is available over the complete speed range.

Matrix converters have been employed increasingly in applications in which wide-range frequency/magnitude control and bidirectional power flow are required. The low voltage gain is one of the problems preventing matrix converters from extensive use in control of ac motors. Voltage ratings of mass-produced motors correspond to the common voltage levels in the existing electrical infrastructure. The torque of an induction motor is proportional to the supply voltage squared, so the 15% reduction in voltage translates into a 28% reduction in the torque developed. Still, in certain "custom" applications, such as dedicated high-performance ac drives, matrix converters seem to have found a market niche.

Another growing application of matrix converters is in low- and medium-power wind-turbine systems, in which the converter serves as an interface between a variable-speed ac generator and the grid. The issue of low voltage gain is unimportant, as the entire system is designed for an optimum match between individual components. Use of matrix converters in electric and hybrid vehicles is also under serious consideration.

5.6 SUMMARY

Ac voltage controllers allow adjustment of the rms values of output voltage and current by means of phase control or pulse width modulation. In phase-controlled controllers, SCRs or triacs are used, while fully controlled switches are employed in PWM controllers (ac choppers). Several types of three-phase ac voltage controllers are available. Ac voltage controllers are often operated as static ac switches.

Three-phase cycloconverters, which are comprised of three dual converters, perform direct ac-to-ac conversion with adjustable output frequency and voltage. However, the output frequency is limited to a small fraction of the input frequency.

Cycloconverters have several advantages, but the device count is high and the control system is complex. Typically, cycloconverters are high-power converters, supplied from dedicated transformers.

Matrix converters, based on bidirectional fully controlled switches that provide direct connections between each input terminal and each output terminal, allow independent sinusoidal modulation of output voltages and input current. As of now, matrix converters remain in the preliminary stage of widespread practical implementation.

Typical applications of ac voltage controllers and static ac switches include in lighting and heating control, in transformer tap changing, as soft starters for induction motors, and in integral-cycle control of ac motors. Cycloconverters are used in high-power low-speed ac adjustable-speed drives. Matrix converters are best suited for applications in which the low voltage gain is of minor importance, such as in electric vehicles or as interfaces between renewable energy sources and the grid.

EXAMPLES

Example 5.1   A single-phase ac voltage controller is used in a movie theater to control incandescent lighting. Neglecting the temperature-related changes of resistance of the lamps, find the reduction in power consumed by the lighting when the firing angle is 60°. What firing angle would result in a 50% reduction of that power?

Solution: The incandescent lighting constitutes a resistive load. Consequently, the ratio of rms values of the output and input voltage, that is, the magnitude control ratio, can be found from Eq. (5.6) as

$$\frac{V_o}{V_i} = \sqrt{\frac{1}{\pi}\left[\pi - \frac{1}{3}\pi + \frac{1}{2}\sin\left(\frac{2}{3}\pi\right)\right]} = 0.897.$$

With a constant resistance, the power is proportional to the squared rms value of voltage, so the resulting power constitutes $0.897^2 = 0.805$ of the full power. Thus, the power consumed is reduced by 19.5%.

The magnitude control ratio for the 50% power reduction is $\sqrt{0.5} = 0.707$. From the control characteristic in Figure 5.3 for $\varphi = 0$, the corresponding value of the firing angle is determined as 90°. Notice that at this firing angle, exactly half of the sinusoidal waveform of the input voltage is passed to the load.

Example 5.2   A single-phase ac chopper is supplied from a 120-V line and operates with a modulation index of 0.75. The load impedance is 2 Ω. Determine the rms value of output voltage and, neglecting the ripple of output current, the rms value of this current.

Solution: The rms value, V_o, of the chopped output voltage is given by Eq. (5.32) as

$$V_o = \sqrt{0.75} \times 120 = 103.9 \text{ V}.$$

However, the rms value, $V_{o,1}$, of the fundamental of this voltage is

$$V_{o,1} = 0.75 \times 120 = 90 \text{ V}$$

and the output current, which is assumed to be purely sinusoidal, has an rms value, I_o, of

$$I_o = I_{o,1} = \frac{V_{o,1}}{Z} = \frac{90}{2} = 45 \text{ A}.$$

Example 5.3   A three-phase six-pulse cycloconverter is supplied from a 460-V 60-Hz line. The load is connected in wye with a resistance of 0.9 Ω/ph and an inductance of 15 mH/ph. The cycloconverter operates with in output frequency of 10 Hz and a magnitude control ratio of 0.7. Neglecting the ripple, find the rms value of the output currents.

Solution: Since the load is connected in wye, the rms value, I_o, of the output currents is given by

$$I_o = \frac{V_{o,\text{LN},1}}{Z}$$

where $V_{o,\text{LN},1}$ denotes the rms value of the fundamental line-to-neutral output voltage and Z is the load impedance, which at the frequency of 10 Hz is

$$Z = \sqrt{0.9^2 + (2\pi \times 10 \times 0.015)^2} = 1.303 \ \Omega/\text{ph}.$$

Using Eq. (5.38), $V_{o,\text{LN},1}$ can be determined as

$$V_{o,\text{LN},1} = \frac{3}{\pi} \times 0.7 \times 460 = 439.3 \text{ V}/\text{ph}.$$

which yields

$$I_o = \frac{439.3}{1.303} = 337.1 \text{ A}/\text{ph}.$$

Example 5.4   A 3Φ–3Φ matrix converter, supplied from a 460-V line, operates with a switching frequency of 5 kHz, a modulation index of 0.5, and a unity power factor. Consider a switching cycle in which the phase angles of the reference input current and output voltage are 135° and 100°, respectively, and determine the switching pattern in that cycle.

Solution: Assuming that $m_{rec} = 1$, the modulation index, m_{inv}, of the virtual inverter can be found from Eq. (5.51) as

$$m_{inv} = \frac{m}{(\sqrt{3}/2)m_{rec}\cos\varphi_i} = \frac{2}{\sqrt{3}}0.5 = 0.577.$$

The phase angles, β_I and β_V, of the reference vectors indicate that the current reference vector, $\vec{i}^*$, is in sector III and the voltage reference vector, $\vec{v}^*$, is in sector II (see Figure 5.30). Thus, $X_I = 0PN$, $Y_I = NP0$, $X_V = PPN$, and $Y_V = NPN$. The in-sector (local) angles, α_I and α_V, are 15° and 10°, respectively. Based on Eqs. (4.77) to (4.79),

$$d_{XI} = 1 \times \sin(60° - 15°) = 0.707$$
$$d_{YI} = 1 \times \sin(15°) = 0.259$$
$$d_{ZI} = 1 - d_{XI} - d_{YI} = 0.034$$

and

$$d_{XV} = 0.577 \times \sin(60° - 10°) = 0.442$$
$$d_{XV} = 0.577 \times \sin(10°) = 0.100$$
$$d_{ZV} = 1 - d_{XV} - d_{YV} = 0.458.$$

The switching period, T_{sw}, which is a reciprocal of the switching frequency, f_{sw}, equals 200 μs. States of the virtual rectifier and inverter and their relative durations are listed in Table 5.2, which is an example-specific version of Table 5.1. Based on the information in Table 5.2, activation of individual switches of the matrix converter can be determined, as shown in Table 5.3 and Figure 5.34. Notice that within the entire switching cycle considered, switches S_{Ab} and S_{Cb} are off and switch S_{Bb} is on. The zero state BBB in the fifth switching subcycle is preferable over the other two zero states, AAA or CCC, as the preceding and following state is BBA, so that only two switches, S_{Ac} and S_{Bc}, must change their states.

Note that the input and output frequencies of the matrix converter have not been specified. This information is not needed, as in space vector PWM methods the required voltage and current vectors are synthesized independently in each switching cycle. Thus, the speed of a vector can be assessed only by comparing its position in two consecutive switching cycles. Here, for example, if the vector of output voltages has moved by 3.6°, the output frequency is 50 Hz, because 5000 switching cycles would produce 18000°, that is, 50 cycles of output voltage.

The modulation index, m, of 0.5 indicates that the line-to-line output voltage is 230 V. As the maximum value of that index is $\sqrt{3}/2$, the magnitude control ratio, M, equals $2m/\sqrt{3} \approx 1.15m$ and the maximum theoretically available output voltage is 398 V.

TABLE 5.2   Switching Pattern for the Matrix Converter in Example 5.4

Switching Subcycle	Rectifier State	Inverter State	t_n / T_{sw}
1	0PN	PPN	0.156
2	0PN	NPN	0.035
3	NP0	NPN	0.009
4	NP0	PPN	0.057
5	Z00 or 0Z0	PPP	0.486
6	NP0	PPN	0.057
7	NP0	NPN	0.009
8	0PN	NPN	0.035
9	0PN	PPN	0.156

TABLE 5.3   Activation of Switches in the Matrix Converter in Example 5.4

Switching Subcycle	State of Matrix Converter	Switches Activated	Duration (μs)
1	BBC	S_{Ba}, S_{Bb}, S_{Cc}	31.2
2	CBC	S_{Ca}, S_{Bb}, S_{Cc}	7.0
3	ABA	S_{Aa}, S_{Bb}, S_{Ac}	1.8
4	BBA	S_{Ba}, S_{Bb}, S_{Ac}	11.4
5	BBB	S_{Ba}, S_{Bb}, S_{Bc}	97.2
6	BBA	S_{Ba}, S_{Bb}, S_{Ac}	11.4
7	ABA	S_{Aa}, S_{Bb}, S_{Ac}	1.8
8	CBC	S_{Ca}, S_{Bb}, S_{Cc}	7.0
9	BBC	S_{Ba}, S_{Bb}, S_{Cc}	31.2

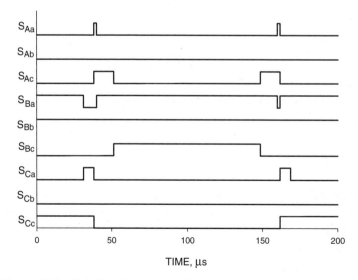

Figure 5.34   Switching signals for individual switches in a matrix converter.

Example 5.5 A three-phase delta-connected PWM ac voltage controller is rated at 10 kVA and 460 V. The supply frequency is 60 Hz, and the modulation index is limited to the range 0.05 to 0.95. Determine the minimum voltage and current ratings of switches, assuming safety margins of 0.4 and 0.2 for the rated voltage and current, respectively. What is the maximum allowable switching frequency for the minimum on- and off-times to be at least 20 μs long?

Solution: The rated voltage, V_{rat}, of switches of the controller is given by Eq. (4.105) as

$$V_{rat} \geq (1 + 0.4) \times \sqrt{2} \times 460 = 911 \, \text{V}.$$

The rated rms value, $I_{o,rat}$, of output currents can be calculated as

$$I_{o,rat} = \frac{10 \times 10^3}{3 \times 460} = 7.25 \, \text{A}$$

and, consequently, the rated current, I_{rat}, of the switches, as required by condition (5.54), must be

$$I_{rat} \geq \frac{\sqrt{2}}{\pi}(1 + 0.2) \times 7.25 = 3.92 \, \text{A}.$$

From Eq. (5.55),

$$T_{sw} \geq \frac{t_{ON(min)}}{m_{min}} = \frac{20}{0.05} = 400 \, \mu s.$$

That is,

$$f_{sw} \leq \frac{1}{400} = 0.0025 \, \text{MHz} = 2.5 \, \text{kHz}.$$

Equation (5.56) would yield the same result, since in the case considered, $m_{max} = 1 - m_{min}$.

PROBLEMS

P5.1 For a single-phase ac voltage controller, sketch the waveforms of output voltage and current for the following values of the firing and extinction angles:

   (a) 60° and 225°
   (b) 90° and 210°
   (c) 120° and 205°
       Assume that the load angle is less than 60°.

P5.2   A single-phase ac voltage controller supplies a load consisting of a 2-Ω resistance and a 5-mH inductance. The controller is fed from a 120-V 60-Hz line, and multipulse gate signals are used to activate the controller's triacs. Estimate the rms output voltage for firing angles of 30°, 90°, and 150°.

P5.3   A single-phase ac voltage controller supplies a resistive load. Use the characteristic in Figure 5.3 to determine the firing angle resulting in the 25% reduction of output voltage

P5.4   A three-phase ac controller in Figure 5.7 is supplied from a 460-V line, and the load is purely resistive. Find the rms output voltage for firing angles of 20°, 90°, and 130°.

P5.5   A single-phase ac chopper supplied from a 115-V line operates with a modulation index of 0.65. Find the rms output voltage.

P5.6   A single-phase ac chopper is supplied from a 120-V 60-Hz line and operates with a constant modulation index of 0.7 and 20 switching intervals per cycle. Find the fundamental output voltage and durations of the pulses and notches of the output voltage.

P5.7   A three-phase delta-connected ac chopper supplied from a 230-V line operates with modulation index of 0.4. Find the rms output voltage and fundamental output voltage.

P5.8   A three-phase six-pulse cycloconverter is supplied from a 460-V line and operates with the magnitude control ratio of 0.8. Find the rms values of the line-to-neutral and line-to-line output voltages of the cycloconverter.

P5.9   List all allowable states of the 3Φ–3Φ matrix converter using a three-letter designation (e.g., AAB).

P5.10   A 3Φ–3Φ matrix converter is supplied from a 460-V line with positive phase sequence. The 10-Ω resistive load is connected in wye. The phase A input voltage can be expressed as $v_A = V_{i,p} \sin \omega t$ Find the instantaneous line-to-line output voltages and input currents at $\omega t = 137°$ if switches S_{Ab}, S_{Ac}, and S_{Ca} are closed.

P5.11   Repeat Example 5.4 with the following data: $f_{sw} = 4$ kHz, $m = 0.35$, $\beta_I = 237°$, $\beta_V = 12°$. What is the output line-to-neutral voltage if the matrix converter in question is supplied from a 230-V line?

P5.12   Determine the minimum required voltage and current ratings of SCRs in a 10-kVA 460-V three-phase fully controlled ac voltage controller. Assume safety margins of 0.4 for the voltage rating and 0.2 for the current rating.

P5.13   An ac voltage controller with the same ratings as those in Problem 5.10 has the SCRs connected in delta after the load. Assuming the same safety margins as in Problems 5.10, determine the minimum required voltage and current ratings of the SCRs.

P5.14 A 5-kVA 120-V 60-Hz single-phase ac chopper operates with 30 switching intervals per cycle. The minimum and maximum values of the modulation index for the PWM operation of the chopper are 0.04 and 0.96, respectively. Assuming the same safety margins as in Problem 5.12, determine the minimum required voltage and current ratings of the main and freewheeling switches. Find the minimum on- and off-times of the switches.

COMPUTER ASSIGNMENTS

*CA5.1 Run PSpice program *AC_Volt_Contr_1ph.cir* for a single-phase ac voltage controller. For firing angles of 35° and 75°, find for the output voltage and current:

(a) The rms value

(b) The rms value of the fundamental

(c) The total harmonic distortion

*CA5.2 Run PSpice program *AC_Volt_Contr_1ph.cir* for a single-phase ac voltage controller. For firing angles of 45° and 105°, obtain the harmonic spectra of the output voltage and current. For the 10 most prominent harmonics, determine the harmonic number and amplitude as a fraction of amplitude of the fundamental.

*CA5.3 Run PSpice program *AC_Volt_Contr_3ph.cir* for a three-phase ac voltage controller. For firing angles of 35° and 75°, find for the output voltage and current:

(a) The rms value

(b) The rms value of the fundamental

(c) The total harmonic distortion

*CA5.4 Run PSpice program *AC_Volt_Contr_3ph.cir* for a three-phase ac voltage controller. For firing angles of 45° and 105°, obtain the harmonic spectra of the output voltage and current. For the 10 most prominent harmonics, determine the harmonic number and amplitude as a fraction of amplitude of the fundamental.

*CA5.5 Run PSpice program *AC_Chopp.cir* for a single-phase ac chopper with an input filter. For $N = 10$ switching intervals per cycle and a magnitude control ratio, M, of 0.4, find for the output voltage and current:

(a) The rms value

(b) The rms value of the fundamental

(c) The total harmonic distortion

   Repeat the assignment for $N = 20$ and $M = 0.6$.

*CA5.6   Run PSpice program *AC_Chopp.cir* for a single-phase ac chopper with an input filter. For $N = 10$ switching intervals per cycle and a magnitude control ratio, M, of 0.4, find:

(a) The total harmonic distortion of the current drawn from the supply source

(b) The real input power to the filter (not equal to the output power because of the nonideal switches employed)

(c) The apparent input power to the filter

(d) The input power factor

Repeat the assignment for $N = 20$ and $M = 0.6$. Observe the oscillograms of the current drawn from the supply source and current drawn by the chopper from the filter. Also, compare the voltage waveform at the input terminals of the chopper (after the filter) with that of the supply source.

*CA5.7   Run PSpice program *Cyclocon.cir* for a single-phase six-pulse cyclo-converter. Observe the oscillograms of the output voltages of the cyclo-converter and constituent rectifiers, output current, and currents in the separating inductors.

LITERATURE

[1] Huber, L., and Borojevic, D., Space vector modulated three-phase to three-phase matrix converter with input power factor correction, *IEEE Transactions on Industry Applications*, vol. 31, no. 6, pp. 1234–1246, 1995.

[2] Rashid, M. H., *Power Electronics Handbook*, 2nd ed., Academic Press, San Diego, CA, 2007, Chap. 18.

[3] Rombaut, C., Seguier, G., and Bausiere, R., *Power Electronic Converters: AC/AC Conversion*, McGraw-Hill, New York, 1987.

6 DC-to-DC Converters

Power electronic converters for the dc-to-dc conversion are described in this chapter. Static dc switches with fully controlled semiconductor power switches and forced-commutated SCRs are then explained. Single-, two-, and four-quadrant step-down choppers and a step-up chopper are analyzed. Guidelines for device selection are provided, and applications of dc-to-dc converters are presented.

6.1 STATIC DC SWITCHES

The simplest type of dc-to-dc power conversion involves connection and disconnection of a dc supply source to a load, with no control of the voltage supplied. Although incapable of physical isolation of the source from the load, power electronic *static dc switches*, find a variety of applications in industrial and consumer electronics. The term *static* implies that the switch changes its state infrequently. In this respect, static dc switches replace traditional electromechanical switches. The latter have a limited life span, due to the wear of contacts and other mechanical parts, and they usually require considerable activating power. In contrast, power electronic switches can, theoretically, operate for an infinite amount of time, and their *power gain*, the ratio of connected power to activating power, is much higher than that in electromechanical switches.

Figure 6.1 is a circuit diagram of a static dc switch based on a fully controlled semiconductor power switch. The semiconductor switch, S, is connected in series with the load. The freewheeling diode, D, parallel to the load, provides a path for the lingering load current when the switch is off. In the on-state of the switch, the output voltage, v_o, equals the fixed dc input voltage, V_i, and the output current, i_o, equals the input current, i_i. The diode is reverse biased and its current, i_D, is zero.

In all subsequent considerations, an RLE load is assumed, making a freewheeling diode necessary. When the switch is turned off, the output current is forced to decrease. The negative rate of change of the current produces a negative voltage across the load inductance. The diode becomes forward biased and starts conducting the output current. Now, $i_o = i_D$, $v_o \approx 0$, and $i_i = 0$. After a short period of time the load current dies out. The voltage drop across switch S equals the input voltage, so the switch is forward biased and ready for the next turn-on. Waveforms of the gate signal,

Introduction to Modern Power Electronics, Second Edition, by Andrzej M. Trzynadlowski
Copyright © 2010 John Wiley & Sons, Inc.

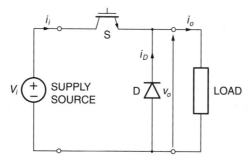

Figure 6.1   Static dc switch based on a fully controlled semiconductor power switch.

g (current or voltage, depending on the type of semiconductor switch), output voltage and current, input current and diode current are shown in Figure 6.2. For simplicity, instantaneous turn-on and turn-off are assumed.

A fully controlled semiconductor power switch can be turned off by an appropriate gate signal, but an SCR operating in a dc circuit requires an auxiliary *commutating circuit* for forced commutation of the device. Various similar circuits have been developed. However, the currently available fully controlled switches have phased out SCRs from most contemporary dc-input converters. Therefore, only an example of in SCR-based static dc switch, which employs a typical, resonant commutating circuit, is described below, to give the reader a clearer view of the issue.

The switch is shown in Figure 6.3. The commutating circuit comprises an auxiliary SCR, T2; diode, D2; inductor, L; and capacitor, C. The polarity of the voltage, v_C, across the capacitor is shown for the on-state of the switch, that is, when the main SCR, T1, is conducting and T2 is off.

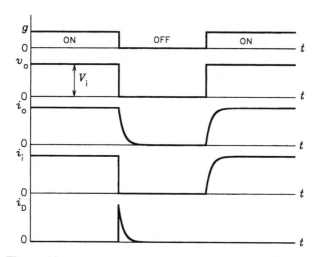

Figure 6.2   Voltage and current waveforms in a static dc switch.

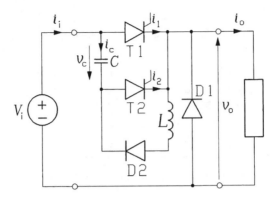

Figure 6.3   SCR-based static dc switch with a resonant commutating circuit.

To turn T1 off, the forward-biased T2 is fired. To better explain the mechanism of forced commutation, the T1−T2−C subcircuit is shown in Figure 6.4 at the instant when T2 starts conducting and therefore can be represented by a closed switch. It can be seen that the capacitor has been connected across T1 and the capacitor voltage imposes reverse bias on that SCR. Since T1 is still conducting, the capacitor is shorted and its discharge current, i_C, flows into the cathode of T1. As a result, a reverse recovery current in T1 is enforced and the SCR turns off, which is tantamount to turning off the entire dc switch being considered.

The same mechanism could be used to turn off the auxiliary SCR, provided that the capacitor had been charged to a voltage of the polarity opposite to that indicated in Figures 6.3 and 6.4. When T1 is off, T2 closes the source−C−T2−load loop. Thus, indeed, following turn-off of T1, capacitor C is charged, via the load, to a voltage of $-V_i$. However, forced commutation of T2 may not be needed if the off-time of T1 is sufficiently long. Then T2 turns off by itself because of the decay of current in the dc-supplied capacitor circuit in question.

When the capacitor is fully charged, T1 can be turned on again, closing the T1−L−D2−C loop. This is a resonant circuit in which, if T1 and D2 were absent, the capacitor voltage and current would have oscillating sinusoidal waveforms (assuming that no significant damping resistance exists in the circuit). However, the diode blocks

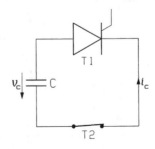

Figure 6.4   Subcircuit of an SCR-based static dc switch.

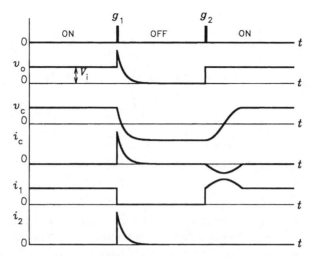

Figure 6.5   Voltage and current waveforms in an SCR-based static dc switch.

a positive current and the resonant process is interrupted after a half-period of the oscillation. At this instant, the capacitor has been recharged to $v_c = V_i$, that is, to the voltage required for a successful turn-on of the main SCR. Pertinent waveforms are shown in Figure 6.5, where g_1 and g_2 denote gate signals of SCRs T1 and T2, respectively, while i_1 and i_2 are currents conducted by these SCRs. For simplicity, instantaneous switching transients of the SCRs and a purely resistive load have been assumed.

A commutating circuit increases the size, weight, and cost of a static dc switch and affects its reliability adversely. Moreover, the forced commutation reduces the maximum available operating frequency of the switch because of the definite duration of the transient conditions illustrated in Figure 6.5. The output voltage spike following turn-on of the auxiliary SCR constitutes an additional disadvantage, as it may damage the load.

6.2 STEP-DOWN CHOPPERS

Power electronic choppers are dc-to-dc converters with adjustable average output voltage and current. As explained in Section 1.5, the principle of operation of a chopper consists of high-frequency on–off switching. The dc supply source is alternately connected with the load and disconnected from it. As a result, the output voltage waveform constitutes a train of short pulses interspersed with short notches. The output current in the inductive load assumed does not have enough time to change significantly within the duration of a single pulse or notch, which makes for low current ripple.

Most practical choppers are of the *step-down* type. Assuming negligible voltage drops across conducting switches, the magnitude of pulses of the output voltage

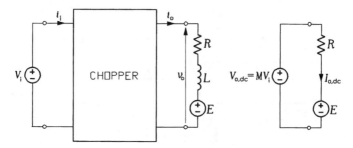

Figure 6.6 Step-down chopper: (a) block diagram; (b) equivalent circuit for the dc output voltage and current.

equals the input voltage, V_i. The average output voltage, $V_{o,\text{dc}}$, depends linearly on the duty ratios of the chopper switches and, generally, can be controlled in the range $-V_i$ to $+V_i$. Consequently, the magnitude control ratio, M, of a chopper can be taken as

$$M = \frac{V_{o,\text{dc}}}{V_i}. \tag{6.1}$$

Figure 6.6 shows a general block diagram of a chopper and the corresponding equivalent circuit for dc components of the output voltage and current. The RLE load may represent a dc motor or a battery charged through an inductive filter. It must be stressed that in this and subsequent figures, the arrows of the voltages and currents indicate the *assumed positive polarities* of these quantities. As is the case of rectifiers covered in Chapter 4, the polarity of the load EMF is the same as that of the average output voltage. In the first and third quadrants, the EMF can be zero since the power flows from the source to the load.

The differential equation of the load is

$$L\frac{di_o}{dt} + Ri_o + E = v_o \tag{6.2}$$

where the output voltage, v_o, can assume values of zero and, depending on the type of chopper, $-V_i$ and/or $+V_i$. Therefore, analyzing a given state of chopper, v_o can be treated as a constant, which makes the solution of Eq. (6.2) be

$$i_o(t) = \frac{v_o - E}{R} + \left[\frac{E - v_o}{R} + i_o(t_0)\right]e^{-(R/L)(t - t_0)} \tag{6.3}$$

where $i_o(t_0)$ is the output current at the initial instant, t_0, of the state considered. The average output current, $I_{o,\text{dc}}$, can be found from the circuit in Figure 6.6a as

$$I_{o,\text{dc}} = \frac{V_{o,\text{dc}} - E}{R} = \frac{MV_i - E}{R}. \tag{6.4}$$

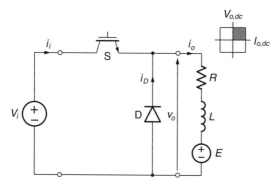

Figure 6.7   First-quadrant chopper.

Depending on the circuit configuration, step-down choppers can operate in one, two, or four quadrants of the operation plane. In subsequent sections we provide detailed descriptions of these converters.

6.2.1   First-Quadrant Chopper

The first-quadrant chopper can only produce positive dc output voltage and current, and the average power always flows from the source to the load. The circuit diagram of the chopper shown in Figure 6.7 is identical to that of the static dc switch in Figure 6.1. A switching variable, x, can be assigned to the fully controlled switch S, and the output voltage of the chopper can be expressed as

$$v_o = x V_i. \tag{6.5}$$

The value of the switching variable indicates the state of a chopper. Equivalent circuits of a chopper in states 1 and 0 are shown in Figure 6.8a and b, respectively. In this and subsequent equivalent circuits, the switch and diode currents are shown with their *actual polarities,* while, as already mentioned, the polarities of the input and output quantities indicate the assumed positive directions of these quantities.

In state 1, the switch is on, the input voltage appears across the load, and the load inductance is charged with electromagnetic energy. The output current waveform follows a growth function expressed by Eq. (6.3) with $v_o = V_i$. In state 0, the switch is off, the freewheeling diode shorts the output terminals, and the discharge of energy stored in the load inductance results in the output current decaying according to Eq. (6.3) with $v_o = 0$.

Denoting by t_{ON} and t_{OFF} the times during which switch S is on and off, that is, the chopper is in states 1 and 0, respectively, the dc output voltage, $V_{o,dc}$, can be calculated as a weighted average of output voltages in the equivalent circuits in

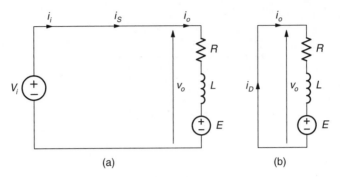

Figure 6.8   Equivalent circuits of the first-quadrant chopper: (a) state 1; (b) state 0.

Figure 6.8. Specifically,

$$V_{o,\mathrm{dc}} = \frac{V_i \times t_{\mathrm{ON}} + 0 \times t_{\mathrm{OFF}}}{t_{\mathrm{ON}} + t_{\mathrm{OFF}}} = \frac{t_{\mathrm{ON}}}{t_{\mathrm{ON}} + t_{\mathrm{OFF}}} V_i = d V_i \qquad (6.6)$$

where d is the duty ratio of switch S (not to be confused with the differentiation operator). Note that Eq. (6.6) could be derived by averaging both sides of Eq. (6.5) and taking into account the fact that the average value of a switching variable equals the duty ratio of the corresponding switch.

From Eqs. (6.1) and (6.6),

$$M = d \qquad (6.7)$$

and as the average output current given by Eq. (6.4) cannot be negative, the range of magnitude control of the output voltage that is available is

$$\frac{E}{V_i} < M \le 1. \qquad (6.8)$$

According to Eq. (6.7), the E/V_i ratio in condition (6.8) represents the minimum duty ratio of the switch allowable for continuous conduction. If operation with a lower value of d is attempted, the output current becomes discontinuous, which, as in rectifiers, should possibly be avoided.

Waveforms of the output voltage and current of a first-quadrant chopper with an E/V_i ratio of 0.25 are shown in Figure 6.9. The magnitude control ratio, M, had initially been 0.50, then changed to 0.75. The output voltage responds instantaneously to the change in M, while the current response is inertial, with the time constant, τ, imposed by the load. Note that waveforms of other voltages and currents in the chopper are easy to determine from the output current and voltage. For example, the voltage across the switch is $V_i - v_o$ and the switch current is $x i_o$.

In practice, output current waveforms in choppers can be assumed to be piecewise linear, thanks to the high switching frequencies employed. This approximation

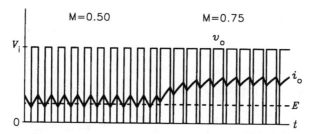

Figure 6.9 Waveforms of output voltage and current in a first-quadrant chopper.

greatly simplifies analysis of choppers, as demonstrated by the following derivation of formulas for the ac and dc components of the output current.

A single cycle of the output voltage and current in the steady state of a first-quadrant chopper is shown in Figure 6.10. The output current, i_o, increases by Δi_o during the t_{ON} time, and decreases by the same amount during the t_{OFF} time. The current increment, Δi_o, can be thought of as a doubled amplitude of the ac component (ripple) of the output current. The waveform of this component is triangular and it can easily be shown that the rms value, $I_{o,ac}$, of such a waveform is $\sqrt{3}$ times lower than that of the amplitude. Thus,

$$I_{o,ac} = \frac{\Delta i_o}{2\sqrt{3}}. \tag{6.9}$$

From Eq. (6.2),

$$di_o = \frac{1}{L}(v_o - E - Ri_o)\,dt. \tag{6.10}$$

During the on-time, $di_o \approx \Delta i_o$, $v_o = V_i$, $i_o \approx I_{o,dc}$, and $dt \approx t_{ON}$. Consequently,

$$\Delta i_o = \frac{1}{L}(V_i - E - RI_{o,dc})t_{ON}. \tag{6.11}$$

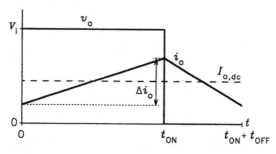

Figure 6.10 Single cycle of the output voltage and current in a first-quadrant chopper.

Analogously, during the off-time, $di_o \approx -\Delta i_o$, $v_o = 0$, $i_o \approx I_{o,dc}$, $dt \approx t_{OFF}$, and

$$\Delta i_o = \frac{1}{L}(E - RI_{o,dc})t_{OFF}. \tag{6.12}$$

Comparing Eqs. (6.11) and (6.12) and solving for $I_{o,dc}$, yields

$$I_{o,dc} = \frac{1}{R}\left(\frac{t_{ON}}{t_{ON} + t_{OFF}}V_i - E\right) = \frac{d_1 V_i - E}{R} = \frac{M V_i - E}{R} \tag{6.13}$$

that is, Eq. (6.4), determined in the preceding section.
   Substituting Eq. (6.13) in Eq. (6.12) gives

$$\Delta i_o = \frac{M V_i}{L} t_{OFF} \tag{6.14}$$

Note that t_{OFF} can be expressed as

$$t_{OFF} = (1 - d)(t_{ON} + t_{OFF}) = \frac{1 - M}{f_{sw}} \tag{6.15}$$

where $f_{sw} \equiv 1/(t_{ON} + t_{OFF})$ denotes the switching frequency of a chopper. Also, $L = \tau R$, where $\tau \equiv L/R$ is the time constant of the load. Thus, Eq. (6.14) can be rearranged to

$$\Delta i_o = \frac{V_i}{R}\frac{M(1 - M)}{\tau f_{sw}}. \tag{6.16}$$

   For generality, to accommodate choppers operating in the third and fourth quadrants, that is, with negative magnitude control ratios, M in Eq. (6.16) can be replaced by $|M|$. Then, Eqs. (6.9) and (6.16) yield

$$I_{o,ac(pu)} = \frac{|M|(1 - |M|)}{2\sqrt{3}\, f_{sw(pu)}} \tag{6.17}$$

where $I_{o,ac(pu)}$ denotes a per-unit rms value of the output ripple current, defined as

$$I_{o,ac(pu)} \equiv \frac{I_{o,ac}}{V_i/R} \tag{6.18}$$

and $f_{sw(pu)}$ is the per-unit switching frequency, defined as

$$f_{sw(pu)} \equiv \tau f_{sw}. \tag{6.19}$$

Relation (6.17) is illustrated in Figure 6.11 in the form of a three-dimensional graph. It can be seen that the maximum ripple occurs at $|M| = 0.5$. The switching frequency

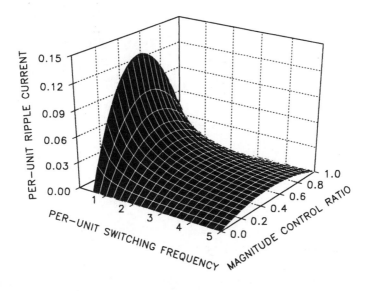

Figure 6.11   Current ripple in a chopper as a function of the magnitude control ratio and switching frequency.

has a strong impact on the ripple current, particularly at low values of this frequency when the switching times of the chopper are comparable with the time constant of the load. Excessively high switching frequencies are impractical, as small improvements in the quality of output current are obtained at the expense of high switching losses. The per-unit switching frequency of about 3 offers the best trade-off between the quality and efficiency of chopper operation.

6.2.2   Second-Quadrant Chopper

The second-quadrant chopper, depicted in Figure 6.12, operates with a positive output voltage and a negative average output current. The average power always flows from the load to the source. Equivalent circuits of the chopper in states 1 and 0 are shown in Figure 6.13. Comparing these circuits with those in Figure 6.8 for a first-quadrant chopper, it can be seen that the circuits have simply been interchanged. Clearly, the configuration of the chopper prevents the input and output currents from flowing in the positive direction.

In state 1, switch S shorts the load and the load EMF, E, supplies the resulting, circuit, which includes the load inductance, L. In state 0, the energy stored in this inductance maintains the current, which is now forced to flow through diode D to the supply source. The instantaneous output voltage is given by

$$v_o = (1 - x)V_i \qquad (6.20)$$

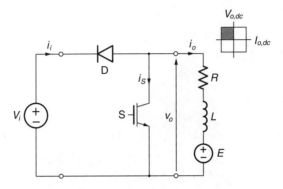

Figure 6.12   Second-quadrant chopper.

and the average output voltage can be determined as

$$V_{o,dc} = \frac{0 \times t_{ON} + V_i \times t_{OFF}}{t_{ON} + t_{OFF}} = \frac{t_{OFF}}{t_{ON} + t_{OFF}} V_i = (1 - d_2)V_i \qquad (6.21)$$

that is,

$$M = 1 - d_2 \qquad (6.22)$$

where d_2 denotes the duty ratio of switch S. For the output current to be continuous and its dc component, $I_{o,dc}$, to be negative, the control range of the chopper must be limited to

$$0 \leq M \leq \frac{E}{V_i} \qquad (6.23)$$

if $E < V_i$. Otherwise, M can be controlled in the full range 0 to 1.

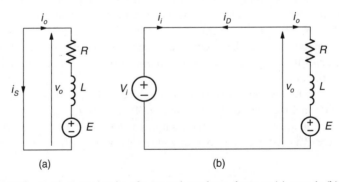

Figure 6.13   Equivalent circuits of a second-quadrant chopper: (a) state 1; (b) state 0.

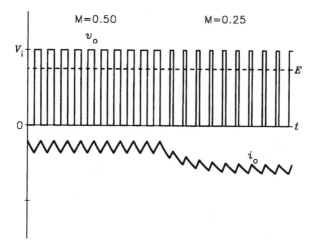

Figure 6.14   Waveforms of output voltage and current in a second-quadrant chopper.

Operation of the second-quadrant chopper is illustrated in Figure 6.14. The E/V_i ratio is 0.75, while the magnitude control ratio has initially been 0.50 and then changed to 0.25. This results in the output current waveform being a mirror image of that in Figure 6.9.

6.2.3   First-and-Second-Quadrant Chopper

While the single-quadrant choppers described in Sections 6.2.1 and 6.2.2 can transmit energy in only one direction, two-quadrant choppers are capable of bidirectional power flow. Figure 6.15 is a circuit diagram of a first-and-second-quadrant chopper, that is, one operating with a positive output voltage and an output current of either polarity. Topologically, the chopper is a combination of first- and second-quadrant choppers. Indeed, if branch S2–D2 were removed, the remaining circuit would be identical with that of a first-quadrant chopper; and vice versa, removal of branch S1–D1 would result in a second-quadrant chopper.

Denoting by x_1 and x_2 switching variables of switches S1 and S2, respectively, the state of the chopper can be designated as $(x_2 x_1)_2$. If, for example, switch S1 is on, that is, $x_1 = 1$, and S2 is off, that is, $x_2 = 0$, the chopper is said to be in state 2 since $10_2 = 2$. Theoretically, a total of four states is possible. However, state 3 is forbidden, as it would create a short circuit involving the supply source and both switches.

In the first quadrant of operation, only switch S1 is controlled, while S2 is turned off. Thus, the chopper alternates between states 0 and 2. Diode D2 is permanently reverse biased, so the entire branch S2–D2 is inactive. Conversely, when the chopper operates in the second quadrant, it is only switch S2 that performs the chopping, the chopper alternates between states 0 and 1, and branch S1–D1 is inactive. Therefore, equations derived in the preceding two sections for first- and second-quadrant

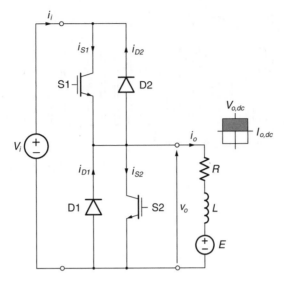

Figure 6.15 First-and-second-quadrant chopper.

choppers can easily be adapted here. In particular, in the first quadrant of operation,

$$v_o = x_1 V_i, \quad x_2 = 0 \tag{6.24}$$

and

$$M = d_1 \tag{6.25}$$

where d_1 denotes the duty ratio of switch S1. In the second quadrant,

$$v_o = (1 - x_2)V_i, \quad x_1 = 0 \tag{6.26}$$

and

$$M = 1 - d_2 \tag{6.27}$$

where d_2 is the duty ratio of switch S2. Note that the designations of switches agree with the quadrants of operation; that is, switch S1 is associated with the first quadrant and S2 with the second quadrant. This convention is observed in the description of all multiquadrant choppers covered in this chapter. The available ranges of the magnitude control of the output voltage are given by Eq. (6.8) for the first quadrant and Eq. (6.23) for the second quadrant.

An example of operation of the chopper is illustrated in Figure 6.16. The E/V_i ratio is 0.5. At the beginning, the chopper operates in the first quadrant with a magnitude control ratio of 0.75 and in the second quadrant with $M = 0.25$. The average output

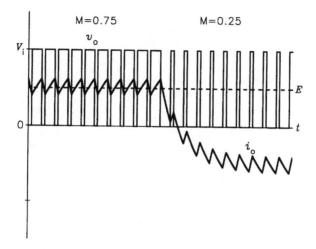

Figure 6.16   Waveforms of output voltage and current in a first-and-second-quadrant chopper.

voltage is always positive, whereas according to Eq. (6.4), the output current, initially positive, changes its polarity to negative but with the same absolute value of the dc component.

6.2.4   First-and-Fourth-Quadrant Chopper

The first-and-fourth-quadrant chopper shown in Figure 6.17 allows bidirectional power flow with a positive output current. The load EMF must be positive for first-quadrant operation and negative when the chopper is to operate in the fourth quadrant. The state of the chopper is designated as $(x_1 x_4)_2$, where x_1 and x_4 denote switching variables of switches S1 and S4, respectively.

For first-quadrant operation, switch S4 must be turned on permanently to provide a path for the output current. Switch S1 performs the chopping, with the duty ratio d_1, so the chopper operates alternately in states 2 and 3. In the fourth quadrant, switch S1 is off, S4 operates with the duty ratio d_4, and the chopper alternates between states 0 and 1.

Operation in the first quadrant was described in Section 6.2.3, although Eq. (6.24) must be modified to

$$v_o = x_1 V_i, \quad x_4 = 1. \tag{6.28}$$

Equivalent circuits of the chopper in the fourth quadrant of operation are shown in Figure 6.18. Inspection of these circuits yields

$$v_o = (x_4 - 1)V_i, \quad x_1 = 0 \tag{6.29}$$

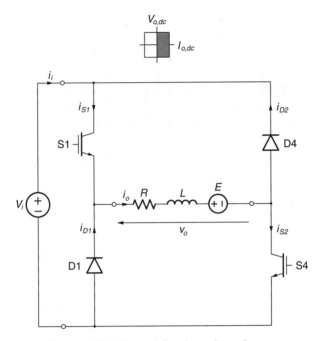

Figure 6.17   First-and-fourth-quadrant chopper.

and

$$V_{o,dc} = \frac{0 \times t_{ON,4} - V_i \times t_{OFF,4}}{t_{ON,4} + t_{OFF,4}} = -\frac{t_{OFF,4}}{t_{ON,4} + t_{OFF,4}} V_i = (d_4 - 1)V_i \qquad (6.30)$$

where $t_{ON,4}$ and $t_{OFF,4}$ denote the on and off-times of switch S4, respectively. Consequently,

$$M = d_4 - 1. \qquad (6.31)$$

For the output current to be continuous and positive, the allowable control range is

$$\frac{E}{V_i} < M \leq 0. \qquad (6.32)$$

Operation of a first-and-fourth-quadrant chopper is illustrated in Figure 6.19, where initially, for first-quadrant operation, $E/V_i = 0.5$ and $M = 0.75$ (see Figure 6.16). The dc output current is maintained constant when the load EMF changes its polarity, so that $E/V_i = -0.5$. It can easily be found from Eq. (6.4) that the magnitude control ratio in the fourth quadrant must be -0.25.

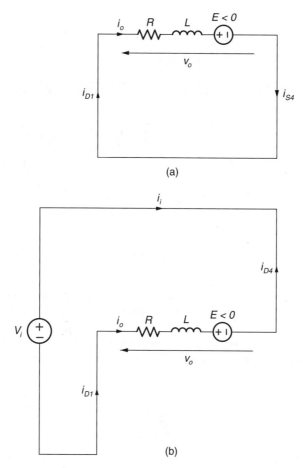

Figure 6.18 Equivalent circuits of a first-and-fourth-quadrant chopper operating in the fourth quadrant: (a) state 2; (b) state 0.

It is worth pointing out that first-quadrant operation could also be performed using switch S4 for chopping, with S1 on continuously. Then, when S4 is off, the output current would close through diode D4. However, this operating mode is not recommended, as it violates the control symmetry, which requires that in each quadrant of operation a different switch–diode pair is employed.

6.2.5 Four-Quadrant Chopper

The four-quadrant chopper is the most versatile of all step-down choppers. Its power circuit, shown in Figure 6.20, comprises four power switches, S1 through S4, and four freewheeling diodes, D1 through D4. Designating a state of the chopper as $(x_1 x_2 x_3 x_4)_2$, where x_1 through x_4 are switching variables of the switches, a theoretical

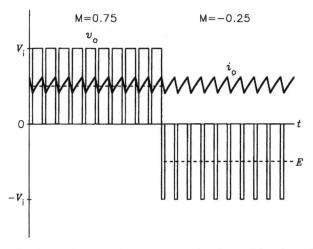

Figure 6.19 Waveforms of output voltage and current in a first-and-fourth-quadrant chopper.

total of 16 states is obtained. However, only states 0, 1, 4, 6, and 9 are utilized. The remaining states would either short-circuit the supply source or spoil the symmetry of control.

Instead of describing operation of the chopper separately for each quadrant, a summary of operating conditions is provided in Table 6.1. "ON state" and "OFF state"

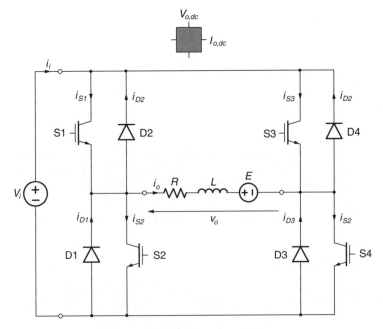

Figure 6.20 Four-quadrant chopper.

TABLE 6.1   Operating Features of the Four-Quadrant Chopper

Quadrant:	I	II	III	IV
E	≥ 0	> 0	≤ 0	< 0
x_1	$0, 1, 0, 1, \ldots$	0	0	0
x_2	0	$0, 1, 0, 1, \ldots$	1	0
x_3	0	0	$0, 1, 0, 1, \ldots$	0
x_4	1	0	0	$0, 1, 0, 1, \ldots$
ON state	9	4	6	1
OFF state	1	0	4	0
ON circuit	S1−D4	S2−D3	S2−S3	S4−D1
OFF circuit	S4−D1	D2−D3	S2−D3	D1−D4
v_o	$x_1 V_i$	$(1 - x_1)V_i$	$-x_3 V_i$	$(x_4 - 1)V_i$
M	d_1	$1 - d_2$	$-d_3$	$d_4 - 1$
M range	E/V_i to 1	0 to E/V_i	-1 to E/V_i	E/V_i to 0

denote states of the chopper corresponding to states of the chopping switch. The "ON circuit" and "OFF circuit" indicate switches and diodes carrying the output current. Duty ratios of the individual switches are denoted by d_1 through d_4. As mentioned earlier, designations of the switches correspond to respective operating quadrants of the chopper. For examples in the third quadrant, it is switch S3 that acts as the modulating (chopping) switch. The M ranges listed are those that satisfy the conditions for a continuous output current. The same magnitude control ranges, valid for all step-down choppers, are illustrated in Figure 6.21.

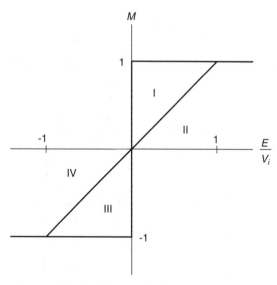

Figure 6.21   Allowable ranges of the magnitude control ratio in step-down choppers.

6.3   STEP-UP CHOPPER

The pulsed output voltage in step-down choppers described in Section 6.2 has a fixed amplitude, practically equal to the input voltage, and an adjustable average value which, taking into account voltage drops in the chopper, is *lower* than the input voltage. In contrast, the step-up chopper produces an output voltage with a fixed average value equal to the input voltage, and with an adjustable amplitude of the pulses, which is *higher* than the input voltage.

The step-up chopper shown in Figure 6.22 consists of a fully controlled switch, a diode, and an inductor. The diode prevents reversal of the output current when the switch is closed. This chopper cannot be emulated by a generic power converter since the inductor constitutes a crucial element for the power conversion performed. Notice the topological similarity of the step-up chopper with the second-quadrant chopper in Figure 6.12, with the load EMF considered a supply source. Indeed, the energy can be transferred from an active load to the input source even if the input voltage is higher than the load EMF.

When the switch is on for the time, t_{ON}, the output voltage, v_o, is zero and the voltage drop, v_L, across the chopper's inductor, conducting the input current, i_i, equals the input voltage, V_i. Thus,

$$v_L = L_c \frac{di_i}{dt} = V_i \qquad (6.33)$$

where L_c denotes the coefficient of inductance of the inductor. Consequently,

$$\frac{di_i}{dt} = \frac{V_i}{L_c} \qquad (6.34)$$

that is, the current in the inductor increases linearly in time. The increment, Δi_i, of

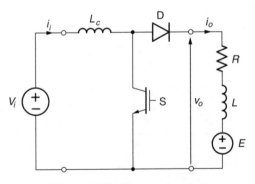

Figure 6.22   Step-up chopper.

the current by the end of the interval is

$$\Delta i_i = \frac{V_i}{L_c} t_{ON}. \tag{6.35}$$

During the following time interval, t_{OFF}, the switch is off and the input current decreases, as the load is no longer shorted by the switch. In the steady state of operation of the chopper, the average input current is constant, which implies that the current drops by the same amount, Δi_i, as it has increased during the preceding time interval, t_{ON}. Assuming that the current changes linearly in time, its derivative can be expressed as

$$\frac{di_i}{dt} = -\frac{\Delta i_i}{t_{OFF}} = -\frac{V_i}{L_c} \frac{t_{ON}}{t_{OFF}}. \tag{6.36}$$

Consequently, the output voltage during the off-time of the switch is given by

$$v_o = V_i - v_L = V_i - L_c \frac{di_i}{dt} = V_i + L_c \frac{V_i}{L_c} \frac{t_{ON}}{t_{OFF}} = V_i \left(1 + \frac{t_{ON}}{t_{OFF}} \right). \tag{6.37}$$

According to Eq. (6.37), the output voltage is constant and equal to its peak value, $V_{o,p}$, so

$$V_{o,p} = V_i \left(1 + \frac{t_{ON}}{t_{OFF}} \right) = \frac{V_i}{1-d} \tag{6.38}$$

where d denotes the duty ratio of switch S.

The average output voltage, $V_{o,dc}$, can be determined as

$$V_{o,dc} = \frac{0 \times t_{ON} + V_i/(1-d) \times t_{OFF}}{t_{ON} + t_{OFF}} = \frac{V_i}{1-d} \frac{t_{OFF}}{t_{ON} + t_{OFF}} = \frac{V_i}{1-d}(1-d) = V_i \tag{6.39}$$

that is, as mentioned before, it is equal to the input voltage. Since $0 \leq d \leq 1$, the amplitude of pulses of the output voltage is higher than that of the input voltage. The magnitude control ratio, M, as defined in Eq. (6.1), is unity, independent of the duty ratio, d.

Operation of a step-up chopper is illustrated in Figure 6.23. For simplicity, a purely resistive load has been assumed. The duty ratio is 0.75, so the pulses of the output voltage have amplitude four times higher than that of the input voltage. The pulses are not exactly rectangular because the current in the chopper inductance decays exponentially, not linearly.

To obtain a continuous output voltage, a large capacitor must be connected in parallel with the load. The diode, D, protects the capacitor from discharging through switch S. If the capacitance is sufficiently high, the voltage is maintained at an average

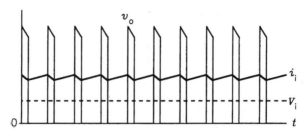

Figure 6.23 Waveforms of the output voltage and input current in a step-up chopper ($d = 0.75$).

level equal to the input voltage. Then

$$M = \frac{1}{1 - d}.$$ (6.40)

It must be pointed out that the parasitic resistances, mostly those of the source and inductor, affect the performance of the chopper. In particular, Eq. (6.38) remains valid only for low values of the duty ratio, d, that is, for low voltage gains $V_{o,p}/V_i$. The maximum available voltage gain, theoretically approaching infinity when d approaches unity, is actually quite limited (see Section 8.2.2). Therefore, in the design of step-up choppers, particularly those with smoothing capacitors and/or when a truly high output voltage is desired, circuit simulation programs, such as PSpice, should be used, taking the parasitic resistances into account.

6.4 CURRENT CONTROL IN CHOPPERS

Control of the output voltage by means of pulse width modulation is most commonly used in choppers. However, in certain applications, such as the torque control of dc motors, it is the output current that is controlled directly. This requires a closed-loop control system that imposes such switching patterns in the chopper that the actual current follows the reference signal desired. If the current is to change the polarity, the control system must cause a transition to an appropriate operating quadrant.

In practice, current-controlled inverters are more common than current-controlled choppers, and their current control systems are very similar to those of choppers. Both these converter types are supplied from dc sources. Therefore, detailed coverage of the current control is delayed until Section 7.2.2.

6.5 DEVICE SELECTION FOR CHOPPERS

Clearly, a static dc switch represents a first-quadrant chopper operating with a low switching frequency. Thus, with regard to the selection of semiconductor devices, there is no need for different treatment of the dc switches and step-down choppers. On the other hand, the rules for device selection for step-down choppers differ from those for the step-up choppers. Therefore, these two types of choppers are considered separately.

Determination of the rated voltage required in semiconductor devices (switches and diodes) used in static dc switches and step-down choppers is straightforward, since the highest voltage across a device equals, at worst, the peak value, $V_{i,p}$, of the input voltage. Although assumed constant (ideal dc), the input voltage may in practice contain a certain ripple, so that the peak value of this voltage is somewhat greater than the average value. Consequently, taking into account the voltage safety margin, s_V, the rated voltage, V_{rat}, of the devices must satisfy the condition

$$V_{rat} \geq (1 + s_V)V_{i,p}. \tag{6.41}$$

If, for example, a chopper is fed by a rectifier supplied from a 460-V line, the peak input voltage, $V_{i,p}$, should be taken as $\sqrt{2} \times 460 = 651$ V.

The contribution of a switch operating with the duty ratio d to the output current, $I_{o,dc}$, is $I_{o,dc}d$, and that of the freewheeling diode is $I_{o,dc}(1-d)$. As d can vary from 0 to 1, the rated current, I_{rat}, of the semiconductor devices must be at least as high as the rated dc output current, $I_{o,dc(rat)}$, of the chopper. Thus, employing the current safety margin, s_I, the condition to be met is

$$I_{rat} \geq (1 + S_I)I_{o,dc(rat)}. \tag{6.42}$$

In a step-up chopper, the rated voltage of the switch and diode must exceed the peak output voltage, $V_{o,p}$. The maximum allowed value, $V_{o,p(max)}$, of the instantaneous output voltage must be specified as a basis for device selection. Then the maximum allowable value, d_{max}, of the duty ratio of the switch is

$$d_{max} = 1 - \frac{V_i}{V_{o,p(max)}} \tag{6.43}$$

and the rated voltage, V_{rat}, of the switch and diode must satisfy the condition

$$V_{rat} \geq (1 + s_V)V_{o,p(max)}. \tag{6.44}$$

The diode carries the entire output current. Therefore, its current rating, $I_{D(rat)}$, must at least equal the rated average output current, $I_{o,dc(rat)}$, of the chopper. Thus,

$$I_{D(rat)} \geq (1 + s_I)I_{o,dc(rat)}. \tag{6.45}$$

To determine the required rated current, $I_{S(rat)}$, of the switch, a chopper with a resistive load and no smoothing capacitor is considered. If the switching frequency is sufficiently high, the output current consists of rectangular pulses whose amplitude is denoted by $I_{o,p}$, while the input current is practically constant and equal to $I_{o,p}$.

The switch carries the input current during the on-time, so that the average value, $I_{S,dc}$, of the switch current is

$$I_{S,dc} = I_{o,p}d. \tag{6.46}$$

The average output current, $I_{o,dc}$, which flows within the off-time, is given by

$$I_{o,dc} = I_{o,p}(1 - d). \tag{6.47}$$

Solving Eqs. (6.46) and (6.47) for $I_{S,dc}$ yields

$$I_{S,dc} = \frac{d}{1 - d} I_{o,dc} \tag{6.48}$$

which leads to the condition

$$I_{S,(rat)} \geq \frac{d_{max}}{1 - d_{max}}(1 + s_I)I_{o,dc(rat)}. \tag{6.49}$$

The same condition is valid for a chopper with a smoothing capacitor. Assuming that the input and output voltages and currents have the dc quality, that is, $v_o = V_{o,dc}$, $i_o = I_{o,dc}$, and neglecting power losses in the chopper,

$$V_i I_i = V_{o,dc} I_{o,dc} = \frac{V_i}{1 - d} I_{o,dc}. \tag{6.50}$$

Thus,

$$I_i = \frac{I_{o,dc}}{1 - d} \tag{6.51}$$

and as

$$I_{S,dc} = I_i d = \frac{d}{1 - d} I_{o,dc} \tag{6.52}$$

the same relation as that of Eq. (6.48) has been obtained.

In any PWM converter, the pulse width modulation cannot be performed within the entire range 0 to 1 of the duty ratio, d, of switches, since at d close to zero or unity, the required on-time, t_{ON}, or off-time, t_{OFF}, would be too short in comparison with the feasible switching times of the switches. These times are related to the minimum and maximum values, d_{min} and d_{max}, of the duty ratio as

$$t_{ON(min)} = \frac{d_{min}}{f_{sw}} \tag{6.53}$$

and

$$t_{\text{OFF(min)}} = \frac{1 - d_{\max}}{f_{\text{sw}}}. \tag{6.54}$$

The last two equations facilitate selection of the switching frequency, f_{sw}, that is appropriate for a given type of semiconductor power switch.

6.6   COMMON APPLICATIONS OF CHOPPERS

Step-down choppers produce high-quality output currents that can be adjusted within a wide range. Voltage control characteristics of choppers are linear. Therefore, step-down choppers are employed primarily in high-performance dc drive systems such as in machine tools and electrical traction.

If a chopper is supplied from an electric battery, operation with the reversed power flow poses no technical problem. However, if a chopper is fed from an ac line, a diode rectifier with a *dc link* must be used, as shown in Figure 6.24. An inductor, which prevents the chopper from drawing a high-frequency current from the ac line, is optional. In certain situations, the line inductance alone can suffice. A capacitor is necessary, however, to absorb the negative input current, which cannot close through the rectifier. In practice, in choppers supplied from a battery, a dc link is often used too, to avoid extra losses in the battery due to the ac component of the input current.

Sustained operation of a chopper in the second and fourth quadrants with a supply arrangement as in Figure 6.24 is not feasible, because the rectifier is not capable of transferring power to the supply line. Attempting to operate the chopper with a negative power flow would charge up the filter capacitor, increasing the input voltage to the chopper. This would cause either a disruption in operation of the chopper due to the increasingly limited control range (see Figure 6.21) or overvoltage damage to the semiconductor devices or the capacitor. One way to solve this problem is to replace the diode rectifier with a controlled rectifier. This solution is preferable when a number of choppers are supplied from a single dc network. A dc mass transit system is a typical example.

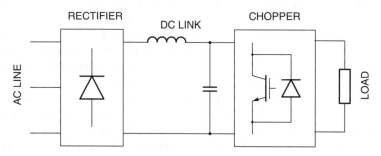

Figure 6.24   Typical supply system of a chopper.

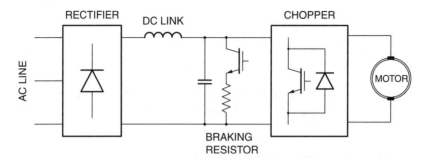

Figure 6.25   Dc drive in a chopper with a breaking resistor.

In the case of a single chopper feeding a dc motor and supplied through a diode rectifier, a *braking resistor* is usually employed to dissipate energy generated by the motor in the second and fourth quadrants of operation. This arrangement is shown in Figure 6.25. When an overvoltage across the dc-link capacitor is detected, the semiconductor switch in series with the braking resistor turns on, bleeding the capacitor of the excessive charge.

Second-quadrant choppers are utilized in autonomous power supply systems based on solar panels charging a battery pack. The battery pack is connected to the input terminals, while the panel, whose voltage is fluctuating, is connected through a reactor to the output terminals of the chopper. Even when the source's voltage is lower than that of the battery pack, the pack can be charged from the source.

Step-up choppers are used primarily as high-voltage sources in certain radar and ignition systems. As explained in Chapter 8, operating principles of step-down and step-up choppers are utilized in switching power supplies.

6.7   SUMMARY

Dc-to-dc power conversion is performed using choppers. In step-down choppers, the fixed dc input voltage is "chopped" into a train of rectangular pulses whose average value is adjustable by means of the width modulation. The load inductance attenuates the ac component of the output current so that the current is practically of dc quality. The higher the switching frequency is, the lower the current ripple becomes.

Step-down choppers can be structured to operate in one, two, or all four quadrants. The number of switch−diode pairs equals that of the operating quadrants. When operated intermittently with a duty ratio of zero or unity, a first-quadrant chopper serves as a static dc switch connecting a load to a dc source.

A step-up chopper generates a train of voltage pulses whose average value is constant, but the amplitude is higher than that of the input voltage. A smoothing capacitor connected in parallel with the load makes it possible to obtain a stepped-up dc output voltage. The voltage gain of a step-up chopper is limited by parasitic resistances in the chopper.

EXAMPLE   263

Step-down choppers are employed primarily in high-performance dc drive systems. Step-up choppers find application in radar and ignition systems.

EXAMPLE

Example 6.1   A step-down chopper fed from a 120-V source operates a dc motor whose armature EMF is 100 V and armature resistance is 0.5 Ω. With a magnitude control ratio of 0.6, find the average output voltage and current of the chopper and determine the quadrant of operation.

Solution: From Eq. (6.1), the average output voltage of the chopper is

$$V_{o,dc} = 0.6 \times 120 = 72 \, V$$

and from Eq. (6.4), the average output current is

$$I_{o,dc} = \frac{72 - 100}{0.5} = -56 \, A.$$

As $V_{o,dc} > 0$ and $I_{o,dc} < 0$, the chopper operates in the second quadrant.

Example 6.2   The same system is considered as in Example 6.1. The chopper operates with a switching frequency of 1.5 kHz, and the armature inductance of the motor is 12 mH. Find the ac component of the armature current and the ripple factor of the current.

Solution: The time constant, τ, of the load of the chopper is

$$\tau = \frac{12 \times 10^{-3}}{0.5} = 0.024 \, s$$

and from Eqs. (6.17) and (6.19),

$$I_{o,ac(pu)} = \frac{0.6(1 - 0.6)}{2\sqrt{3} \times 0.024 \times 1.5 \times 10^3} = 0.00192.$$

Now, based on Eq. (6.18), the ac component, $I_{o,ac}$, of the armature current of the motor supplied by the chopper can be calculated as

$$I_{o,ac} = \frac{120}{0.5} \times 0.00192 = 0.46 \, A.$$

In Example 6.1, the absolute value of the dc component of the output current of the chopper has been found to be 56 A. Hence, the ripple factor, RF_I, equals

$$RF_I = \frac{0.46}{56} = 0.008 = 0.8\%.$$

Example 6.3   A four-quadrant chopper operates with a switching frequency of 800 Hz and supplies power to a dc motor whose armature EMF is -200 V. The input voltage of the chopper is 250 V. Find:

(a) The voltage control ranges for the third and fourth quadrants of operation of the chopper
(b) The on- and off-times of the modulating switch when the chopper operates in the third quadrant with the magnitude control ratio of -0.9
(c) The on- and off-times of the modulating switch when the chopper operates in the fourth quadrant with the magnitude control ratio of -0.5

Solution:

(a) The load EMF/input voltage ratio, E/V_i, is $-200/250 = -0.8$. Therefore, according to Table 6.1, the magnitude control ratio, M, when the chopper operates in the third quadrant must be within the range -1 to -0.8. The allowable range of M for the fourth quadrant is -0.8 to 0.
(b) Considering the circuit diagram of the chopper in Figure 6.19, the modulating switch in the third quadrant is S3. From Table 6.1, the duty ratio, d_3, of this switch is

$$d_3 = -M = 0.9.$$

Consequently, the on-time, $t_{ON,3}$, of the switch is

$$t_{ON,3} = \frac{0.9}{800} = 1.125 \times 10^{-3}\,\text{s} = 1.125\,\text{ms}$$

and the off-time, $t_{OFF,3}$, is

$$t_{OFF,3} = \frac{0.1}{800} = 1.25 \times 10^{-4}\,\text{s} = 0.125\,\text{ms}.$$

(c) In this case, it is the switch S4 that performs the modulation. Since its duty ratio, d_4, is

$$d_4 = M + 1 = -0.5 + 1 = 0.5$$

then

$$t_{ON,4} = t_{OFF,4} = \frac{0.5}{800} = 6.25 \times 10^{-4}\,\text{s} = 0.625\,\text{ms}$$

Example 6.4 A step-up chopper is supplied from a 240-V source. Find the duty ratio of the chopper switch such that the output voltage has a peak value of 510 V. What is the duration of output voltage pulses if the switching frequency is 1 kHz?

Solution: From Eq. (6.38), the required duty ratio, d, is

$$d = 1 - \frac{V_i}{V_{o,p}} = 1 - \frac{240}{510} = 0.53.$$

As the pulses of the output voltage appear when the switch is off, then their duration, t_{OFF}, is

$$t_{OFF} = \frac{1 - 0.53}{10^3} = 4.7 \times 10^{-4}\,\text{s} = 0.47\,\text{ms}.$$

Example 6.5 A two-quadrant chopper, rated at 15 kVA and 200 V, is supplied from a diode rectifier with a low-pass filter such that the peak supply voltage does not exceed 250 V. Find the minimum voltage and current ratings of switches and diodes of the chopper. Use safety margins of 0.4 for the rated voltage and 0.2 for the rated current.

Solution: According to Eq. (6.41), the rated voltage of the two semiconductor power switches and two freewheeling diodes of the chopper must be at least $1.4 \times 250\,\text{V} = 350\,\text{V}$. The rated current, $I_{o,dc(rat)}$, of the chopper is $(15{,}000\,\text{VA})/(200\,\text{V}) = 75\,\text{A}$. From Eq. (6.42), the rated current of the switches and diodes may not be less than $1.2 \times 75\,\text{A} = 90\,\text{A}$.

PROBLEMS

P6.1 A chopper is supplied from a 240-V dc source. The load of the chopper consists of a 10-Ω resistance in series with a 20-mH inductance. The chopper operates with a magnitude control ratio of 0.7 and a switching frequency of 1.2 kHz. Find:

(a) The quadrant of operation of the chopper

(b) The average output voltage

(c) The average output current

P6.2 Can the chopper in Problem 6.1 operate in the second quadrant? Justify the answer.

P6.3 For the chopper in Problem 6.1, find:
 (a) The on-time of the modulating switch
 (b) The off-time of the switch
 (c) The output ripple current
 (d) The ripple factor of the output current

P6.4 For the chopper in Problem 6.1, find:
 (a) The per-unit switching frequency
 (b) The per-unit output ripple current

P6.5 A two-quadrant chopper is supplied from a 150-V dc source and operates with a magnitude control ratio of –0.7 and a switching frequency of 900 Hz. The chopper supplies power to a dc motor whose armature resistance is 0.5 Ω and armature inductance is 1.7 mH. The motor rotates with a constant speed such that the armature EMF is –120 V. Find:
 (a) The operating quadrant of the chopper
 (b) The average armature voltage of the motor
 (c) The average armature current

P6.6 For the chopper in Problem 6.5, find:
 (a) The duty ratio of the modulating switch
 (b) The on-time of the switch
 (c) The off-time of the switch
 (d) The ac component of the output current

P6.7 The same chopper and motor as in Problem 6.5 are considered, but the armature voltage and EMF are 125 and 100 V, respectively, and the armature current is positive. Find:
 (a) The operating quadrant of the chopper
 (b) The average armature current of the motor
 (c) The magnitude control ratio of the chopper
 (d) The duty ratio of the modulating switch

P6.8 A two-quadrant chopper fed from a 300-V dc source supplies power to a dc motor whose armature resistance is 0.1 Ω. The armature current is to be maintained constant at 150 A, independent of the motor speed, which affects the armature EMF. Find the magnitude control ratio of the chopper and the duty ratio of the modulating switch when the armature EMF is:
 (a) 210 V
 (b) −210 V

P6.9 For the chopper in Problem 6.8, find the average output current when:
 (a) The armature EMF is 100 V and the magnitude control ratio of the chopper is 0.5
 (b) The armature EMF is −200 V and the magnitude control ratio is –0.8

P6.10   The armature inductance of the motor in Problem 6.9 is 0.5 mH. Find the switching frequency of the chopper such that the per-unit ripple current does not exceed 0.02 pu.

P6.11   A four-quadrant chopper is supplied from a 400-V dc source. The absolute value of the load EMF is 240 V. For each operating quadrant of the chopper, find the magnitude control range and the range of the duty ratio of the modulating switch.

P6.12   A step-up chopper is supplied from a 12-V battery and operates with a switching frequency of 2 kHz. Find the amplitude of pulses of the output voltage and their duration when the duty ratio of the chopper switch is 0.9.

P6.13   A single-quadrant chopper rated at 150 VA is to be supplied from a 12-V battery. Determine the minimum required voltage and current ratings of the semiconductor devices of the chopper.

P6.14   A two-quadrant chopper is rated at 3 kVA and 300 V. A six-pulse diode rectifier fed from a three-phase 230-V line is to be used as the supply source. Determine the minimum required voltage and current ratings of the semiconductor devices of the chopper.

P6.15   Repeat Problem 6.14 for a chopper rated at 10 kVA and 600 V and a rectifier fed from a three-phase 460-V line.

P6.16   A step-up chopper is supplied from a 6-V battery and rated at 240 VA. The highest allowable duty ratio of the chopper switch is 0.9. Determine the minimum required rated currents of the switch and diode of the chopper.

COMPUTER ASSIGNMENTS

*CA6.1   Run PSpice program *DC_Switch.cir* for an SCR-based static dc switch. Observe the oscillograms of the voltages and currents illustrated in Figure 6.5.

*CA6.2   Run PSpice program *Chopp_1Q.cir* for a first-quadrant chopper. For a magnitude control ratio of 0.8, find for both the output voltage and current of the chopper:

(a) The dc component

(b) The rms value

(c) The rms value of the ac component

(d) The ripple factor

*CA6.3   Run PSpice program *Chopp_2Q.cir* for a second-quadrant chopper. For a magnitude control ratio of 0.4, find for both the output voltage and current of the chopper:

(a) The dc component

(b) The rms value

(c) The rms value of the ac component

(d) The ripple factor

*CA6.4   Run PSpice program *Chopp_12Q.cir* for a first-and-second-quadrant chopper. With an arbitrarily set value of the magnitude control ratio, observe oscillograms of the voltages and currents in both quadrants of operation of the chopper.

CA6.5   Develop a PSpice circuit file for a first-and-fourth-quadrant chopper. With an arbitrarily set magnitude control ratio, observe oscillograms of the voltages and currents in both quadrants of operation of the chopper.

CA6.6   Develop a PSpice circuit file for a four-quadrant chopper. For each quadrant of operation, write a separate set of lines of program for switching signals of individual switches of the chopper (when running the program for a specific quadrant, lines for the other three quadrants will be "commented out" using asterisks).

*CA6.7   Run PSpice program *Step_Up_Chopp.cir* for a step-up chopper. Set the duty ratio of the switch to 0.75. Compare the voltage and current waveforms with those in Figure 6.22. Repeat the simulation for a duty ratio of 0.5.

*CA6.8   Run PSpice program *Step_Up_Chopp.cir* for a step-up chopper. With a duty ratio for the switch of 0.8, compare the dc component of the current conducted by the switch with that of the output current.

CA6.9   Based on Eqs. (6.5) through (6.7), develop a computer program for computation of the ripple current and ripple factor in a chopper. Use the program to verify the results of computer Assignments 6.2(c) and 6.3(c).

LITERATURE

[1] Abraham, L., Power electronics in German railway propulsion, *Proceedings of the IEEE*, Apr. 1988, pp. 472–480.

[2] Dewan, S. B., Slemon, G. R., and Straughen, A., *Power Semiconductor Drives*, Wiley, New York, 1984, Chapt. 5.

[3] Ohno, E., *Introduction to Power Electronics*, Clarendon Press, Oxford, UK, 1988, Sect. 5.1.

[4] Rashid, M. H., *Power Electronics Handbook*, 2nd Ed., Academic Press, San Diego, CA, 2007, Chapt. 13.

7 DC-to-AC Converters

Principles of dc-to-ac power conversion using power electronic inverters are presented in this chapter. Voltage- and current-source inverters in the square-wave and PWM operation modes are also discussed, and voltage and current control techniques are reviewed. Multilevel and soft-switching inverters are introduced. Device selection for inverters is outlined and major inverter applications are described.

7.1 VOLTAGE-SOURCE INVERTERS

As mentioned in Chapter 1, dc-to-ac power conversion is performed by inverters. An inverter is supplied from a dc source, while ac output voltage has a strong fundamental component with adjustable frequency and amplitude. Depending on the type of source, *voltage-source inverters* (VSIs) and *current-source inverters* (CSIs) can be distinguished. Voltage-source inverters are the second most common power electronic converters after rectifiers.

The dc input voltage for a voltage-source inverter can be obtained from a rectifier, usually of the uncontrolled diode type, or from another dc source, such as a battery or a photovoltaic array. As shown in Figure 7.1, if a rectifier is used, the inverter is supplied via an LC dc link, similar to that used for choppers (see Figure 6.24). The link capacitor can be thought of as a voltage source, since voltage across it cannot change instantly. The main task of the inductor is to isolate the supplying rectifier and power system from the high-frequency component of an inverter input current. In contrast to a capacitor, a link inductor is not inherently necessary. Indeed, in some practical inverters, the inductor is disposed of to reduce the size and cost of a converter.

Inverters can be built with any number of output phases. In practice, single- and three-phase inverters are most common. However, construction of ac motors with more than three phases has recently been proposed, to increase their reliability in certain critical applications. Such motors have to be supplied from appropriate multiphase inverters.

In the past, SCRs were used in high- and medium-power inverters. SCR-based inverters required commutating circuits, similar to those described in Section 6.1, to turn the SCRs off. Commutating circuits increase inverter size and cost and reduce reliability and switching frequency. Presently, fully controlled semiconductor power switches, mostly IGBTs (in medium-power inverters) and GTOs or IGCTs

Introduction to Modern Power Electronics, Second Edition, by Andrzej M. Trzynadlowski
Copyright © 2010 John Wiley & Sons, Inc.

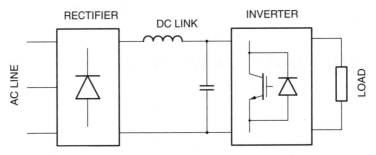

Figure 7.1   Voltage-source inverter supplied from a diode rectifier.

(in high-power inverters), are employed. Therefore, fully controlled switches are assumed in all subsequent considerations in this chapter.

7.1.1   Single-Phase Voltage-Source Inverter

The identity of the block diagrams in Figures 6.24 and 7.1 is not coincidental. Indeed, when controlled appropriately, a four-quadrant chopper becomes a single-phase inverter (analogous to a dual converter, which can also serve as a single-phase cycloconverter). The inverter is so controlled that the time-averaged output voltage changes sinusoidally in time. As the output current is usually shifted in phase with respect to the fundamental output voltage, the converter operates in all four quadrants within each cycle of operation.

Figure 7.2 is a circuit diagram of a single-phase voltage-source inverter based on fully controlled semiconductor power switches. As already mentioned, the bridge topology of the inverter is the same as that of the four-quadrant chopper in Figure 6.20,

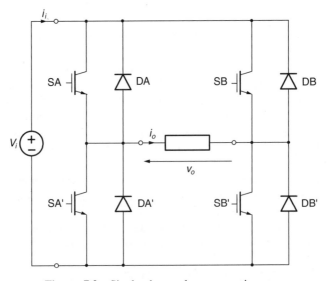

Figure 7.2   Single-phase voltage-source inverter.

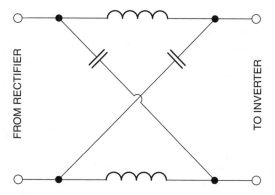

Figure 7.3   Circuit diagram of a Z-source.

although drawn here with a slightly different arrangement and designation of elements. An ideal voltage source, producing a dc input voltage, V_i, is assumed.

Switches in any leg of an inverter may not be on simultaneously, as they would short the supply source. This condition, called *a shot-through*, was covered in Section 3.3. Semiconductor switches, even fast switches, require finite transition times from one conduction state to another. Therefore, in practice, to avoid a shot-through, a switch is turned off shortly before turn-on of the second switch in the same leg. The interval between the turn-off and turn-on signals is called *dead time* or *blanking time*.

Considering the issue of dead time, it is worth mentioning the *Z-source inverter*, recently proposed in the literature. The power circuit of the inverter is identical to that of the VSI, but the dc-link capacitor is replaced by the inductive–capacitive circuit shown in Figure 7.3. Other configurations are also possible, as is application of the idea to other power electronic converters. Thanks to the presence of inductors in the dc link, the need for dead time is eliminated, as turning on both switches in a given leg of the inverter is no longer dangerous. Even more important, a Z-source circuit makes it possible to both step down and step up (boost) the output voltage. Z-source inverters have not yet reached the stage of full commercialization, and interested readers are referred to the many papers on the subject.

For simplicity in subsequent considerations, zero dead time will be assumed. If so, each leg of the inverter can assume two states only: either the upper (common-anode) switch is on and the lower (common-cathode) switch is off or the other way around. Thus, switching variables a and b can be assigned to individual legs of the inverter and defined as

$$a = \begin{cases} 0 & \text{if SA is OFF and SA}' \text{ is ON} \\ 1 & \text{if SA is ON and SA}' \text{ is OFF} \end{cases}$$

$$b = \begin{cases} 0 & \text{if SB is OFF and SB}' \text{ is ON} \\ 1 & \text{if SB is ON and SB}' \text{ is OFF} \end{cases}$$

(7.1)

An inverter state is designated as ab_2. If, for example, $a = 1$ and $b = 1$, the inverter is said to be in state 3, as $11_2 = 3$. Clearly, four states are possible, states 0 through 3.

When a given switching variable assumes a value of 1, the positive terminal of the supply source is connected to the corresponding output terminal of the inverter. Vice versa, a value of 0 indicates connection of the negative terminal of the source to the output terminal of the inverter. Consequently, the output voltage, v_o, of the inverter can be expressed as

$$v_o = V_i(a - b) \tag{7.2}$$

and the voltage can assume three values only: 0, $+V_i$, and $-V_i$, corresponding to states 0 or 3, state 2, and state 1, respectively.

When the output current, i_o, is of such polarity that it cannot flow through a switch that is turned on, the freewheeling diode parallel to the switch provides a path for the current. If, for example, the output current is positive, as shown in Figure 7.2, and switch SB is on, the current is forced to close through diode DB because switch SB′ must be off. In a similar way, it is easy to determine conditions under which each of the other three diodes becomes necessary for continuous conduction of the output current. The freewheeling diodes would not be needed if the load were purely resistive. Otherwise, however, interrupting a current in the load inductance would cause dangerous overvoltage.

The basic version of the *square-wave* operation mode of the inverter is described by the following control law:

$$a = \begin{cases} 1 & \text{for } 0 < \omega t \leq \pi \\ 0 & \text{otherwise} \end{cases}$$

$$b = \begin{cases} 1 & \text{for } \pi < \omega t \leq 2\pi \\ 0 & \text{otherwise} \end{cases} \tag{7.3}$$

where ω denotes the fundamental output radian frequency of the inverter. Only states 1 and 2 are utilized. Waveforms of the output voltage and current in an RL load of an inverter in the basic square-wave mode are shown in Figure 7.4. The rms fundamental output voltage, $V_{o,1}$, and harmonic content, $V_{o,h}$, of this voltage equal $0.9V_i$ and $0.435V_i$, respectively, so the total harmonic distortion, THD, is 0.483.

The total harmonic distortion of the output voltage can be minimized by interspersing states 1 and 2 with states 0 and 3, lasting 0.81 rad (46.5°) each in the ωt domain, as shown in Figure 7.5. Then, in comparison with basic square-wave operation, the fundamental output voltage decreases by 8%, to $0.828V_i$, but the total harmonic distortion is reduced by as much as 40%, to 0.29. The control laws yielding the optimal square-wave mode are

$$a = \begin{cases} 1 & \text{for } \alpha_d < \omega t \leq \pi + \alpha_d \\ 0 & \text{otherwise} \end{cases}$$

$$b = \begin{cases} 1 & \text{for } \pi + \alpha_d < \omega t \leq 2\pi + \alpha_d \\ 0 & \text{otherwise} \end{cases} \tag{7.4}$$

where $\alpha_d = 0.405$ rad (23.2°).

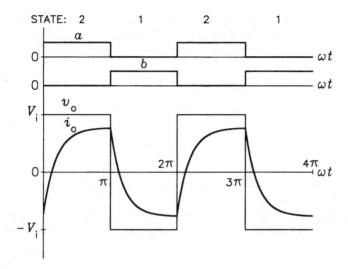

Figure 7.4 States, switching variables, and waveforms of output voltage and current in a single-phase VSI in the basic square-wave mode.

The quality of operation of the inverter can be improved further by pulse width modulation. The single-phase inverter has only one output voltage, so the space vector PWM technique is not applicable. Here, for illustration purposes, a simple PWM strategy based on a sinusoidal modulating function, $F(m, \omega t) = m \sin \omega t$, is assumed. The concept of modulating function was introduced in Section 5.1.3.

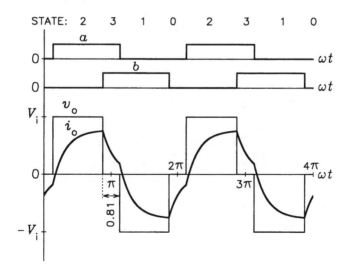

Figure 7.5 States, switching variables, and waveforms of output voltage and current in a single-phase VSI in the optimal square-wave mode.

As usual in PWM converters, the cycle of the output voltage is divided into a number of switching intervals. Denoting the duty ratios (average values) of switching variables a and b in the nth switching interval by d_{an} and d_{bn}, the control law for the inverter is

$$d_{an} = \frac{1}{2}[1 + F(m, \alpha_n)]$$

$$d_{bn} = \frac{1}{2}[1 - F(m, \alpha_n)]$$

(7.5)

where m denotes the modulation index and α_n is the phase angle of the output voltage in the center of the switching interval.

Waveforms of the output voltage and current in an RL load, when the inverter operates in the PWM mode, are shown in Figure 7.6 for $N = 10$ switching intervals

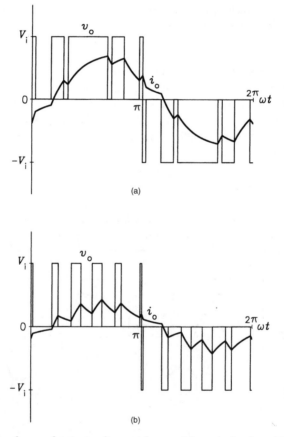

Figure 7.6 Waveforms of output voltage and current in a single-phase VSI in the PWM mode: (a) $m = 1$; (b) $m = 0.5$ ($N = 10$).

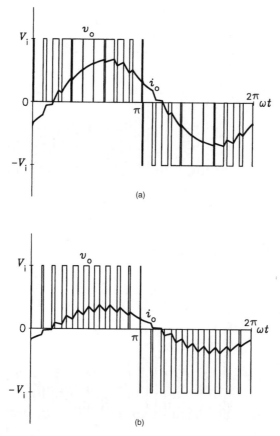

Figure 7.7 Waveforms of output voltage and current in a single-phase VSI in the PWM mode: (a) $m = 1$; (b) $m = 0.5$ ($N = 20$).

per cycle of output voltage, and in Figure 7.7 for $N = 20$. Unsurprisingly, as

$$N = \frac{f_{sw}}{f_{o,1}} \tag{7.6}$$

where f_{sw} and $f_{o,1}$ denote the switching frequency and fundamental output frequency, respectively, the quality of output current increases with the switching frequency. Unfortunately, so do switching losses in the inverter switches; thus, a reasonable trade-off between the quality and efficiency of operation must be sought. With respect to Eq. (7.6), it must be stressed that the frequency ratio N (not necessarily an integer) has been introduced here to facilitate comparison of waveforms and spectra. In practice, it is the switching frequency, usually constant, that is the major defining parameter of operation of a PWM inverter (and any other PWM power converter).

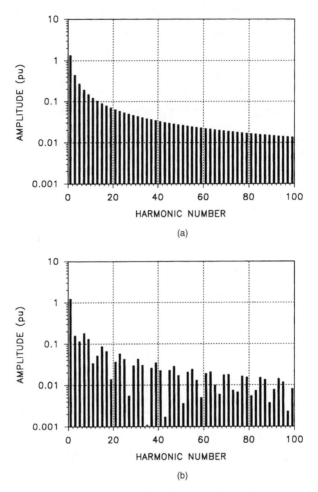

Figure 7.8 Harmonic spectra of output voltage in a single-phase VSI: (a) square-wave mode; (b) optimal square-wave mode.

The progression in improvement of operating characteristics of the inverter from the basic square-wave mode, through the optimal square-wave mode, to the PWM mode can best be explained using harmonic spectra of the output voltage and current. All the following spectra pertain to an inverter supplied from a 1 per-unit dc voltage source and supplying an RL load with a per-unit impedance of 1 and a load angle of 30°.

The voltage spectra in the basic and optimal square-wave modes are shown in Figure 7.8a and b, respectively. Thanks to the half-wave symmetry of the voltage waveforms, only odd harmonics are present. As pointed out on other occasions, the load inductance acts as a low-pass filter of the output current. Hence, for the high quality of that current, low-order harmonics of the output voltage should have

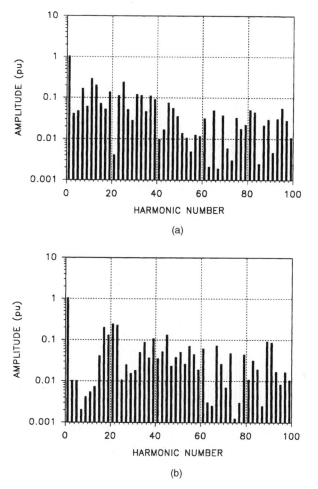

Figure 7.9 Harmonic spectra of output voltage in a single-phase VSI in the PWM mode: (a) $N = 10$; (b) $N = 20$ ($m = 1$).

possibly low amplitudes. Indeed, the third and fifth harmonics resulting from the optimal distribution of inverter states over the cycle of output voltage are distinctly lower than those in the basic square-wave mode.

Further reduction of the low-order voltage harmonics is obtained in the PWM mode, as illustrated in Figure 7.9 for $m = 1$, especially when the number of switching intervals is high. Notice, for example, that the lowest-order harmonic whose amplitude is higher than 10% of the fundamental amplitude is the ninth for $N = 10$ (Figure 7.9a) and the seventeenth for $N = 20$ (Figure 7.9b). However, the maximum available fundamental output voltage has a peak value of 1, so the maximum available rms value, $V_{o,1(\text{max})}$, of this voltage is $0.707V_i$ only. This is about 21% lower than the rms voltage in the basic square-wave operation mode and about 15% lower than that in the optimal square-wave mode.

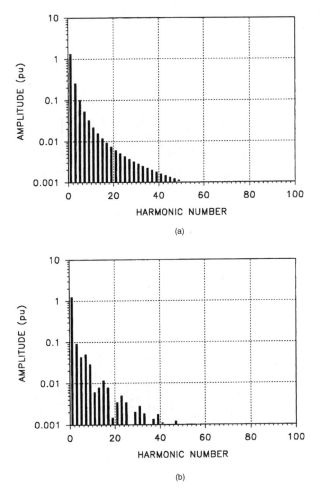

Figure 7.10 Harmonic spectra of output current in a single-phase VSI: (a) basic square-wave mode; (b) optimal square-wave mode.

Harmonic spectra of the output currents produced by voltages whose spectra are shown in Figures 7.8a through 7.9b are depicted in Figures 7.10a through 7.11b, respectively. The highest-amplitude current harmonics in an inverter operating in either of the square-wave modes are the low-order harmonics, such as the third, fifth, and seventh. In contrast, in a PWM inverter, the highest amplitudes belong to harmonics whose harmonic numbers are close to the number of switching intervals per cycle of the output voltage. These are the seventh, eleventh, and thirteenth when $N = 10$ (Figure 7.11a) and the seventeenth, twenty-first, and twenty-third when $N = 20$ (Figure 7.11b), all of them with amplitudes more than one order of magnitude lower than that of the fundamental.

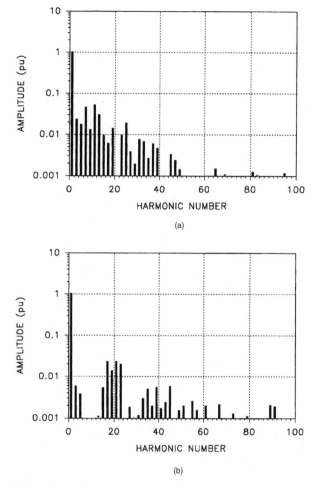

Figure 7.11 Harmonic spectra of output current in a single-phase VSI in the PWM mode: (a) $N = 10$; (b) $N = 20$ ($m = 1$).

The input current, i_i, of the inverter, given by

$$i_i = (a - b)i_o \tag{7.7}$$

has a large dc component, $I_{i,\text{dc}}$. The product of V_i and $I_{i,\text{dc}}$ constitutes the average input power to the inverter. The waveform of the input current in the optimal square-wave mode is shown in Figure 7.12a and that in the PWM mode, with $m = 1$ and $N = 20$, in Figure 7.12b. Harmonic spectra of these currents are shown in Figure 7.13a and b, respectively. The fundamental frequency of the currents is twice the output frequency, and the first harmonic is comparable with the dc component. Other than that, the harmonic spectra in the low-frequency region are similar to those of the respective output currents. Note that apart from the fundamental, the most

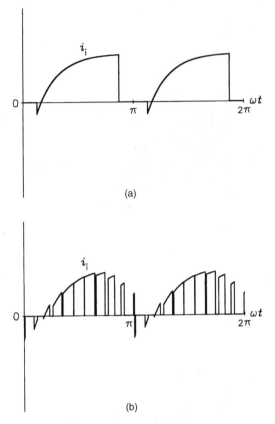

Figure 7.12 Waveforms of input current in a single-phase VSI: (a) optimal square-wave mode; (b) PWM mode ($m = 1, N = 20$).

prominent harmonics of the input current in a PWM inverter are clustered about the $N/2$ value of the harmonic order. Thus, if the switching frequency is sufficiently high, a small dc link can prevent most of the high-frequency content of the input current from reaching the supply source (power system or battery).

The practically instant changes of the input current in Figure 7.11b confirm the necessity for a dc-link capacitor. Such current could not be drawn directly from an ac grid, via the line-side rectifier, because of all the inductances in the path of that current.

The output frequency in all types of inverters is controlled by the appropriate timing of switching instants. However, control of the magnitude of the output voltage in the square-wave mode can only be realized outside the inverter, by adjusting the dc input voltage. Specifically, a controlled rectifier must be used in place of the diode rectifier shown in Figures 7.1 and 7.2. In contrast, the PWM mode of operation allows the voltage control by adjusting the modulation index, m, which equals the magnitude control ratio, M. Here, the latter parameter represents the ratio of rms

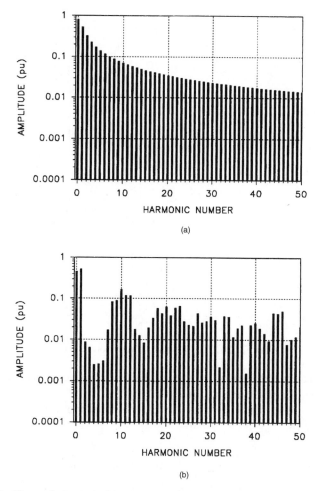

Figure 7.13 Harmonic spectra of input current in a single-phase VSI: (a) optimal square-wave mode; (b) PWM mode ($m = 1, N = 20$).

fundamental output voltage, $V_{o,1}$, to the maximum value, $V_{o,1(max)}$, of this voltage available using a given PWM strategy.

7.1.2 Three-Phase Voltage-Source Inverter

Most practical inverters, especially medium- and high-power inverters, are of the three-phase type. Shown in Figure 7.14, the power circuit of a three-phase voltage-source inverter is obtained by adding a third leg to the single-phase inverter. Assuming, as before, that of the two power switches in each leg (phase) of the inverter, one and only one is always on [i.e., neglecting the short time intervals when both switches are off (dead time)], three switching variables, a, b, and c, can be assigned to the inverter. A state of an inverter is designated as abc_2, making for a total of eight states, from

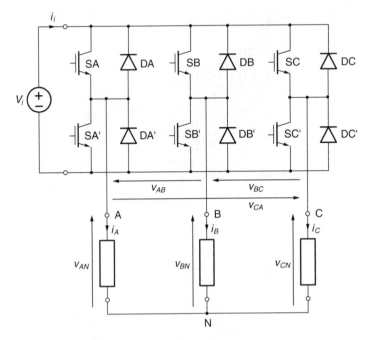

Figure 7.14   Three-phase voltage-source inverter.

state 0, when all output terminals are clamped to the negative dc bus, through state 7, when they are clamped to the positive bus.

It is easy to show that the instantaneous line to line output voltages, v_{AB}, v_{BC}, and v_{CA}, are given by

$$\begin{bmatrix} v_{AB} \\ v_{BC} \\ v_{CA} \end{bmatrix} = V_i \begin{bmatrix} 0 & -1 & 0 \\ 0 & 1 & -1 \\ -1 & 0 & 1 \end{bmatrix} \begin{bmatrix} a \\ b \\ c \end{bmatrix} \tag{7.8}$$

In a balanced three-phase system, the instantaneous line-to-neutral output voltages, v_{AN}, v_{BN}, and v_{CN}, can be expressed as

$$\begin{bmatrix} v_{AN} \\ v_{BN} \\ v_{CN} \end{bmatrix} = \frac{1}{3} \begin{bmatrix} 1 & 0 & -1 \\ -1 & 1 & 0 \\ 0 & -1 & 1 \end{bmatrix} \begin{bmatrix} v_{AB} \\ v_{BC} \\ v_{CA} \end{bmatrix} \tag{7.9}$$

which, when combined with Eq. (7.8), yields

$$\begin{bmatrix} v_{AN} \\ v_{BN} \\ v_{CN} \end{bmatrix} = \frac{V_i}{3} \begin{bmatrix} 2 & -1 & -1 \\ -1 & 2 & -1 \\ -1 & -1 & 2 \end{bmatrix} \begin{bmatrix} a \\ b \\ c \end{bmatrix} . \tag{7.10}$$

Equations (7.8) and (7.10) allow determination of the line-to-line and line-to-neutral output voltages for all states of an inverter. The results are assembled in

TABLE 7.1   States and Voltages of the Three-Phase Voltage-Source Inverter

State	abc	v_{AB}/V_i	v_{BC}/V_i	v_{CA}/V_i	v_{AN}/V_i	v_{BN}/V_i	v_{CN}/V_i
0	000	0	0	0	0	0	0
1	001	0	-1	1	$-\frac{1}{3}$	$-\frac{1}{3}$	$\frac{2}{3}$
2	010	-1	1	0	$-\frac{1}{3}$	$\frac{2}{3}$	$-\frac{1}{3}$
3	011	-1	0	1	$-\frac{2}{3}$	$\frac{1}{3}$	$\frac{1}{3}$
4	100	1	0	-1	$\frac{2}{3}$	$-\frac{1}{3}$	$-\frac{1}{3}$
5	101	1	-1	0	$\frac{1}{3}$	$-\frac{2}{3}$	$\frac{1}{3}$
6	110	0	1	-1	$\frac{1}{3}$	$\frac{1}{3}$	$-\frac{2}{3}$
7	111	0	0	0	0	0	0

Table 7.1. The line-to-line voltages can assume only three values, 0 and $\pm V_i$, while the line-to-neutral voltages can assume five values, 0, $\pm V_i/3$, and $\pm 2V_i/3$.

If the 5–4–6–2–3–1 ... sequence of states is imposed in an inverter, each state lasting one-sixth of the desired period of the output voltage, the individual line-to-line and line-to-neutral voltages have the waveforms shown in Figure 7.15. Notice that

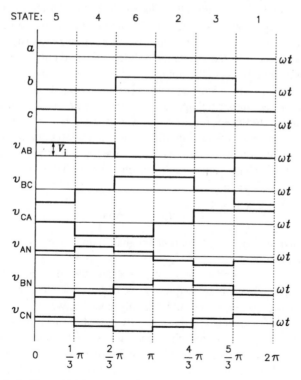

Figure 7.15 Switching variables and waveforms of output voltages in a three-phase VSI in the square-wave mode.

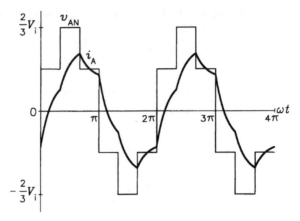

Figure 7.16   Waveforms of output voltage (line-to-neutral) and current in a three-phase VSI in the square-wave mode (RL load).

the voltages, although not sinusoidal, are balanced. This is the square-wave mode of operation, in which each switch of the inverter is turned on and off once within the cycle of output voltage. No wonder that most of the SCR-based inverters of the past were operated in this mode only. The peak value, $V_{LL,1,p}$, of the fundamental line-to-line output voltage is approximately $1.1V_i$ and that of the line-to-neutral voltage, $V_{LN,1,p}$ is $0.64V_i$. Both voltages have the same total harmonic distortion, THD, of 0.31. The ratio $V_{LL,1,p}/V_i$ represents the voltage gain of the inverter, which in this case is greater than unity. As in the square-wave single-phase inverter, the magnitude control of the output voltage must be realized on the dc supply side.

Waveforms of the output voltage and current for one phase of the inverter are shown in Figure 7.16. Except for the triple wave forms, they are similar to those in a single-phase inverter in the same mode and rich in low-order harmonics. These, as in the case of input currents in three-phase rectifiers (see Section 4.1.2), are absent from balanced three-phase voltages and currents. The input current, i_i, is related to the output currents, i_A, i_B, and i_C, as

$$i_i = ai_A + bi_B + ci_C \tag{7.11}$$

and is illustrated in Figure 7.17. Its fundamental frequency is six times higher than that of the output currents.

Apart from the high fundamental output voltage, the advantage of square-wave operation is the low switching frequency, as within a single cycle of the output voltage each switch is turned on and off only once. However, output currents in PWM inverters are of distinctly higher quality than those in inverters operating in the square-wave mode. Therefore, this mode is reserved primarily for high-power inverters with slow switches, such as GTOs or IGCTs. It is employed sporadically in PWM inverters when the maximum possible output voltage is needed.

Pulse trains of the switching variables and waveforms of output voltages in a three-phase PWM VSI are shown in Figure 7.18. To illustrate the impact of the load

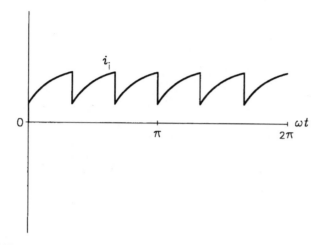

Figure 7.17   Waveform of input current in a three-phase VSI in the square-wave mode.

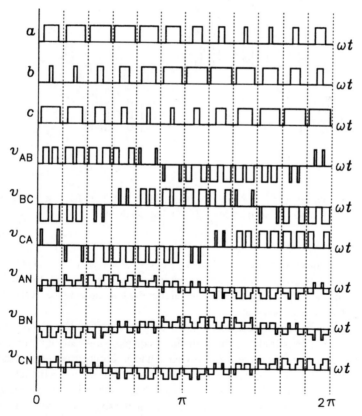

Figure 7.18   Switching variables and waveforms of output voltages in a three-phase VSI in the PWM mode.

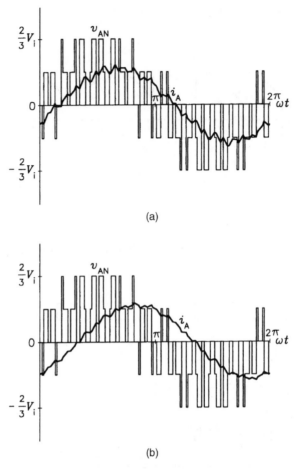

Figure 7.19 Waveforms of the output voltage and current in an RL load of a three-phase VSI in the PWM mode: (a) load angle of 30°; (b) load angle of 60°.

angle on the output current, waveforms of one of the line-to-neutral voltages and the corresponding output current are shown in Figure 7.19. The load angle is 30° in Figure 7.19a and 60° in Figure 7.19b. The higher the load inductance is, the smoother the current waveform becomes.

Input current waveforms corresponding to output currents in Figure 7.19 are shown in Figure 7.20. Interestingly, the fundamental frequency of the input current in three-phase PWM inverters is three times higher than the output frequency, whereas as noted earlier, in inverters operating in the square wave it is six times higher. Note that the possibility of a negative input current in voltage-source inverters necessitates a capacitor in the dc link, as semiconductor devices in the supply rectifier could not pass such current. Similar to the single-phase inverter analyzed in the preceding section, the input current in the PWM mode of a three-phase inverter contains, apart from the

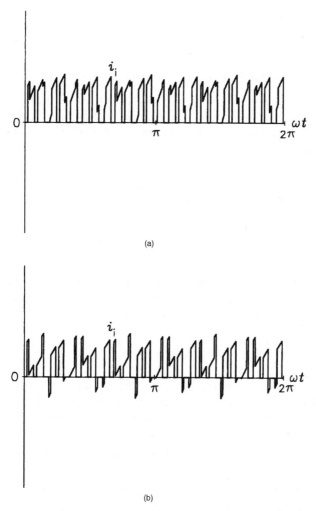

Figure 7.20   Waveforms of input current in a three-phase VSI in the PWM mode: (a) load angle of 30°; (b) load angle of 60°.

dc component, consists mostly of high-order harmonics. Therefore, to attenuate the high-frequency harmonic content of the current drawn from the supply source, the components of the dc link can be significantly smaller than those needed for a comparable square-wave mode inverter.

When supplied from a diode rectifier, a voltage-source inverter has no means of reversing the direction of power flow. This can be inconvenient in certain applications, such as adjustable-speed ac drives, in which the load is capable of power generation. Then, either a braking resistor in the dc link, such as that used in chopper systems (see Figure 6.25), or a controlled rectifier must be employed. Use of the IGBT power module shown in Figure 2.24 in an inverter-fed ac drive system capable of four-quadrant operation is described in Section 7.7.

7.1.3 Voltage Control Techniques for Voltage-Source Inverters

Scores of PWM techniques for voltage control of VSIs have been proposed, and even a brief review of them would be a daunting task. Therefore, only three of the most popular methods for three-phase inverters are described. Characteristics desired in a PWM technique include:

1. Good utilization of the dc supply voltage, that is, a possibly high value of the maximum voltage gain, $K_{V(\max)}$, defined here as

$$K_{V(\max)} \equiv \frac{V_{LL,1,p(\max)}}{V_i} \qquad (7.12)$$

   where $V_{LL,1,p(\max)}$ denotes the maximum peak value of the fundamental line-to-line output voltage available using the technique under consideration.

2. Linearity of the voltage control, that is,

$$V_{LL,1,p}(m) = m V_{LL,1,p(\max)} \qquad (7.13)$$

   where m denotes the modulation index, which in inverters is identical to the magnitude control ratio. As usual, the magnitude control ratio is defined as the ratio of the actual output voltage (line-to-line or line-to-neutral, peak or rms value) to the maximum available value of this voltage.

3. Low amplitudes of low-order harmonics of the output voltage, to minimize the harmonic content of the output current.

4. Low switching losses in the inverter switches.

5. Sufficient time allowance for proper operation of the inverter switches and control system.

A good trade-off between conflicting requirements 3 and 4 is particularly important. As demonstrated earlier, the quality of output current increases with an increase in the number of switchings per cycle of the output voltage (compare Figures 7.5 and 7.6). On the other hand, each switching is associated with an energy loss in the switch. Thus, the quality of operation of a PWM inverter can be increased at the expense of power efficiency, and vice versa.

Carrier-Comparison PWM Technique   In the 1960s, when the first PWM VSIs were built, only analog control systems were available. The simple *carrier-comparison (triangulation)* technique developed for those inverters has survived today in an integrated-circuit version. The magnitude and frequency of the output

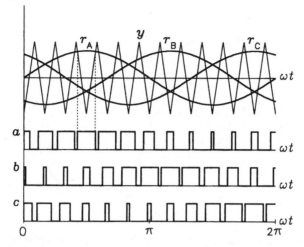

Figure 7.21   Carrier-comparison PWM technique ($N = 12, m = 0.75$).

voltage are the controlled variables. The principle of carrier-comparison methods is illustrated in Figure 7.21. Reference waveforms, r_A, r_B, and r_C, given by

$$r_A(\omega t) = F(m, \omega t)$$

$$r_B(\omega t) = F(m, \omega t - \tfrac{2}{3}\pi) \qquad (7.14)$$

$$r_C(\omega t) = F(m, \omega t - \tfrac{4}{3}\pi)$$

where $F(m, \omega t)$ denotes the modulating function employed, are compared with a unity-amplitude triangular waveform y. Values of the switching variables, a, b, and c, change from 0 to 1 and from 1 to 0 at every sequential intersection of the carrier and respective reference waveforms.

The sinusoidal modulating function, $F(m, \omega t) = m \sin \omega t$ is simple, but the voltage gain of the inverter can be increased significantly using one of the many nonsinusoidal modulating functions developed over the years. All those functions consist of a fundamental and triple harmonics, which are not reflected in the three-phase output voltages and currents of the inverter. For example, the *third-harmonic modulating function*, shown in Figure 7.22 with $m = 1$, is given by

$$F(m, \omega t) = \frac{2}{\sqrt{3}}m \left(\sin \omega t + \frac{1}{6} \sin 3\omega t \right) \qquad (7.15)$$

having thus only the fundamental and third harmonics. At $m = 1$, the fundamental equals $2 \big/ \sqrt{3} \approx 1.15$ which represents a 15% increase in the voltage gain at no additional expense to the inverter.

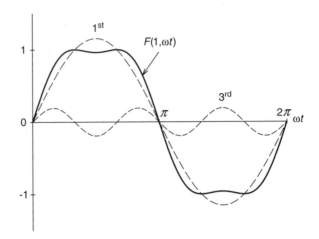

Figure 7.22 Third-harmonic modulating function and its components at $m = 1$.

Space Vector PWM Techniques Comparing Figures 4.53 and 7.14, it can be seen that the voltage-type three-phase PWM rectifier and the voltage-source three-phase PWM inverter have identical topologies of the power circuit. Consequently, the space vector approach to pulse width modulation described in various parts of Section 4.3 can be employed in the VSI control as well. Definitions of switching variables and states of the inverter are identical with those of the voltage-type PWM rectifier (see Section 4.3.2).

The voltage space vector plane with the reference vector $\vec{v}^*$ is shown in Figure 7.23 in the per-unit format, with the maximum available magnitude of that vector as the base. Clearly, Figure 7.23 is just a minor modification of Figure 4.56. The modulation index, m, constitutes the magnitude of $\vec{v}^*$. Assuming negligible voltage drops in the inverter, the highest available peak value of the output line-to-line voltage equals the dc input voltage, V_i. Thus,

$$m = \frac{V_{LL,p}^*}{V_i} \qquad (7.16)$$

where $V_{LL,p}^*$ denotes the reference peak line-to-line voltage. Equation (7.16) allows determination of the required value of m for specified output voltages of the inverter.

In the steady state, when the fundamental output voltage and current maintain fixed magnitude and frequency, m is constant and $\vec{v}^*$ rotates with a constant speed. However, the space vector PWM technique allows synthesis of an instantaneous voltage vector, which may change in magnitude and speed from one switching cycle to another. This is required, for example, in the vector control methods of high-performance ac drives fed from VSIs.

As an alternative, the reference voltage vector, $\vec{v}^*$, can be expressed in volts and represent the actual output line-to-neutral voltages of the inverter. In that case, based on the Park transformation (4.74), the magnitude of $\vec{v}^*$ equals $1.5 V_{LN,p}^*$, that is, one

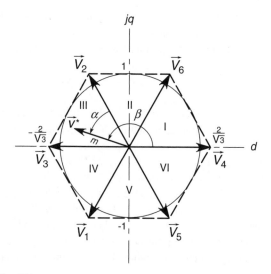

Figure 7.23 Voltage space vector plane of a three-phase VSI (per-unit format).

and a half of the reference peak value of the line-to-neutral voltage. This corresponds to $\sqrt{3}/2$ of the reference peak line-to-line voltage, $V_{LN,p}^*$. Thus, according to Eq. (7.16), the maximum magnitude of $\vec{v}^*$, which corresponds to $m = 1$, equals $\sqrt{3}V_i/2$ (see Figure 4.55b for an analogous diagram of the voltage-type PWM rectifier). As already stated, it is this value that constitutes the base of the per-unit voltage vectors in Figure 7.23.

The angular position, β, of the reference vector allows determination of the sector of the complex plane in which the vector is located within the given sampling cycle of the digital modulator. Specifically,

$$S = \mathrm{int}\left(\frac{3}{\pi}\beta\right) + 1 \tag{7.17}$$

where β is expressed in radians and S is the sector number (1 to 6). The in-sector position, α, of $\vec{v}^*$ is then given by

$$\alpha = \beta - \frac{\pi}{3}(S - 1). \tag{7.18}$$

In a manner analogous to the PWM rectifier control, the revolving reference voltage vector is synthesized from stationary vectors, $\vec{V}_X$ and $\vec{V}_Y$, framing the sector in question, and a zero vector, $\vec{V}_0$ or $\vec{V}_7$. Durations T_X, T_Y, and T_Z of states generating those vectors are given by Eqs. (4.77) to (4.79), in the per-unit format with T_{sw} as the base, or, directly, by Eqs. (4.90) to (4.92).

Times T_X, T_Y, and T_Z indicate only how long a given inverter state should last in the switching cycle under consideration; how the cycle is divided between the employed states must also be specified. The two most commonly used state sequences can be called a *high-quality sequence* and a *high-efficiency sequence*. The high-quality sequence, so called because for a given number of switching intervals per cycle it tends to result in the lowest total harmonic distortion of output currents, is

$$X - Y - Z_1 - Y - X - Z_2 \cdots \tag{7.19}$$

where each state in the sequence lasts half of the allotted time. States Z_1 and Z_2, complementarily 0 and 7, are placed in such order that a transition from one state to another involves switching in one inverter leg only. In other words, only one switching variable changes its value. For example, in sector II, where $X = 6$ and $Y = 2$ (see Figure 7.22), the high-quality state sequence is 6–2–0–2–6–7$\cdots$ or, in binary notation, 110–010–000–010–110–111. This sequence results in such a switching pattern that each switching variable has N pulses per cycle of output voltage (see Eq. (7.6)). It means that each inverter switch is turned on and off once per switching cycle.

The number of switchings can be further reduced at the expense of the somewhat increased distortion of output currents, when the high-efficiency state sequence

$$X - Y - Z - Y - X \cdots \tag{7.20}$$

is employed. Here, sequential states X and Y last $T_X/2$ and $T_Y/2$ seconds, respectively, and state Z lasts T_Z seconds. Moreover, $Z = 0$ in the even sectors (II, IV, and VI) and $Z = 7$ in the odd sectors (I, III, and V). If sector II is again considered as an example, the high-efficiency state sequence is 6–2–0–2–6$\cdots$, or 110–010–000–010–110. Note that variable c is zero throughout all switching cycles in the sector in question. With this state sequence, the average number of pulses of a switching variable per cycle of output voltage is $2N/3 + 1 \approx 2N/3$. As a result, the switching losses decrease by about 30% compared with the high-quality state sequence, but the ripple of output current of the inverter increases slightly. Switching patterns generated by the two variants of the space vector PWM technique described are illustrated in Figures 7.24 and 7.25. In both cases, the reference vector is assumed to be in sector I.

Programmed PWM Techniques   The best compromise between efficiency and quality of inverter operation is achieved in *programmed*, or *optimal switching pattern*, PWM techniques. To understand their principle, consider the switching pattern for phase A of an inverter shown in Figure 7.26. The $a(\omega t)$ waveform has both half-wave and quarter-wave symmetry (see Appendix C). Consequently, the full-cycle switching pattern is determined uniquely by switching angles α_1 through α_4 in the first quarter-cycle. These angles, whose number, K, here 4, is arbitrary, will subsequently be called *primary switching angles*. Switchings in the remaining three quarter-cycles and in phases B and C occur at *secondary switching angles*, which are simple linear functions of the primary angles. Switching patterns $b(\omega t)$ and $c(\omega t)$ differ from $a(\omega t)$ by the appropriate phase shifts only; that is,

$$\alpha(\omega t) = b\left(\omega t + \frac{2}{3}\pi\right) = c\left(\omega t + \frac{4}{3}\pi\right). \tag{7.21}$$

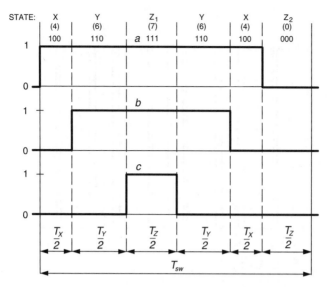

Figure 7.24   High-quality state sequence.

Condition (7.21) implies that waveforms of all three switching variables have identical harmonics.

According to Eqs. (7.8) and (7.10), output voltages of an inverter depend linearly on the switching variables. Therefore, all these voltages have the same harmonic spectrum as $a(\omega t)$, except for triple harmonics, which thanks to condition (7.21) are absent in voltage waveforms. The half-wave symmetry of waveforms of switching

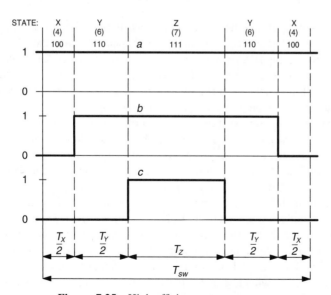

Figure 7.25   High-efficiency state sequence.

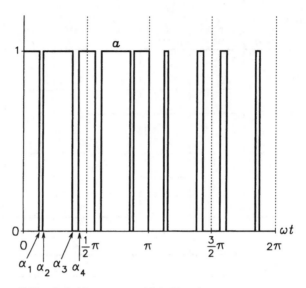

Figure 7.26 Switching pattern with half- and quarter-wave symmetries.

variables results in the absence of even harmonics as well, while the quarter-wave symmetry allows expression of the amplitude of the kth harmonic, $A_{k,p}$, of $a(\omega t)$ as

$$A_{k,p} = \frac{4}{k\pi} \left[\sum_{i=1}^{K} (-1)^{i-1} \cos k\alpha_i - \frac{1}{2} \right]. \tag{7.22}$$

It can be seen that the amplitude of each harmonic depends on all the primary switching angles. Therefore, the K angles can be set to such values that the fundamental and $K - 1$ higher harmonics have the desired (if feasible) amplitudes. This requires solving K equations (7.22) for $\alpha_1, \alpha_2, \ldots, \alpha_K$, with given K values of $A_{k,p}$, including $k = 1$, and with the constraint $0 < \alpha_1 < \alpha_2 < \cdots < \alpha_K$. The equations are nonlinear and solvable only by numerical computations.

The most common approach to a definition of optimal values of primary switching angles is the *harmonic-elimination* PWM technique. It consists of setting $A_{1,p}$ to $MA_{1,p(\max)}$, where M is the magnitude control ratio and $A_{1,p(\max)}$ denotes the maximum amplitude of the fundamental available with the given number of primary switching angles. Amplitudes, $A_{5,p}, A_{7,p}, A_{11,p}, \ldots$, of the $(K - 1)$th lowest-order odd and nontriple harmonics are set to zero. For example, if $K = 5$, the maximum available amplitude of the fundamental, $A_{1,p(\max)}$, of $a(\omega t)$ is 0.583 and the fifth, seventh, eleventh, and thirteenth harmonics can be eliminated, leaving the seventeenth as the lowest-order harmonic. For comparison, in space vector PWM techniques, $A_{1,p(\max)}$ reaches a value of $1/\sqrt{3} \approx 0.577$, that is, about 1% less.

Optimal primary switching angles for various values of K and M may be found in the technical literature of the subject. The optimal angles in the case $K = 5$ are

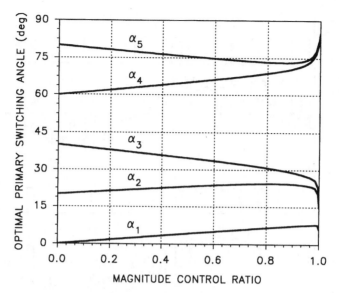

Figure 7.27 Optimal primary switching angles as functions of the magnitude control ratio ($K = 5$).

shown in Figure 7.27 as functions of the magnitude control ratio, M. A harmonic spectrum of the output line-to-neutral voltage in Figure 7.28, determined for $M = 1$, demonstrates the desired elimination of low-order, even, and triple harmonics.

The number of pulses of a switching variable in programmed PWM techniques is $2K + 1$ per cycle of the output voltage. With the same number of pulses per cycle, these techniques tend to produce currents of higher quality than do other PWM

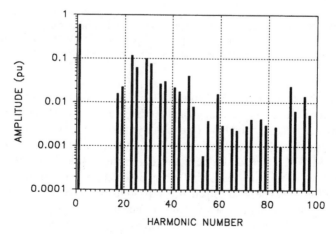

Figure 7.28 Harmonic spectrum of line-to-neutral voltage with the harmonic-elimination technique ($K = 5$, $M = 1$).

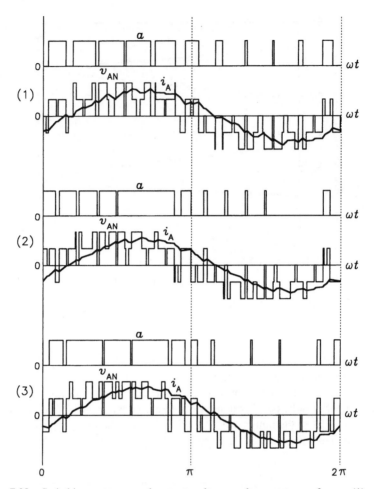

Figure 7.29 Switching patterns and output voltage and current waveforms: (1) carrier-comparison PWM with sinusoidal reference; (b) space vector PWM with high-efficiency state sequence; (c) programmed PWM with harmonic elimination.

techniques. This is illustrated in Figure 7.29, which shows three switching patterns and the corresponding output voltage and current waveforms. These were obtained using (1) a carrier-comparison PWM technique with sinusoidal reference waveforms, (2) a space vector PWM technique with a high-efficiency state sequence, and (3) a programmed PWM technique for harmonic elimination ($K = 5$). In all three cases, an RL load with the load angle of $30°$ was assumed, and the magnitude control ratio was set to 0.9. The load impedance was adjusted to compensate for unequal fundamental voltages and to obtain identical fundamental currents. The increasing quality of the current from case (1) through (2) to (3) can be observed and confirmed by determination of the total harmonic distortion, THD. The respective values of the THD are 11.6, 9.2, and 5.2%.

Computation of optimal switching angles is so time consuming that it cannot be done in real time. Therefore, sets of values of switching angles for each value of the magnitude control ratio must be stored in the memory of a digital control system. Besides the need for a large memory, the disadvantage of programmed PWM techniques manifests itself in disruption of the optimal switching patterns with rapid changes in the reference frequency and/or magnitude of the output voltage.

The programmed PWM techniques allow generation of high-quality sinusoidal currents when the vectors of output voltage and current maintain a fixed magnitude and revolve with a constant speed. However, they are incapable of following a varying reference voltage vector, which the space vector PWM strategies (and, to a certain extent, the carrier-comparison method) can easily do. Therefore, practical applications of the programmed PWM techniques are generally limited to ac power sources (e.g., uninterruptible power supplies), which operate with a constant output frequency and a narrow range of voltage control.

Random PWM Techniques Subsequent considerations pertain to space vector PWM techniques, which dominate modern power electronic converters. In practice, the switching cycles are usually made to coincide with the sampling cycles, so that the switching frequency, f_{sw}, of the inverter equals the fixed sampling frequency, f_{smp}, of the digital modulator. The constant switching frequency results in clusters of higher harmonics in the frequency spectra of output voltage and current. The harmonic clusters are centered about multiples of f_{sw}.

By randomly varying (dithering) consecutive switching periods, aperiodic switching patterns are generated, and part of the harmonic power is shifted to the continuous spectrum of the output voltages. The random PWM (RPWM) techniques have been found to alleviate undesirable acoustic, vibration, and electromagnetic interference (EMI) effects in electromechanical systems, typically ac drives, fed from PWM inverters. For example, the typical unpleasant whine of an ac motor supplied from a deterministically modulated inverter can be changed to soothing "static" by employing an RPWM method in the inverter.

In the classic RPWM strategy, consecutive switching (and sampling) periods are drawn randomly from a uniform-probability pool of values ranging from $xT_{sw,ave}$ to $(2 - x)T_{sw,ave}$ where $T_{sw,ave}$ denotes the average switching period desired and the coefficient x (usually, 0.5) determines the shortest switching period as a fraction of the average period. The results of such period dithering applied to the popular space vector strategy are illustrated in Figure 7.30. Output currents of an inverter controlled using the regular version of this strategy with $N = T_o/T_{sw} = 48$, where T_o denotes the period of output voltage, are shown in Figure 7.30a. Because of the balanced load, all current waveforms are rippled identically. In the random mode, when N is varied randomly in the range 24 to 72, the current waveforms are as in Figure 7.30b. Their shapes change from cycle to cycle and phase to phase. Note that the average number of switching intervals per cycle is the same in both cases, so that the average amount of the current ripple remains unchanged. However, there is a big difference in the power spectra of output voltages and currents.

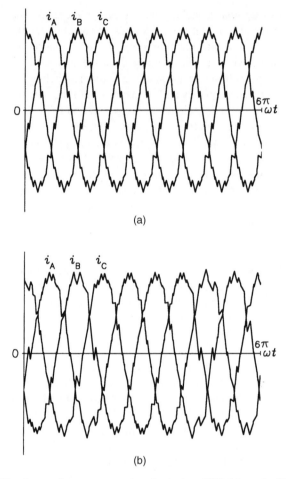

(a)

(b)

Figure 7.30   Waveforms of output current in a three-phase VSI: (a) regular PWM; (2) random PWM.

The frequency spectrum of line-to-neutral output voltage shown in Figure 7.31a for the regular PWM is purely discrete. Clusters of higher harmonics appear about multiples of N, that is, multiples of f_{sw}. In contrast, apart from the fundamental harmonic, which has not changed, the spectrum of the same voltage of the randomly modulated inverter, taken over 75 cycles of output voltage and shown in Figure 7.31b, displays continuous power density. All higher harmonics have disappeared. If the spectrum is measured over a much higher number of cycles, the practically white-noise quality can be observed.

Variations of the sampling frequency in the classic RPWM method are disadvantageous. The sampling frequency in digital systems is normally so selected as to represent the best compromise between various operational requirements, especially when the pulse width modulator is part of a larger control scheme. To maintain

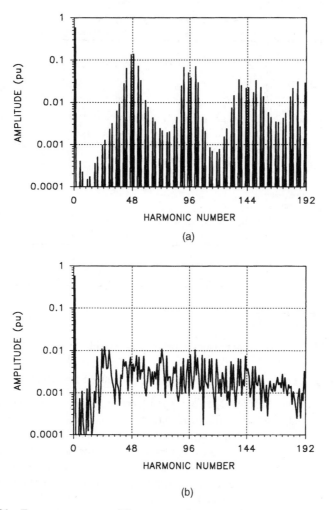

Figure 7.31 Frequency spectra of line-to-neutral output voltage in a three-phase VSI: (a) regular PWM; (2) random PWM.

a fixed sampling frequency, variable-delay random pulse width modulation (VD-RPWM) was developed. The switching periods vary, and their average equals the constant sampling period. This is realized by delaying consecutive switching cycles with respect to the corresponding sampling cycles. The delay time, t_{del}, changes from cycle to cycle, being calculated as $r\,T_{\mathrm{smp}}$, where r is a random number from the range 0 to 1. It may happen that a given switching period is too short to be feasible. In that case it is set to the minimum allowable value, $T_{\mathrm{sw,min}}$, and as a result, the lengths of switching cycles vary from $T_{\mathrm{sw,min}}$ to $2T_{\mathrm{smp}}$. Principles of the two random PWM techniques described are illustrated in Figure 7.32 together with regular PWM with fixed switching frequency.

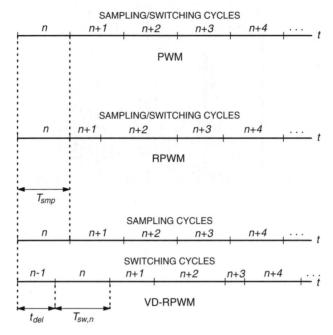

Figure 7.32 Comparison of random PWM techniques with the regular PWM method.

7.1.4 Current Control Techniques for Voltage-Source Inverters

The PWM techniques described perform open-loop, or *feedforward*, control of output voltages in a voltage-source inverter. If output currents, which depend not only on the voltages but also on the load, are to be controlled instead, feedback from current sensors must be provided. A control system compares the actual currents with the reference currents and generates appropriate switching variables, a, b, and c, for individual phases of the inverter. As a result, the switching variables and output voltages are pulse width modulated such that the output current waveforms follow the reference waveforms.

High-performance current control is a challenging task because in most practical cases the inverter load is unknown and varying. A successful current control technique should ensure:

1. Good utilization of the dc supply voltage, meant here as the feasibility of producing a possibly high current in a given load
2. Low static and dynamic control errors
3. Low switching losses in the inverter, that is, possibly infrequent switching
4. Sufficient time allowance for proper operation of the inverter switches and control system

It can be seen that requirements 1, 3, and 4 match closely those for the voltage control PWM techniques. As is the case with those techniques, many current control methods have been proposed in the technical literature. The concepts range from very simple to quite complex, the latter involving machine-intelligence systems such as neural networks and fuzzy-logic controllers. Here, four classic approaches to the current control in three-phase inverters are discussed.

Hysteresis Current Control A voltage-source inverter with the simplest version of *hysteresis control* of output currents is shown in Figure 7.33. The output currents, i_A, i_B, and i_C, of an inverter are sensed and compared with the respective reference current waveforms, i_A^*, i_B^*, and i_C^*. Current errors, Δi_A, Δi_B, and Δi_C, are applied to current controllers that produce switching variables, a, b, and c, for the inverter.

The $a = f(\Delta i_A)$ characteristic of a current controller for phase A of the inverter is shown in Figure 7.34. The characteristic constitutes a hysteresis loop, described as

$$\alpha = \begin{cases} 0 & \text{if } \Delta i_A < -\frac{h}{2} \\ 1 & \text{if } \Delta i_A > \frac{h}{2} \end{cases} \tag{7.23}$$

where h denotes the width of the loop. If $-h/2 \leq \Delta i_A \leq h/2$, the value of variable a remains unchanged. Characteristics of controllers for the other two phases are the same and therefore all considerations can be limited to phase A only. The loop width, h, can be thought of as the width of a tolerance band for the controlled current, i_A, since as long as the current error, Δi_A, remains within this band, no action is taken by the controller. If the error is too high, that is, the actual current is lower than its reference value by more than $h/2$, variable a assumes a value of 1. According to Eq. (7.10), this makes voltage v_{AN} equal to or greater than zero, which is a necessary condition for current i_A to increase. Analogously, switching a to zero when the output

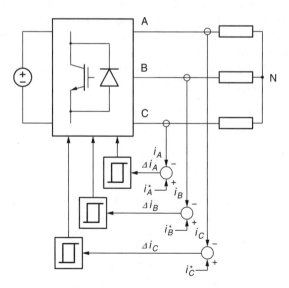

Figure 7.33 Hysteresis current control scheme.

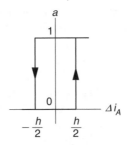

Figure 7.34 Characteristic of the hysteresis current controller.

current is too high causes v_{AN} to be equal to or less than zero, which is conducive for i_A to decrease.

Interaction between phases of an inverter is a disadvantage of the scheme described. Note that if, for example, variable a is switched from zero to 1 to increase current i_A, the voltage v_{AN} increases as required, but voltages v_{BN} and v_{CN} decrease, which may be detrimental for control of currents i_B or i_C. As a result, in certain loads, such as ac motors, current errors tend sporadically to exceed a tolerance of $\pm h/2$. The value of h affects the average switching frequency, which increases with a decrease in the width of the tolerance band. This is illustrated in Figure 7.35, which shows waveforms of output currents in an inverter with an RL load and with h equal to 20% (Figure 7.35a) and 10% (Figure 7.35b) of the peak value of the sinusoidal reference currents.

The hysteresis control is an excellent choice when a fast response to rapid changes in the reference current is required, because the current controllers have negligible inertia and delays. An example situation, when the magnitude, frequency, and phase of reference currents are changed simultaneously, is illustrated in Figure 7.36 for current i_A. It is only the time constant of the load that limits the speed of transition of the current to the new waveform.

In practice, the width, h, of the tolerance band is often made proportional to the magnitude control ratio, M, defined here as

$$M \equiv \frac{I_{o,p}^*}{I_{o,p(\text{max})}^*} \tag{7.24}$$

where $I_{o,p}^*$ denotes the peak value of the reference output current and $I_{o,p(\text{max})}^*$ is the maximum available peak value of that current. If $h(m) = Mh(1)$, the ripple factor of the current is maintained at an approximately constant level, independent of the magnitude control ratio. It must be pointed out that when the peak value of reference currents is increased beyond $M = 1$, an inverter transits from the PWM mode to the square-wave mode of operation.

Instead of using three current controllers, one for each phase of an inverter, two controllers can be employed in the space vector scheme depicted in Figure 7.37. Based on the Park transformation given by Eq. (4.73) and applied to the output currents of

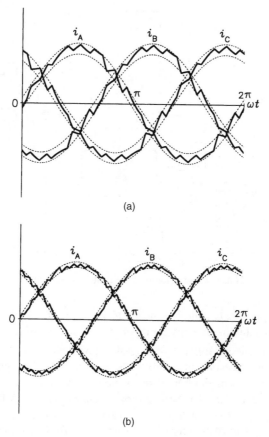

(a)

(b)

Figure 7.35 Waveforms of output currents in a VSI with hysteresis current control: (a) 20% tolerance; (b) 10% tolerance.

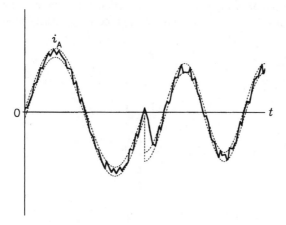

Figure 7.36 Waveform of output current in a VSI with hysteresis current control at a rapid change in the magnitude, frequency, and phase of the reference current.

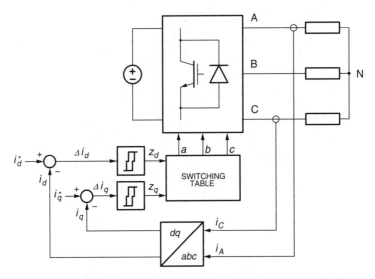

Figure 7.37 Space vector version of the hysteresis current control scheme.

an inverter, components i_d and i_q of the space vector, $\vec{i}$, of output currents are computed and compared with respective components, i_d^* and i_q^*, of the reference current vector, $\vec{i}^*$. The current controllers have three-level outputs, z_d and z_q, as illustrated in Figure 7.38 for the d-component controller. Signals z_d and z_q are applied to a switching table that for each of the nine distinct pairs of values of the signals selects a specific state of the inverter. The best control effects are obtained when state 0 or 7 is imposed for $(z_d, z_q) = (0,0)$, state 1 for $(0,1)$ and $(1,1)$, state 2 for $(1,-1)$, state 3 for $(1,0)$, state 4 for $(-1,0)$, state 5 for $(-1,1)$, and state 6 for $(-1,-1)$ and $(0,-1)$.

Ramp-Comparison Current Control The hysteresis control systems are characterized by unnecessarily high switching frequencies, especially at low values of the magnitude control ratio, since the three current controllers act independent of each

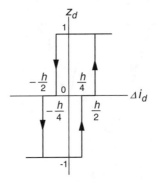

Figure 7.38 Characteristic of a current controller for the space vector version of the hysteresis current control scheme.

RAMP GENERATOR

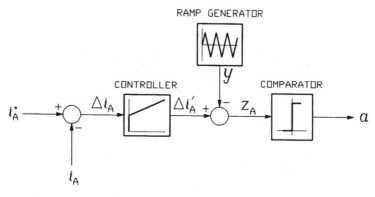

Figure 7.39   Ramp-comparison scheme for a current-controlled VSI.

other. Also, the somewhat chaotic operation of the inverter can be perceived as a disadvantage. To stabilize the switching frequency, *ramp-comparison control* can be employed. A block diagram of this control scheme is shown in Figure 7.39 for phase A of an inverter.

The current error, $\Delta i_A = i_A^* - i_A$, is applied to the input of a linear controller, usually of the proportional-integral (PI) type. The output signal, x_A, of the controller is, in turn, compared with a triangular ramp signal, y, similar to that used in the carrier-comparison PWM technique for voltage-controlled inverters. The difference, z_A, of those two signals activates a comparator, which generates the switching variable a according to the equation

$$a = \begin{cases} 0 & \text{if } Z_A \leq 0 \\ 1 & \text{if } Z_A > 0 \end{cases}. \tag{7.25}$$

Identical control loops are used for the other two phases.

The carrier-comparison method was not mentioned by accident: It is easy to see that the ramp-comparison control can be thought of as the carrier-comparison PWM technique with the processed current error, $\Delta i_A'$, as the modulating function. If, for example, $i_A < i_A^*$ and the voltage v_{AN} should be increased to boost i_A, switching variable a is modulated by signal $\Delta i_A'$ in such a way that wide pulses are interspersed with narrow notches. For an illustration, imagine that signal r_A in Figure 7.21 is replaced by $\Delta i_A'$.

Waveforms of the output current of an inverter with the ramp-comparison current control are shown in Figure 7.40 for two values of the ratio of the ramp signal frequency, f_r, to the fundamental output frequency, f_1, of the inverter. The frequency ratio, f_r/f_1, is 10 in Figure 7.40a and 20 in Figure 7.40b, and the same RL load as before is employed. Comparing the waveforms with those in Figure 7.35 for the hysteresis current control, greater regularity of the switching pattern and stability of the switching frequency can be discerned.

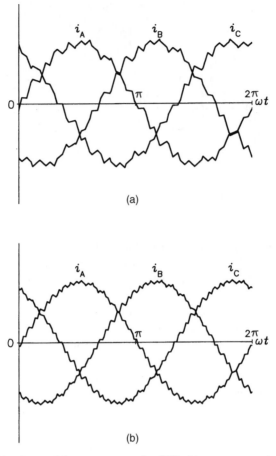

Figure 7.40   Waveforms of the output current in a VSI with ramp-comparison current control: (a) $f_r/f_1 = 10$; (b) $f_r/f_1 = 20$.

A significant advantage of this control scheme lies in the increased number of adjustable parameters. In contrast to hysteresis control, in which only the width of the hysteresis band of the controllers can be adjusted, the ramp-comparison system allows tuning the settings of the linear controller as well as of the amplitude and frequency of the ramp signal. Disadvantages of the ramp-comparison control involve somewhat reduced speed of response to rapid changes in reference currents (this can be improved by using a high-gain proportional controller) and, generally, an increased average switching frequency. The latter characteristic results from the absence of zero states (state 0 or state 7) of the inverter. Indeed, since all three current errors cannot simultaneously be of the same polarity, the values of switching variables imposed by the comparators are never all zero or one.

The ramp-comparison technique is usually implemented in an analog system. A related discrete PWM scheme, the *current-regulated delta modulator*, is shown in

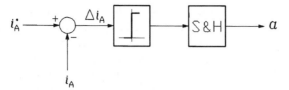

Figure 7.41   Current-regulated delta modulation scheme for a current-controlled VSI.

Figure 7.41. In the digital delta modulator, the ramp signal generator is replaced by a sample-and-hold circuit, which allows fixing the switching frequency of an inverter. Typically, no linear controller is used, and the comparator acts as an infinite-gain proportional (P) controller. Again, detrimentally to the efficiency of operation of the modulator, no zero states are imposed.

Predictive Current Control   An optimal switching pattern for a given set of values of the reference currents could be determined if the parameters of the load were known. This assumption underlies the principle of *predictive current controllers*. Two basic types of these controllers minimize either the average switching frequency, f_{sw}, or the total harmonic distortion of the controlled currents at a fixed value of f_{sw}. In each step of the operation, based on the response of the load to known voltage changes, a predictive controller estimates of the load parameters to optimize selection of the next state of the inverter. Microprocessors or digital signal processors are used to handle the computations involved.

Linear Current Control   In the three control schemes described so far, current feedback directly enforces switching patterns in the inverter. A different, indirect approach, illustrated in Figure 7.42, involves traditional linear controllers of the PI

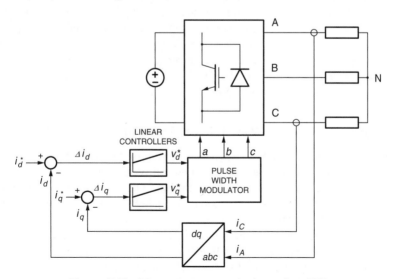

Figure 7.42   Linear current control scheme for a VSI.

(proportional-plus-integral) type. Similar to the space vector version of hysteresis control, the reference current is expressed as a space vector $\vec{\imath}^*$, whose components, i_d^* and i_q^*, serve as reference signals for the respective components, i_d and i_q, of the actual vector, $\vec{\imath}$, of output currents. Control errors, Δi_d and Δi_q, are converted by the linear controllers into components, v_d^* and v_q^*, of the voltage reference vector, $\vec{v}^*$ (line-to-line or line-to-neutral), to be realized by the inverter using the voltage space vector PWM technique described in Section 7.2.1. Indeed, given the magnitude and phase of $\vec{v}^*$, the corresponding values of m and α required in Eqs. (4.77) through (4.79) can easily be determined.

7.2 CURRENT-SOURCE INVERTERS

The freewheeling diodes typical of voltage-source inverters become redundant if an inverter is supplied from a dc current source. Then the current entering any leg of the inverter cannot change its polarity and therefore can only flow through the semiconductor power switches. Use of the current source prevents overcurrents, even in case of a short circuit in the inverter or load. However, continuity of the current must be preserved during commutation between switches. Therefore, if fully controlled switches are used, switching signals of the outgoing switch and incoming switch overlap a little.

The absence of freewheeling diodes reduces the size and weight of the power circuit and further increases the reliability of the current-source inverter (CSI). The practical current source consists of a controlled rectifier, usually SCR-based, and an inductive dc link. The output current of the rectifier is maintained at a constant level employing a closed control loop. The dc-link inductor attenuates the current ripple. In subsequent considerations, an ideal dc input current, I_i, is assumed.

In practice, current-source inverters are used primarily for control of ac motors. As such, they are invariably of the three-phase type, and therefore only those inverters are covered here. If needed, a single-phase current-source inverter can be obtained by removing one leg from the three-phase inverter. Control strategies for the single-phase current-source inverter can be developed by proper adaptation of those for three-phase inverters, described later in this section.

7.2.1 Three-Phase Square-Wave Current-Source Inverter

Figure 7.43 is a block diagram of a three-phase current-source inverter supplied from a three-phase ac line through a controlled rectifier and a dc link. The rectifier uses a feedback loop to maintain a constant current in the link. The power circuit of the inverter is depicted in Figure 7.44. The input current cannot be reversed, so a power flow from the load to the source requires reversal of the input voltage. To protect asymmetrical power switches, blocking diodes must be connected with them in series. Symmetrical switches, such as non-punch-through IGBTs, do not require blocking diodes.

As simultaneous conduction of both switches in an inverter leg is allowed, then, theoretically, as many as $2^6 = 64$ states of the inverter are possible, and six switching

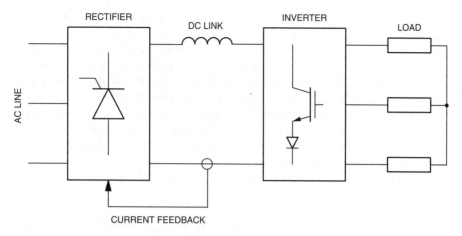

Figure 7.43 Current-source inverter supplied from a controlled rectifier.

variables, one for each of the inverter switches, can be defined. In the subsequent considerations, variables a, b, and c represent switches SA, SB, and SC (e.g., $a = 1$ means that SA is in the on-state); a', b', and c' are assigned to switches SA', SB', and SC'.

The high-power current-source inverters, typically based on relatively slow GTOs, operate in the square-wave mode only. Within the consecutive one-sixths of the period

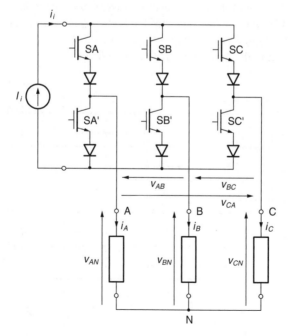

Figure 7.44 Three-phase current-source inverter.

of output voltage, the conducting switch pairs are SA & SB′, SA & SC′, SB & SC′, SB & SA′, SC & SA′, and SC & SB′. Note that the same conduction sequence is employed in a six-pulse rectifier. Indeed, if the inverter is used to feed an ac motor, which generates a three-phase counter-EMF of rotation, it can be thought of as a six-pulse rectifier operating in the inverter mode. The controlled dc voltage source and dc-link inductor supplying the inverter correspond directly to the commonly used RLE load of the rectifier, while the ac load EMF constitutes a counterpart of the rectifier's supply voltage. Vice versa, a rectifier supplied from an ac generator and feeding a dc motor represents an inverse of a current-source inverter fed from a dc generator and supplying an ac motor.

It is easy to show that the output currents, i_A, i_B, and i_C, of the inverter are given by

$$\begin{bmatrix} i_A \\ i_B \\ i_C \end{bmatrix} = \begin{bmatrix} a - a' \\ b - b' \\ c - c' \end{bmatrix} I_i \tag{7.26}$$

with currents i_{AB}, i_{BC}, and i_{CA} in a balanced delta-connected load given by

$$\begin{bmatrix} i_{AB} \\ i_{BC} \\ i_{CA} \end{bmatrix} = \frac{1}{3} \begin{bmatrix} 1 & -1 & 0 \\ 0 & 1 & -1 \\ -1 & 0 & 1 \end{bmatrix} \begin{bmatrix} i_A \\ i_B \\ i_C \end{bmatrix}. \tag{7.27}$$

Waveforms of switching signals for square-wave operation are shown in Figure 7.45, and those of the corresponding currents are shown in a wye-connected load (i_A, i_B, and i_C) and a delta-connected load (i_{AB}, i_{BC}, and i_{CA}) in Figure 7.46. The *current gain*, K_I, defined as the ratio of the peak value, $I_{L,1,p}$, of the fundamental output line current to the input current, I_i, is $2\sqrt{3}/\pi \approx 1.1$.

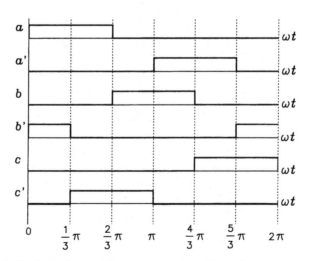

Figure 7.45 Switching variables in a three-phase CSI in the square-wave mode.

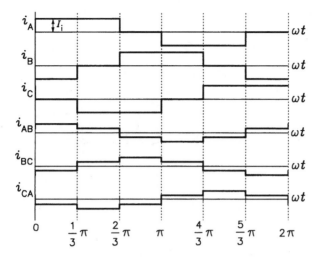

Figure 7.46   Waveforms of output currents in a three-phase CSI in the square-wave mode.

Rapid changes in the output currents during transitions from one state to another would cause high voltage spikes across inductive loads. Therefore, in practice, the commutation between inverter switches is purposefully prolonged, to limit the spikes to an allowable, safe level. Also, if possible, low-reactance loads are selected, such as induction motors with low leakage inductances. The output voltage waveform depends on the load. This is illustrated in Figure 7.47, depicting waveforms of output voltage and current in one phase of an inverter. Figure 7.47a shows the voltage and current with an RL load, with waveforms in Figure 7.47b corresponding to an LE load, that is, a series connection of an inductance and an ac EMF. The later load is commonly used to model ac motors in adjustable-speed drives, which represent an almost exclusive application of current-source inverters.

The square-wave mode of operation does not provide magnitude control of the output current within the inverter; hence, the current must be controlled in the rectifier supplying the inverter. The rectifier also allows bidirectional power flow, which is an important advantage of current-source inverters. As already mentioned, the input current is always positive, so the negative power flow requires negative average output voltage from the rectifier. Apart from the simplicity and reliability of the power circuit, the excellent dynamics of current control, important in high-performance ac drives, represents a distinct advantage of current-source inverters. On the other hand, the highly distorted stepped waveforms of the output current constitute an obvious weakness.

7.2.2   Three-Phase PWM Current-Source Inverter

PWM current-source inverters are feasible, although much less common in practice than are their voltage-source counterparts. Figure 7.48 is a block diagram of such an inverter. Comparing it with the square-wave inverter in Figure 7.43, it can be seen

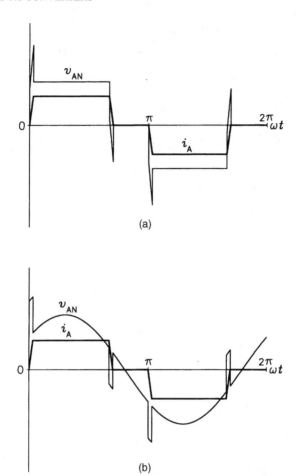

(a)

(b)

Figure 7.47   Waveforms of output voltage and current in a three-phase CSI in the square-wave mode: (a) RL load; (b) LE load.

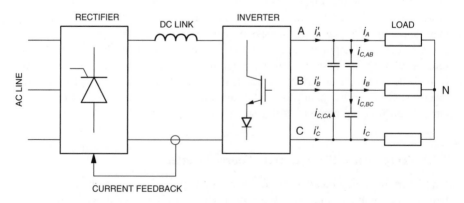

Figure 7.48   Three-phase PWM current-source inverter.

that the only difference in the power circuitry consists in the addition of capacitors between the output terminals. The capacitors act as low-pass filters for output currents, shunting most of the high-frequency harmonic content of pulsed currents produced by the inverter switches. Clearly, a PWM current-source inverter represents an exact inverse of the current-type PWM rectifier depicted in Figure 4.45.

Control strategies for PWM current-source inverters differ from those for voltage-source inverters. The pulsed output currents i'_A, i'_B, and i'_C are generated using switching patterns such that only two switches, one in the common-cathode group and one in the common-anode group, are on simultaneously. Typically, the two conducting switches belong in different legs of the inverter, so that the currents flow in two phases. However, in certain applications, the shot-through situations, when both switches in a given phase are on, are included in the control strategy to ease the voltage stress on the output capacitors.

As a result of a given PWM technique, a number of current pulses appear in each phase of the inverter within a single cycle of output current. This number equals $2P$, where P denotes the always-odd number of pulses of each switching variable. A fixed switching pattern is generated, while the magnitude control of the output currents is realized in the supplying rectifier. Similar to PWM voltage-source inverters, the waveforms of output currents, i_A, i_B, and i_C, of a PWM current-source inverter are rippled sinusoids. Both open- and closed-loop control of output currents are feasible.

Two PWM techniques have found widespread application in current-source inverters. The first resembles the classic carrier-comparison method for voltage-source inverters; however, both the carrier and modulating function are different here. Generation of a switching pattern with $P = 5$, which results in 10 pulses of currents i'_A, i'_B, and i'_C per cycle, is illustrated in Figure 7.49. It shows waveforms of the triangular carrier, y, trapezoidal reference signal, r, and switching variable a.

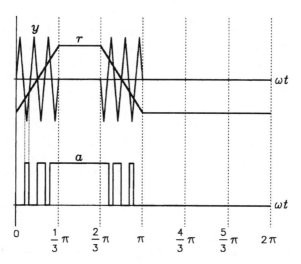

Figure 7.49 Carrier-comparison method for the PWM CSI ($P = 5$).

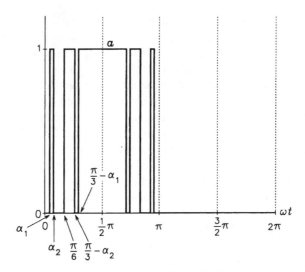

Figure 7.50   Optimal switching pattern for the PWM CSI with two primary switching angles.

Waveforms of other switching variables are identical, but shifted in phase. Specifically, those of b and c are delayed by 120° and 240° with respect to $a(\omega t)$, while waveforms of variables a', b', and c' are shifted with respect to those of a, b, and c, by a half-cycle. Note that the switching pattern for a square-wave inverter displays the same property (see Figure 7.45). The best attenuation of low-order harmonics in the output currents is achieved when the ratio of peak values of the reference and carrier signals is 0.82.

The second technique is a programmed, harmonic-elimination PWM method. By proper selection of primary switching angles, specific harmonics of the output currents can be canceled. A switching pattern, again with $P = 5$, is shown in Figure 7.50. Two optimal primary switching angles, α_1 and α_2, 7.93° and 13.75°, respectively, allow elimination of the fifth and seventh harmonics. If the eleventh harmonic was to be canceled as well, 14-pulse ($P = 7$) currents would have to be produced using three optimal switching angles of 2.24°, 5.60°, and 21.26°. An optimal switching pattern, $a(\omega t)$, must have quarter- and half-wave symmetry and must be symmetrical about 30° and 150°. No switching is permitted in the interval 60° to 120°.

Waveforms of the output currents, i_A and i'_A, capacitor current, $i_{C,AB}$, and output voltage, v_{AN}, in an inverter with a wye-connected RL load and a carrier-comparison pulse width modulation with $P = 9$ are shown in Figure 7.51. The ripple of the output current could be reduced further by employing larger output capacitors or increasing the number of switching pulses, P. The output voltage waveform depends on the load. Interestingly, the current gain, K_I, for the pulsed currents, i'_A, i'_B, and i'_C, may exceed unity. For example, with the programmed method for elimination of the fifth and seventh harmonics, $K_I = 1.029$, higher than that of 0.955 for the square-wave inverter. However, the output capacitors divert a part of the fundamentals of the pulsed currents from the output, so that the current gain for the output currents i_A, i_B, and i_C is lower than unity.

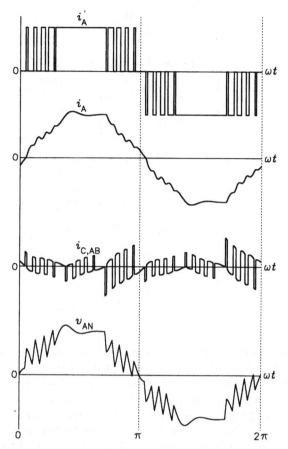

Figure 7.51 Waveforms of the output current, capacitor current, and output voltage in a three-phase PWM CSI (wye-connected RL load, $P = 9$).

7.3 MULTILEVEL INVERTERS

The three-phase voltage-source inverters described in Section 7.1 are by far the most common dc-to-ac power converters encountered in practice. As explained later, those inverters can be classified as *two-level* inverters. Recently, *multilevel* voltage-source inverters have excited widespread interest. Multilevel inverters offer better performance than two-level inverters, but they are more complex and costly and are employed primarily in high-voltage applications.

The "two-level" adjective describing the voltage-source inverters presented so far arises from the fact that the potential of any output terminal can assume two values only. Indeed, the switches of an inverter connect each terminal to either the positive or negative dc bus. Consequently, the line-to-line voltage can assume three values, and the line-to-neutral voltage, five values. Particularly in the square-wave mode, this limits the options available for minimizing the distortion of the voltage waveform.

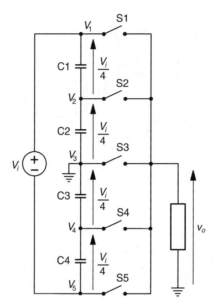

Figure 7.52   Generic five-level inverter.

If more than two voltage levels were obtainable at the output terminals, the output voltage waveforms could be made to better resemble the sine waves.

The idea of a single-phase multilevel inverter, specifically a five-level inverter, is illustrated in Figure 7.52. The four capacitors C1 through C4 make up a voltage divider, which constitutes a dc link for the inverter. The center node of the divider and one terminal of the load are grounded. Five switches, S1 through S5, of which one and only one is assumed to be on at any time, allow applying any of the five fractional voltages, V_1 through V_5, to the nongrounded load terminal. Thus, the output voltage, v_o, could assume any of these five values, including $v_o = V_3 = 0$.

Note that even if the number of levels of the generic multilevel inverter were limited to two (by elimination of switches S2 through S4 and capacitors C2 and C3 between them), the resulting topology would not be equivalent to that of the single-phase inverter in Figure 7.2. The power circuit of that inverter is of the bridge type, while the inverter in Figure 7.52 has the *half-bridge* topology. A practical (although of marginal usefulness) half-bridge single-phase inverter is shown in Figure 7.53. Clearly, the inverter is capable of a two-level output voltage only (i.e., $v_o = \pm V_i/2$), while the full-bridge inverter of Figure 7.2 may have an output voltage of either $-V_i$, 0, or V_i.

A full-bridge single-phase l-level generic inverter could be built by disconnecting the grounded terminal of the load from the ground and connecting it to another half-bridge inverter, supplied from the same dc source and capacitive dc link. The output voltage could assume $l(l - 1) + 1$ levels. Analogously, a three-phase multilevel generic inverter could be made of three half-bridges connected in parallel to the source and dc link. Sketching these two converters is left to the reader.

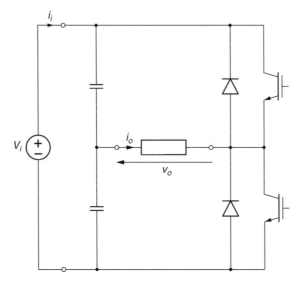

Figure 7.53   Half-bridge voltage-source inverter.

The unidirectional semiconductor devices in practical multilevel inverters require somewhat more complicated topologies. The power circuit of the *three-level neutral-clamped* (or *diode-clamped*) inverter is shown in Figure 7.54. Each of the three legs of the inverter is composed of four semiconductor power switches, S1 through S4; four freewheeling diodes, D1 through D4; and two clamping diodes, D5 and D6. The necessity of clamping diodes is easy to demonstrate by considering what would happen if, for example, diode D5 were missing and switch S1 were turned on. Clearly, capacitor C1, serving as a voltage source, would then be shorted. Similarly, diode D6 prevents shorting capacitor C2 by switch S4.

Theoretically, the four switches in each inverter leg imply the possibility of 2^4 states of a leg, and 2^{12} (i.e., 4096!) states of the entire inverter. In practice, only three states of a leg are used, which makes a total of 27 states of the inverter. A ternary switching variable can thus be assigned to each inverter phase and, for phase A, defined as

$$a = \begin{cases} 0 & \text{if S1, S2 are OFF and S3, S4 are ON} \\ 1 & \text{if S1, S4 are OFF and S2, S3 are ON} \\ 2 & \text{if S1, S2 are ON and S3, S4 are OFF.} \end{cases} \qquad (7.28)$$

Switching variables b and c for the other two phases are defined analogously. It is easy to see that the voltage of a given output terminal of the inverter with respect to the "ground" (inverter's neutral), G, can be expressed in terms of the associated switching variable and input voltage. For example, the voltage, v_A, at terminal A is

$$v_A = \frac{a-1}{2} V_i. \qquad (7.29)$$

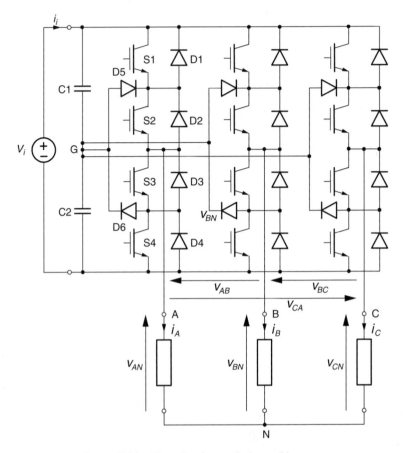

Figure 7.54 Three-level neutral-clamped inverter.

Consequently, the output line-to-line voltages are given by

$$\begin{bmatrix} v_{AB} \\ v_{BC} \\ v_{CA} \end{bmatrix} = \frac{V_i}{2} \begin{bmatrix} 1 & -1 & 0 \\ 0 & 1 & -1 \\ -1 & 0 & 1 \end{bmatrix} \begin{bmatrix} a \\ b \\ c \end{bmatrix} \qquad (7.30)$$

and, based on Eq. (7.9), the line-to-neutral voltages are given by

$$\begin{bmatrix} v_{AN} \\ v_{BN} \\ v_{CN} \end{bmatrix} = \frac{V_i}{6} \begin{bmatrix} 2 & -1 & -1 \\ -1 & 2 & -1 \\ -1 & -1 & 2 \end{bmatrix} \begin{bmatrix} a \\ b \\ c \end{bmatrix} \qquad (7.31)$$

Control methods for the three-level inverter allow changes from only 0 to 1, 1 to 2, and vice versa for each switching variable, while transitions from 0 to 2 and 2 to 0 are forbidden. This ensures smooth commutation and reduces the chances for a

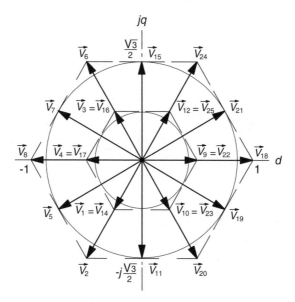

Figure 7.55 Voltage space vectors of a three-level neutral-clamped inverter.

shoot-through, since of the four switches in a leg, only two change their conduction states simultaneously. The dc voltage is shared by at least two switches, so their voltage ratings can be lower than those of switches in a regular, two-level inverter supplied from the same source.

Analogous to the three-phase two-level inverter, a state of the three-level inverter can be designated as abc_3. For example, an inverter is said to be in state 14 if $a = 1$, $b = 1$, and $c = 2$, since $112_3 = 14$. Of the 27 states, states 0, 13, and 26 are zero states, yielding zero voltages at all three output terminals. Space vectors of the line-to-neutral voltages across a wye-connected load, corresponding to individual states of the inverter, are shown in Figure 7.55 in the per-unit format, with V_i as the base voltage. It can be seen the active voltage vectors can be divided into three groups. High-voltage vectors such as $\vec{V}_{18}$ have a magnitude of V_i; medium-voltage vectors such as $\vec{V}_{21}$ are $(\sqrt{3}/2)V_i$ strong; and the magnitude of low-voltage vectors such as $\vec{V}_9$ is $V_i/2$. In comparison with classic two-level inverters, this diversity allows much more freedom in developing a variety modulation strategies.

Comparing the vector diagram with that in Figure 7.23 for the two-level inverter, it is easy to guess that the 18–21–24–15–6–7–8–5–2–11–20–19··· state sequence, each state maintained for one-twelfth of the desired cycle of the output voltage, represents the square-wave operation mode. The corresponding waveforms of switching variables and output voltages of the inverter are shown in Figure 7.56. Although the voltage gain is 1.065, that is, slightly lower than the 1.1 value for a two-level inverter, the higher quality of the output voltage waveforms in the three-level inverter is obvious (compare Figure 7.16). The low-distortion waveform of output current i_A, shown in Figure 7.57 with that of voltage v_{AN}, confirms this conclusion.

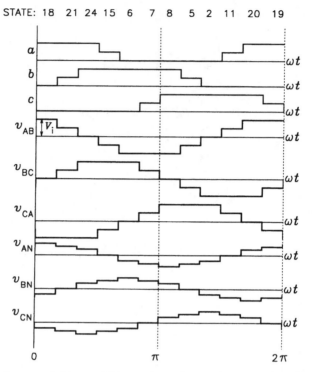

Figure 7.56 States, switching variables, and waveforms of output voltage in a three-level neutral-clamped inverter in the square-wave mode.

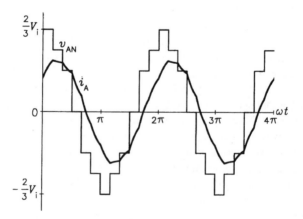

Figure 7.57 Waveforms of output voltage and current in a three-level neutral-clamped inverter in the square-wave mode.

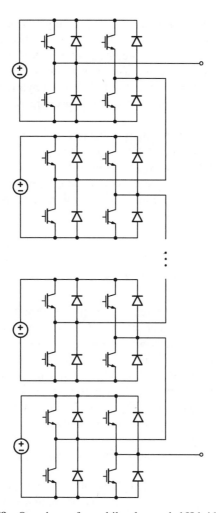

Figure 7.58   One phase of a multilevel cascaded H-bridge inverter.

An even higher quality of operation can be achieved by pulse width modulation, with significantly lower switching frequencies than those typical for two-level
inverters. Several PWM techniques based on the space vector approach have been
developed. Note that the number of stationary voltage vectors to be used for synthesis
of the revolving reference vector is much higher here than in the classic two-level
voltage-source inverter. This allows more freedom in designing effective space vector
PWM strategies.

Other topologies of multilevel inverters are also possible. For example, in a flying-
capacitor inverter, the clamping diodes are replaced with additional capacitors. In the
cascaded H-bridge inverter shown in Figure 7.58, the multilevel inverter is assembled
from the classic single-phase inverters described in Section 7.11 and depicted in

Figure 7.2, each constituent inverter (H-bridge) supplied from a separate dc source. An H-bridge can generate three voltage levels between its output terminals: $-V_i$, 0, and V_i. Such inverters are particularly convenient in systems supplied from batteries, such as electric vehicles, where battery cells would constitute the sources. The number of H-bridges in an l-level inverter is $(l - 1)/2$. Several hybrid topologies combining subcircuits of two of the types of multilevel inverters described have also been proposed.

Multilevel inverters are particularly suitable for high-power high-voltage applications. Fast voltage-controlled IGBTs can be employed in place of the relatively sluggish GTOs or IGCTs. The high number of semiconductor devices, such as 30 devices for the three-phase three-level neutral-clamped inverter versus 12 devices for the two-level voltage-source inverter, is an obvious disadvantage of these otherwise promising dc-to-ac converters.

7.4 SOFT-SWITCHING INVERTERS

All the inverters and, for that matter, all other power converters with fully controlled switches presented so far are characterized by *hard switching*. This means that the transition of their switching devices from one conduction state to another occurs in the presence of nonzero voltages and currents. Consider, for example, phase A of the voltage-source inverter in Figure 7.14. When switch SA is off and diode DA is not conducting, the full input dc voltage, V_i, appears across them. Conversely, if switch SA is on and diode DA$'$ is not conducting, current i_A flows through the switch. Thus, a switch turns on with a nonzero voltage across it and turns off when carrying a nonzero current. As a result, as already explained in Chapter 2, each switching is associated with a certain amount of energy loss, which depends on the values of current and voltage in question.

If the switching frequency were high enough, currents practically free of ripple could be produced in voltage-source inverters. Then, however, the switching losses would increase dramatically. Therefore, in practice, the switching frequency is usually limited not as much by the dynamic properties of the switches as by the thermal effects, the switching losses accounting for up to 50% of total losses. Hard switching is also detrimental from the point of view of the EMI radiated through generation of electromagnetic waves, due to rapid current changes.

If the voltage across a device transitioning from the off-state to the on-state were zero, or no current flowed through the device to be turned off, the resulting *soft switching* would cause minimum losses and EMI. As described in Chapter 8, two types of soft switching, zero-voltage switching (ZVS) and zero-current switching (ZCS), are employed in *resonant converters* for low-power switching power supplies. The phenomenon of electrical resonance is utilized there to impose lossless zero-voltage or zero-current switching conditions. The significant reduction in switching losses in resonant converters allows for high switching frequencies, leading to minimization of filters, magnetic components, and heat sinks, and, consequently, of an entire converter.

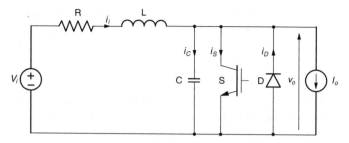

Figure 7.59   Switched network to illustrate the operating principle of a resonant dc link.

A *resonant dc-link inverter* represents an endeavor to implement the ZVS principle in voltage-source power inverters. To explain the basic idea of the resonant dc link, the switched network in Figure 7.59 is considered first. It consists of a supply source producing an ideal dc input voltage, V_i, a resonant LC circuit, a semiconductor switch S, and a freewheeling diode D. The resonant circuit is nonideal, having a resistance of R ohms. The current source at the network's output represents a load whose inductance is assumed to be much higher than that, L, of the resonant circuit. Consequently, the output current, I_o, can be assumed constant during the short cycle of operation of the network.

The operation cycle can be divided into three subcycles. In the first subcycle, the switch is closed for a period of t_1, and the inductance, L, is charged with the electromagnetic energy due to the increased flow of input current, i_i. As explained later, the switch current does not start flowing immediately after the switching signal, g, is applied to the switch at $t = 0$, but after a short delay, t_0. The capacitor, C, has been discharged in the preceding cycle, so its current, i_C, is zero, as are the diode current, i_D, and the output voltage, v_o. Since

$$V_i = Ri_i + L\frac{di_i}{dt} \tag{7.32}$$

and

$$i_s = i_i - I_o \tag{7.33}$$

then

$$i_i = \frac{V_i}{R}\left(1 - e^{(R/L)t}\right) + I_1 e^{(R/L)t} \tag{7.34}$$

and

$$i_s = \frac{V_i}{R}(1 - e^{(R/L)t}) + I_1 e^{(R/L)t} - I_o \tag{7.35}$$

where I_1 denotes the initial input current, $i_i(t_0)$. At the end of the subcycle considered, the input current attains a value of I_2:

$$I_2 = i_i(t_1) = \frac{V_i}{R}(1 - e^{(R/L)t_1}) + I_1 e^{(R/L)t_1}. \tag{7.36}$$

In the second subcycle, the switch opens and the energy stored in the inductance is discharged through the capacitor. Now both the switch and diode currents are zero, while the capacitor current and output voltage are given by

$$i_C = i_i - I_o \tag{7.37}$$

and

$$v_o = \frac{1}{C}\int_0^t i_C dt. \tag{7.38}$$

Kirchhoff's voltage law for the resonant circuit can be written as

$$V_i = Ri_i + L\frac{di_i}{dt} + v_o \tag{7.39}$$

and substituting Eq. (7.37) in Eq. (7.38) and Eq. (7.38) in Eq. (7.39) yields

$$V_i = Ri_i + L\frac{di_i}{dt} + \frac{1}{C}\int_0^t i_i dt - \frac{I_o}{C}t. \tag{7.40}$$

Assuming that the resonant circuit is underdamped, solving Eq. (7.40) for i_i and finding i_C and v_o from Eqs. (7.37) and (7.38) gives

$$i_C = \left\{\left[\frac{V_i - RI_o}{L\omega_r} - \frac{\alpha}{\omega_r}(I_2 - I_o)\right]\sin\omega_r t + (I_2 - I_o)\cos\omega_r t\right\}e^{-\alpha t} \tag{7.41}$$

and

$$v_o = V_i - RI_o + \left\{\left[\frac{\alpha}{\omega_r}(V_i - RI_o) + \frac{I_2 - I_o}{C\omega_r}\right]\sin\omega_r t - (V_i - RI_o)\cos\omega_r t\right\}e^{-\alpha t} \tag{7.42}$$

where

$$\alpha = \frac{R}{2L} \tag{7.43}$$

and

$$\omega_r = \sqrt{\frac{1}{LC} - \alpha^2} \tag{7.44}$$

denotes the damped resonance frequency.

If quality of the resonant circuit is high, $RI_o << V_i$ and $\alpha << \omega_r$, and Eqs. (7.41) and (7.42) can be simplified, to give

$$i_C = Ie^{-\alpha t} \sin(\omega_r t + \varphi_1) \tag{7.45}$$

and

$$v_o = V_i - Ve^{-\alpha t} \cos(\omega_r t + \varphi_2) \tag{7.46}$$

where

$$I = \sqrt{\left(\frac{V_i}{L\omega_r}\right)^2 + (I_2 - I_o)^2} \qquad \varphi_1 = \tan^{-1}\frac{L\omega_r(I_2 - I_o)}{V_i} \tag{7.47}$$

and

$$V = \sqrt{V_i^2 + \left(\frac{I_2 - I_o}{C\omega_r}\right)^2} \qquad \varphi_2 = \tan^{-1}\frac{I_2 - I_o}{C\omega_r V_i}. \tag{7.48}$$

As $L\omega_r \approx 1/C\omega_r$, then $\varphi_1 \approx \varphi_2$. At the beginning of the subcycle in question, the output voltage is zero, then increases to a certain peak value, $V_{o,p}$, and decreases back to zero. If not for the freewheeling diode, the voltage would next become negative, completing the cycle of resonant oscillation. However, once the diode becomes forward biased, it shorts the output of the network, and the output voltage remains at the zero level.

The instant at which the output voltage gets back to zero after half-wave oscillation marks the beginning of the third subcycle of the operating cycle. Denoting the duration of the second subcycle by t_2, the input current at the end of the subcycle assumes the value of I_3:

$$I_3 = i_1(t_2) = I_o + i_C(t_2) = I_o + Ie^{-\alpha t_2} \sin(\omega_r t_2 + \varphi_1). \tag{7.49}$$

In the third subcycle, whose duration is t_3, the switch and capacitor currents are zero, and so is the output voltage, while the diode current, i_D, is given by

$$i_D = I_o - i_i \tag{7.50}$$

where i_i can be found from the again-valid Eq. (7.32). As a result, the input current can be expressed as

$$i_i = \frac{V_i}{R}\left(1 - e^{(R/L)t}\right) + I_3 e^{(R/L)t} \tag{7.51}$$

and the diode current as

$$i_D = I_o - \frac{V_i}{R}(1 - e^{-\alpha t}) - I_3 e^{-\alpha t}. \tag{7.52}$$

Obviously, Eq. (7.52) is valid only as long as it yields a positive value of i_D. When the left-hand side of that equation reaches zero, the diode ceases to conduct.

The next cycle of operation of the network should begin before the diode current reaches zero, so that the switch is turned on under a zero-voltage condition. On the other hand, the flow of the switch current, i_S, cannot begin until the diode has ceased to conduct, since simultaneous conduction of these two antiparallel devices is not possible. This explains the delay, t_0, of the switch current in the first subcycle. In retrospect, the three subcycles can be called *charging*, *resonance*, and *dead time* (this dead time should not to be confused with that for avoidance of the shot-through in hard-switching converters). Waveforms of the voltage and currents considered are illustrated in Figure 7.60. The ZVS conditions for both the switch and diode can easily be observed.

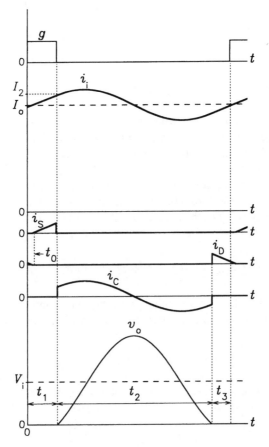

Figure 7.60 Waveforms of voltage and current in the resonant dc link.

Equation (7.48) indicates strong dependence of the output voltage, v_o, on the $I_2 - I_o$ difference. Note that the output current, I_o, has been assumed constant only within a single cycle of operation of the network considered. Therefore, the control system of a resonant dc-link inverter must monitor both the output current and the input current at the t_1 instant of each operating cycle.

In a practical inverter, the output voltage, v_o, of the resonant dc link is fed to a regular voltage-source inverter. The three legs of the power circuit will thus alternately play the role of the switch–diode pair in the hypothetical network analyzed before. The shot-through states of each leg are employed in the charging subcycle. The average value of the pulsed voltage v_o is only slightly lower than the dc supply voltage, V_i, because of the low power loss in the resistance of the dc link. However, the peak value of v_o, which can be up to three times as high as the dc voltage supplied to the resonant circuit, may require the inverter switches and diodes of unacceptably high voltage ratings. Therefore, an additional clamping arrangement is often needed to shave off the peaks of the voltage pulses.

The circuit diagram of a three-phase resonant dc link inverter with an active clamp is shown in Figure 7.61. The active clamp consists of a clamping capacitor, C_{cl}, precharged to $(k_{cl} - 1)V_i$ volts and an antiparallel connection of a switch, S, and a diode, D. The *clamping voltage ratio*, k_{cl}, is usually set to 1.2 to 1.4. As in the hypothetical switching network considered before, the dc bus is shorted in the first subcycle by switches in one leg of the inverter.

In the second subcycle, the link voltage across the resonant capacitor, C, increases toward its natural resonant peak. However, when the voltage reaches the clamping

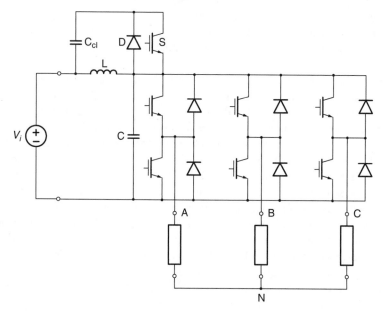

Figure 7.61   Three-phase resonant dc link with an active clamp.

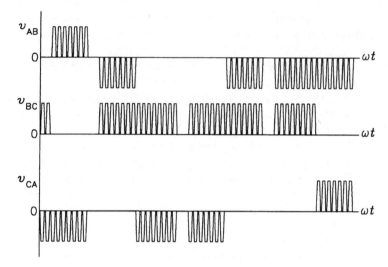

Figure 7.62 Waveforms of line-to-line output voltages in a resonant dc-link inverter.

value, V_{cl}, equal to $k_{cl}V_i$, diode D starts conducting. As a result, the clamping capacitor shunts the resonant inductor, L, and the bus voltage becomes clamped to $V_i + (k_{cl} - 1)$ $V_i = V_{cl}$. The diode current charges the clamping capacitor. Next, switch S is turned on under the ZVS condition, the diode current eventually transfers to the switch, and the clamping capacitor loses its extra charge. The switch is turned off as soon as the net charge received by the clamping capacitor reaches zero, so that the clamping circuit is ready for the next cycle of operation. The dc bus voltage decreases to zero to complete the resonance subcycle, and after the dead time, appropriate switches of the inverter are turned on again, initiating the next operating cycle of the inverter.

The resonance frequency, ω_r, must be at least two orders of magnitude higher than the output frequency, so that the continuous rectangular pulses of the line-to-line voltages of hard-switching voltage-source inverters can be replaced by series of separate clipped or unclipped resonant pulses. This is called a *discrete pulse modulation*, as an integer number of voltage pulses appear within each switching cycle. The discrete pulse modulation in a resonant dc-link inverter with an active clamp is illustrated in Figure 7.62, which shows line-to-line output voltages of the inverter. For good resolution, only a fragment of a cycle is depicted, the resonance frequency here being 100 times higher than the output frequency.

In low-voltage inverters, the unclamped version of the resonant dc link is preferable, because of the higher (approaching unity) voltage gain. Shaving off the peaks of the pulsed voltage produced by the link reduces this factor significantly.

The requirement that all switches must change their states in synchronism with the resonating dc bus constitutes a major disadvantage of the resonant dc link inverters. The resultant discrete pulse width modulation is less precise that that in hard-switching inverters, limiting the quality and control range of output currents.

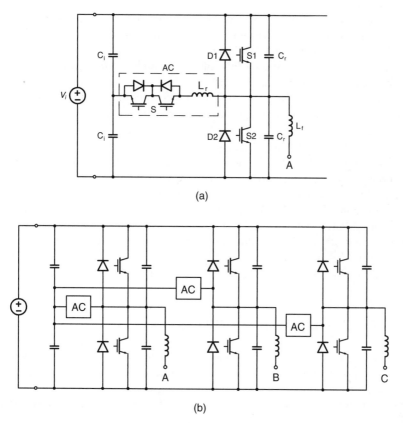

(a)

(b)

Figure 7.63 Auxiliary resonant commutated pole inverter: (a) one phase with the auxiliary circuit; (b) the entire inverter.

It would be more convenient to activate individual switches independently while maintaining the soft-switching conditions, especially in high-power applications. This type of operation has been achieved in an *auxiliary resonant commutated pole inverter*, where the term *pole* refers to the *totem-pole* arrangement of inverter switches (see Figure 3.14).

A power circuit of one phase of the auxiliary resonant commutated pole inverter is shown in Figure 7.63a, and the entire three-phase inverter, in Figure 7.63b. The auxiliary circuit, AC, which triggers the resonances, is connected between the mid-point of two dc-link capacitors, C_i, and the center of the pole. Each inverter pole consists of two main switches, S1 and S2, two freewheeling diodes, D1, and D2, and two resonant capacitors, C_r. The auxiliary circuit is comprised of a bidirectional switch, S, and a resonant inductor, L_r, the switch being made of two semiconductor switches (e.g., IGBTs) and two diodes. The other inductor, L_f, in each phase of the inverter constitutes a part of the output filter and plays a role in shaping the voltage and current transients during the resonance.

The bidirectional switch allows freedom in timing the resonance. The zero crossing of the voltage of a resonant capacitor allows lossless commutation of the parallel main switch. The zero crossing of the resonant inductor allows lossless commutation of the auxiliary switch. The efficiency of the inverter is high and the switching frequency is controllable. However, control is difficult because of the complexity of the three possible commutation modes, two of which require different approaches for low and high currents. Interested readers are directed to the article by De Doncker and Lyons [1] for details.

In comparison with hard-switching inverters, soft-switching inverters are characterized by lower losses and a lower level of parasitic side effects, such as the electromagnetic interference (EMI) associated with switching of large currents. On the other hand, soft-switching inverters suffer from increased complexity of the power circuit and control algorithm, higher voltage stresses on switches, and narrower control ranges. Many topological and control variations have been proposed, but most of the power industry has been faithful to the mature technology of hard-switching inverters. Thus, despite the ingenuity of the operating principles, and apart from certain niche applications, soft-switching power inverters are still relatively rare in practice.

7.5 DEVICE SELECTION FOR INVERTERS

The rated voltage, V_{rat}, of electronic devices in regular two-level hard-switching inverters must be at least equal to the peak value, $V_{i,p}$, of the input voltage, which generally is not ideally constant in time. If an inverter is supplied from a rectifier, it is the peak value of the ac voltage feeding the rectifier, which should be taken as $V_{i,p}$. Thus,

$$V_{rat} \geq (1 + s_V)V_{i,p}. \tag{7.53}$$

As mentioned in Section 7.1.3, the average input voltage, V_i, in PWM voltage-source inverters, equals the maximum available peak value of the fundamental line-to-line output voltage, $V_{LL,1,p(max)}$. In the square-wave operation mode, $V_i \approx 0.9\,V_{LL,1,p}$. These relations facilitate selection of the dc supply source.

Semiconductor power switches in current-source inverters are subjected to switching-generated voltage spikes (see Figure 7.47). Denoting the maximum expected instantaneous value of the line-to-line output voltage by $V_{LL(max)}$, the rated voltage of the switches must satisfy the condition

$$V_{rat} \geq (1 + s_V)V_{LL(max)}. \tag{7.54}$$

When controlled correctly, the maximum voltage stress in three-level inverters across any device does not exceed a half of the input voltage. However, under faulty operating conditions, some devices may be subjected to full input voltage. It is therefore left to the designer's discretion, whether Eq. (7.53) should be used directly or with the right-hand side reduced by half. Often, three-level inverters are selected mostly for their high voltage capability, and correct control is assumed.

In softswitching, the maximum voltage stresses depend on the specific topology. For example, in resonant dc-link inverters, if the resonant pulses of the input voltage are not clamped, their peak value may be up to 2.5 times as high as the dc voltage at the input to the resonant circuit. With clamping, the typical voltage stresses on the devices are on the order of 1.3 to 1.5 of the dc voltage. The right-hand side of condition (7.53) should be modified accordingly.

The highest current stresses in voltage-source inverters occur in the square-wave mode of operation, in which each switch may be forced to conduct the output current for the entire half-cycle of output voltage. This applies also to the freewheeling diodes in inverters employed in systems with a bidirectional power flow. If an inverter is to operate with unidirectional power flow only, the average current of the freewheeling diodes depends on the load angle. As illustrated in Figure 7.64a, with a purely resistive load (R load) the diodes do not conduct at all, and each switch must pass the entire half-wave of the current. Thus, similar to ac voltage controllers [see Eq. (5.54)], the rated current, $I_{S(rat)}$, of the switches should satisfy the condition

$$I_{S(rat)} \geq \frac{\sqrt{2}}{\pi}(1 + s_I)I_{L(rat)} \tag{7.55}$$

where $I_{L(rat)}$ denotes the rated rms value of the line output current of the inverter designed. For the freewheeling diodes, the worst-case scenario, depicted in Figure 7.64b, is a purely inductive load (L load) when each switch and each diode in a given leg of the inverter conduct a quarter-wave of the output line current. As a result, the required rated current, $I_{D(rat)}$, of the diodes can be as low as half of $I_{S(rat)}$.

The pulses of the line output currents in current-source inverters operating in the square-wave mode are one-third of a cycle long (see Figure 7.46). Thus, the average switch current amounts to one-third of the input current, I_i. On the other hand, the easy-to-calculate rms value, $I_{L,1}$, of the fundamental line current equals $(\sqrt{6}/\pi)I_i$. Consequently, the necessary condition for the rated current of switches is

$$I_{S(rat)} \geq \frac{\pi}{3\sqrt{6}}(1 + s_I)I_{L(rat)}. \tag{7.56}$$

Condition (7.55) can be employed for the determination of current ratings of power switches in three-level inverters. Interestingly, even with a purely inductive load, the diodes, both freewheeling and clamping diodes, conduct average currents of not more than one-fourth of the maximum average current that can flow through a switch. It is so because the freewheeling diodes in a given leg of the inverter share the output current with the clamping diodes when the current cannot flow through the switches. Thus, diodes with a rated current, $I_{D(rat)}$, equal to $I_{S(rat)}/4$ can safely be selected.

Resonant circuits in a soft-switching inverter do not significantly affect currents in the main switches. For example, the current conducted by both switches in the leg of

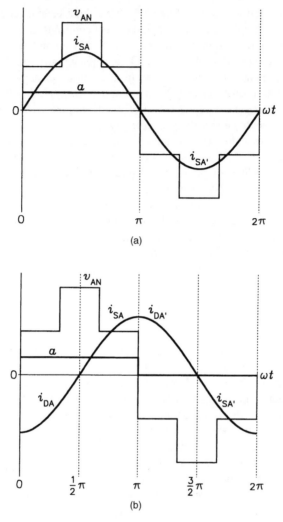

Figure 7.64 Idealized line-to-neutral voltage and line current waveforms in a voltage-source inverter in the square-wave mode: (a) R load; (b) L load.

a resonant dc-link inverter during the charging subcycle is simply too low compared with the output current. Therefore, the same rules as for hard-switching inverters can be employed for the selection of switches and diodes.

To evaluate the shortest on- and off-times of switches in PWM inverters, information about the specific PWM strategy employed is required. Sometimes, as in the case of hysteresis current control, the switching patterns are difficult to predict by theoretical analysis and computer simulations are needed. Therefore, no generalized formulas can be derived. Example calculations of $t_{ON(min)}$ and $t_{OFF(min)}$ are presented in Example 7.3.

7.6 COMMON APPLICATIONS OF INVERTERS

Generally, power inverters are used either in direct dc-to-ac or indirect ac-to-ac power conversion systems. For example, a battery-powered electric vehicle driven by an ac motor employs a direct dc-to-ac conversion scheme in which a voltage-source inverter interfaces the battery with the motor. Another example may involve an off-grid farm supplied (among other sources) from an array of solar cells, where an inverter produces the three-phase voltage for ac machinery used on the farm. Such supply systems usually include a battery that is charged by the array during daylight, and the inverter is fed from the battery rather than from the array.

If, in another case, a photovoltaic array were used to supplement the power supply from an ac system, the system shown in Figure 7.65 could be used. In this photovoltaic utility interface, a PWM voltage-source inverter receives the dc voltage from a photovoltaic array and converts it into high-frequency ac voltage applied to a transformer. The transformer provides electrical isolation and voltage matching between the array and the ac power system. Thanks to the high frequency involved, the transformer is much smaller than a 60-Hz transformer with comparable voltage and power ratings. The secondary voltage of the transformer is converted back to a dc voltage by a diode rectifier, and this voltage is fed to a phase-controlled rectifier, separated from the diode rectifier by an inductive filter. The controlled rectifier, connected to the utility grid, operates in the inverter mode, the filter and the dc voltage representing an RLE load with a negative EMF. Thus, power flows into the grid.

Inverters are widely used in various renewable energy systems, and the photovoltaic utility interface described is only one member of a large family of such interfaces. They are presented in Chapter 9, devoted to "green" applications of power electronics.

Active power filters for elimination of harmonic currents in a power system represent another practical example of the direct dc-to-ac conversion scheme. A block diagram of an active power filter is shown in Figure 7.66. The filter maintains sinusoidal currents in a power line that supplies a nonlinear load, most often a rectifier. Two feedback loops are employed: an outer loop for control of the line current and an inner loop for the current-controlled inverter that produces currents compensating the harmonic currents in the line. Based on the measured line currents and voltages, a control system establishes sinusoidal reference current signals for individual phases

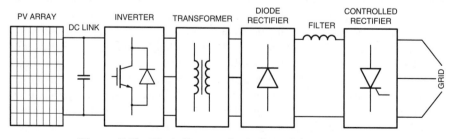

Figure 7.65 Block diagram of a photovoltaic utility interface.

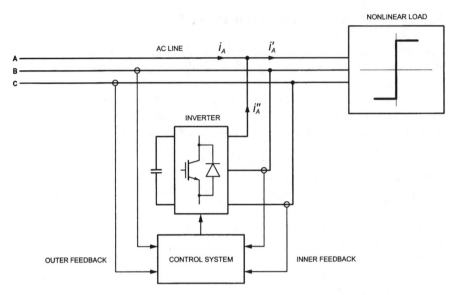

Figure 7.66   Block diagram of an active power filter.

of the line. The reference currents are in phase with the respective phase voltages for a unity power factor. When the reference line current signals are subtracted from actual current signals, reference signals for inverter output currents are obtained. For example, denoting by i_A, i'_A, and i''_A the line, load, and inverter currents in phase A, respectively, the reference inverter current, i''^*_A, is given by

$$i''^*_A = i'_A - i^*_A \tag{7.57}$$

where i^*_A denotes the reference line current.

Voltage and current waveforms in an active power filter connected to an ac line that feeds a three-phase controlled rectifier are shown in Figure 7.67. For clarity, the current waveforms are idealized; that is, the current drawn by the rectifier has a purely square wave and the line current is ideally sinusoidal, assuming that the inverter provides exactly the current required. In reality, the instantaneous changes of inverter current shown cannot be realized in a voltage-source inverter, so that complete elimination of harmonics is not possible. Note the difficult operating conditions of the control system that must compute running waveforms of the reference currents.

Since the load is to be fully powered by the ac line, no real power is required from the inverter. Therefore, only a capacitor is connected to the inverter's input as a dc link, without any supply source, and the losses in the inverter are covered by a low amount of real power drawn from the system. Control of the inverter includes a provision for maintaining constant voltage across the capacitor. Often, to match the voltage and current ratings of the filter with those of the ac line, a transformer is used as an interface between the inverter and the line. A similar arrangement can

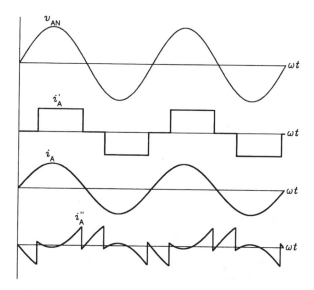

Figure 7.67 Waveforms of voltage and current in an active power filter.

be used for reactive power control in a power system, such as an instantaneous VAr controller. In that case the inverter produces sinusoidal currents that lead respective voltage sine waves in phase, to provide compensation for inductive loads in the system and improve the power factor.

Uninterruptible power supply (UPS) systems are critical in such facilities as hospitals, communication and computer centers, airports, or military installations, which require a constant supply of electric power, even in the case of failure of the grid. These facilities have their own backup battery rooms and diesel-engine or fuel cell–powered ac generators. Two basic types of UPS are standby (off-line) and on-line UPSs. In a standby UPS system, the load is supplied from the grid, while the battery is trickle charged to sustain a full charge. When the supply from the power system is interrupted, a voltage-source inverter is activated automatically to convert the dc voltage from the battery into the regular ac voltage used in the facility. This arrangement is maintained until the normal power supply is restored or the standby generator is brought into operation. Then the inverter returns to its passive status and the battery is charged for fast recovery of full readiness.

Apart from providing backup power when needed, an on-line UPS system isolates a sensitive load from the grid to protect the load from line transients and harmonic pollution. A rectifier–dc link–inverter cascade separates the grid from the load, while the battery connected in parallel to the dc-link transformer maintains full charge. The ac voltage produced by the inverter is then supplied to the load through a low-pass filter, which mitigates the higher harmonics and makes the load voltage sinusoidal. Thus, power quality fed to the load is high and independent of possible disturbances in the grid. A UPS system is shown in Figure 7.68. If the static power switch is normally on, the system operates as an off-line UPS, whereas if it is normally off,

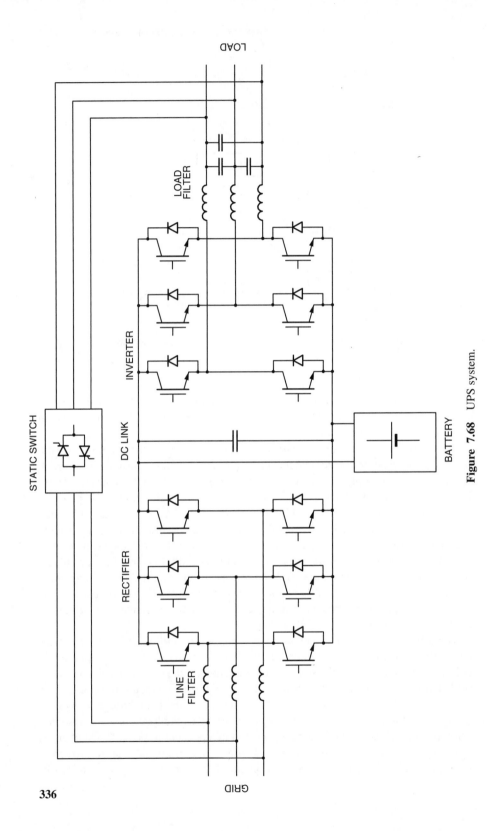

Figure 7.68 UPS system.

it operates as an on-line UPS. An isolation transformer (not shown) is often used between the inverter and the load.

Adjustable-speed ac drives, based primarily on induction motors, constitute the most common use of inverters for indirect ac-to-dc power conversion. Various types of ac motor drives have been developed over the years for the purpose of control of speed, torque, and position. Depending on the quality of control, the drive systems can be classified as low- or high-performance and, considering the control principles, as scalar- or vector-controlled. Generally, the speed of an ac motor depends on the frequency of the stator voltage, while the torque developed is related to the stator current.

In scalar-controlled drive systems, only the frequency and amplitude of the fundamental stator voltage or current are adjusted, which precludes the high performance of the system under transient operating conditions. Therefore, scalar-controlled low-performance drives are employed in such machinery as adjustable-speed pumps, compressors, fans, and grinders, where high control quality would be superfluous. In fact, those drives consume most of the electrical energy generated in developed countries. However, high performance is required from motor drives used in electric traction, hybrid and electric cars, lifts, and elevators, or in adjustable-torque applications such as winders in paper, plastic, and textile factories, and cold-rollers in steel mills. These drives are vector-controlled, which means that those are the instantaneous values of the stator voltage or current that are adjusted continuously. As a result, the transient current waveforms (e.g., during a speed reversal of a drive) often do not resemble the sinusoids typical for steady-state operation. Current-controlled voltage-source inverters are usually employed.

In variable-speed induction motor drives, to maintain the maximum available torque at a constant level, the ratio of stator voltage to frequency should be kept constant, which is known as the *constant volts per hertz* (CVH) control. Above the rated speed, adherence to the CVH principle would mean increasing the stator voltage above the rated value, which is not permitted. Therefore, with frequencies higher than rated, the voltage is maintained constant at the rated level. This area of operation is called *field weakening*, because the intensity of the magnetic field decreases when a frequency increase is not accompanied by a voltage increase. The maximum frequency allowed is that which causes the motor to rotate with the maximum speed permitted. At very low frequencies, the stator voltage must be somewhat higher than that indicated by the CVH rule to compensate for the voltage drop across the stator resistance.

A block diagram of a scalar speed control system with an induction motor is shown in Figure 7.69. The angular velocity, ω_M, of the motor is measured by a speed sensor and compared with the reference velocity, ω^*. The speed error signal, $\Delta\omega_M$, is applied to a slip controller whose output variable, ω_{sl}^*, constitutes the reference slip velocity of the motor, that is, the difference desired between the angular velocity of the revolving magnetic field of the stator and the rotor speed. The slip velocity of a motor must be limited for stability purposes and to protect the motor from too-high currents in the stator and rotor windings. For that purpose the slip controller's input–output characteristic exhibits saturation. When the signal ω_{sl}^* is added to the

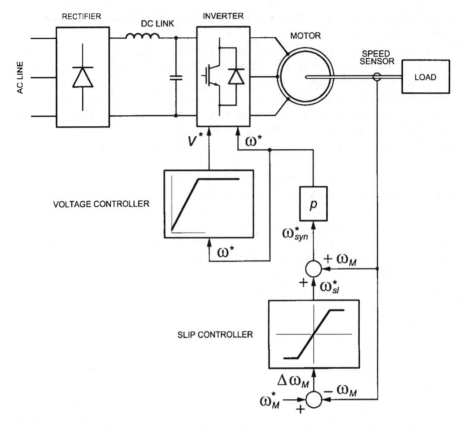

Figure 7.69   Block diagram of an ac drive system with scalar speed control.

motor speed signal, ω_M, a reference signal for synchronous speed, ω^*_{syn}, of the stator field is obtained. The synchronous speed signal is now multiplied by the number, p, of pole pairs of the stator, to result in the reference radian frequency, ω^*, for the stator voltage to be produced by the inverter. The reference magnitude signal, V^*, for the voltage is generated in a voltage controller in dependence on the level of ω^*; that is, the CHV principle is obeyed below the rated frequency and the voltage is made constant in the field weakening area.

In certain applications, such as lift drives or electric ac traction, a drive system is required to have the bidirectional power flow capability to absorb the potential or kinetic energy of the mechanical part of the system during braking. In low-power drives, a braking resistor can be used to dissipate the energy transferred by the inverter from the load. Such an ac drive system, based on the modular frequency changer with IGBTs in Figure 2.24, is shown in Figure 7.70a. The diode in series with the seventh switch serves as a freewheeling diode for the parasitic inductance of the braking resistor circuit. As shown in Figure 7.70b, the same switch and diode, and an external inductor, can serve as a step-up chopper to boost the input voltage to the inverter.

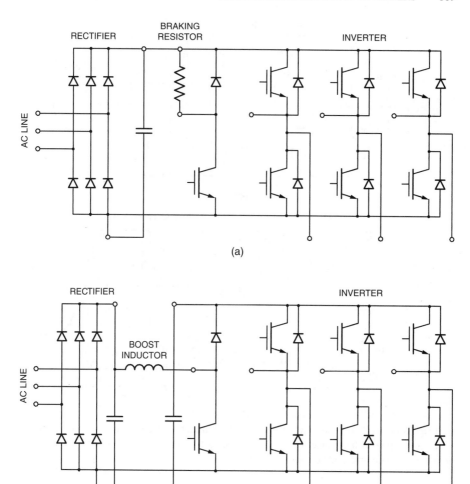

Figure 7.70 Use of the modular frequency changer of Figure 2.24 in an ac drive: (a) a system with a braking resistor; (b) a system with a step-up chopper.

The inverter supply capacitor is charged by the chopper to a voltage higher than that across the dc-link capacitor.

To recover the braking energy and return it to the supply line, a controlled rectifier is required. SCR-based rectifiers are commonly used in practice, but they require large line filters to improve waveforms of the currents drawn from the line. More elegant solutions, shown in Figure 7.71 and allowing four-quadrant operation of the motor, involve a PWM rectifier whose power circuit is a copy of that of the inverter. The current-type rectifier in Figure 7.71a, identical to the current-source inverter in Figure 7.44, has already been covered in Section 4.3.3. The voltage-type rectifier in Figure 7.71b can be thought of as a voltage-source inverter operating with a

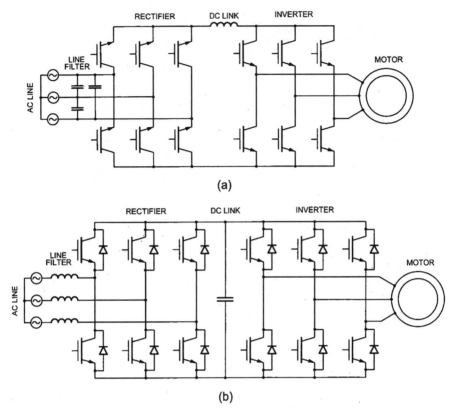

Figure 7.71 PWM rectifier–inverter cascades for bidirectional power flow in ac motor drives: (a) current-type rectifier, inductive dc link, and current-source inverter; (b) voltage-type rectifier, capacitive dc link, and voltage-source inverter.

reversed power flow and an inductive-EMF load (LE load). The cascade of a PWM rectifier, dc link, and a PWM inverter forms a high-performance frequency changer whose applications can be extended beyond the field of adjustable-speed drives (see Figure 7.68).

Single-phase voltage- and current-source inverters are used for induction heating in industry, providing high-frequency currents for the heating coil. Often, a capacitor is connected in series or in parallel with the coil to create a resonant circuit at the inverter output. The output frequencies vary from less than a hundred hertz to a few hundred kilohertz, depending on the application. Inverters are also used in electric arc welding equipment, which requires an isolation transformer between the welding electrodes and the utility supply line. The supply voltage is rectified, converted into a high-frequency ac voltage in the inverter, transformed, rectified again, and applied to the electrodes.

In recent years, low-power sources of electricity utilizing the hydro and wind power have been gaining popularity in the United States, and many small independent operators sell the power to local utilities. The power plant may consist of a synchronous generator, driven by a waterwheel or a wind turbine as the prime mover, and a utility interface. Usually, the speed of the prime mover varies, which causes fluctuations of the frequency of the voltage generated. Also, the magnitude of the voltage may be affected by the speed changes. Therefore, the ac voltage produced by the generator is rectified and fed to a voltage-source inverter that stabilizes the frequency and magnitude of the output voltage, as long as the prime energy does not fall below a specific minimum level. An isolation transformer is often used between the inverter and power system. More details about wind power systems may be found in Chapter 9.

7.7 SUMMARY

The dc-to-ac power conversion is performed by inverters. Depending on the characteristics of the dc source employed, voltage- and current-source inverters can be distinguished. Apart from the dissimilar dc links, these two types of dc-to-ac converters differ by the absence of freewheeling diodes in the current-source inverters. PWM current-source inverters require output capacitors to smooth the output current waveforms. Inverters can be built with any number of phases, three-phase voltage-source inverters being most common in practice.

Inverters can be made to operate in the square-wave or PWM mode. The square-wave mode of operation is simple and characterized by a low number of switchings per cycle of the output voltage. However, a higher quality of the output quantities is obtained in PWM inverters, and a number of PWM techniques have been developed. Feedforward (open-loop) voltage control and feedback (closed-loop) current control are employed in voltage-source inverters, while feedforward current control is typical for current-source inverters.

Multilevel voltage-source inverters provide high-quality output currents in the square-wave mode and are particularly useful in high-voltage applications. Switching losses are minimized in soft-switching inverters, which are also characterized by reduced EMI effects. On the other hand, these converters are more complex than hard-switching converters and have lower resolution of control of output voltage and current.

Inverters have found many applications in direct dc-to-ac and indirect ac-to-ac power conversion schemes. They form a crucial part of adjustable-speed ac drives, improve the quality of currents in power systems, provide an uninterruptible ac power supply, and interface photovoltaic arrays and wind- and water-driven synchronous generators with utility lines. Inverters are also used in such industrial processes as induction heating and electric arc welding.

EXAMPLES

Example 7.1   A voltage-source inverter is supplied from a 620-V dc source and feeds a balanced wye-connected load. At a certain instant, the inverter is in state 3 and the output currents in phases A and B are -72 and 67 A, respectively. Neglect the voltage drops in the inverter and determine all the output voltages (line-to-neutral and line-to-line) and the input current.

Solution: In state 3, the switching variables, a, b, and c, of the inverter are 0, 1, and 1, respectively, since $011_2 = 3$. The line-to-line output voltages can now be calculated from Eq. (7.8) as

$$
\begin{bmatrix} v_{AB} \\ v_{BC} \\ v_{CA} \end{bmatrix} = 620 \begin{bmatrix} 1 & -1 & 0 \\ 0 & 1 & -1 \\ -1 & 0 & 1 \end{bmatrix} \begin{bmatrix} 0 \\ 1 \\ 1 \end{bmatrix} = \begin{bmatrix} -620 \\ 0 \\ 620 \end{bmatrix} \text{V}
$$

and from Eq. (7.10), the line-to-neutral voltages as

$$
\begin{bmatrix} v_{AN} \\ v_{BN} \\ v_{CN} \end{bmatrix} = \frac{620}{3} \begin{bmatrix} 2 & -1 & -1 \\ -1 & 2 & -1 \\ -1 & -1 & 2 \end{bmatrix} \begin{bmatrix} 0 \\ 1 \\ 1 \end{bmatrix} = \begin{bmatrix} -414 \\ 207 \\ 207 \end{bmatrix} \text{V}.
$$

The individual line output currents are $i_A = -72$ A, $i_B = 67$ A, and $i_C = -i_A - i_B = 5$ A. Thus, according to Eq. (7.11),

$$
i_i = 0 \times (-72) + 1 \times 67 + 1 \times 5 = 72 \text{ A}.
$$

Example 7.2 A PWM inverter, supplied from a 310-V dc source, is controlled using the voltage space vector technique with a switching frequency of 4 kHz and a high-efficiency state sequence. In a certain switching interval, the per-unit reference voltage vector is $0.75 \angle 280°$. Determine the switching pattern of the inverter switches in this interval. Assuming that the modulation index does not change, what is the rms fundamental line-to-line output voltage of the inverter?

Solution: The switching period is

$$
T_{SW} = \frac{1}{f_{SW}} = \frac{1}{4 \times 10^3} = 2.5 \times 10^{-4} \text{s} = 250 \ \mu\text{s}
$$

and the reference voltage vector is in sector V, which extends from 240 to 300° and is framed by stationary vectors $\hat{V}_1$ and $\hat{V}_5$ (see Figure 7.22). Thus, the in-sector angle, α, of the vector is $280° - 240° = 40°$, and the durations of the involved inverter states, 1, 5, and 7, are

$$
T_X = T_1 = 0.75 \times 250 \times \sin(60° - 40°) = 64 \ \mu\text{s}
$$

$$
T_Y = T_5 = 0.75 \times 250 \times \sin(40°) = 121 \ \mu\text{s}
$$

$$
T_Z = T_7 = 250 - 64 - 121 = 65 \ \mu\text{s}.
$$

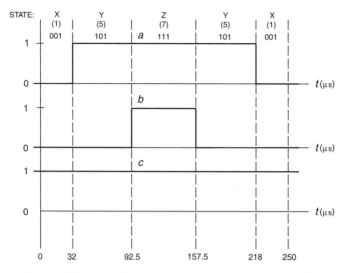

Figure 7.72   Switching pattern of the inverter in Example 7.2.

Thus, the sequence of states required is: state 1 for 32 μs, state 5 for 60.5 μs, state 7 for 65 μs, state 5 for 60.5 μs, and state 1 for 32 μs.

The switching pattern described is illustrated in Figure 7.72. It can be seen that no switching occurs in phase C. Also, as long as the reference voltage vector remains in the same sector, the next switching cycle begins with the same state 1 that concluded the previous cycle. Consequently, no switching takes place in the inverter during cycle-to-cycle transition. As a result, the switching losses are reduced by a third compared with the high-quality state sequence.

As the maximum available peak fundamental line-to-line output voltage, $V_{LL,1p(max)}$, equals, ideally, the supply dc voltage, here 310 V, then

$$V_{LL,1p} = 0.75 \times 310 = 232.5 \text{ V}$$

and

$$V_{LL,1} = \frac{232.5}{\sqrt{2}} = 164 \text{ V}.$$

Example 7.3   A certain space vector PWM technique (a number of them have been proposed) for a three-level neutral-clamped inverter requires that for a modulation index of less than 0.5, low-voltage vectors such as $\vec{V}_3$ or $\vec{V}_4$ be used for synthesis of the reference voltage vector, $\vec{v}^*$. Find the three vectors and duty ratios of the corresponding states of the inverter which will produce a reference vector whose magnitude corresponds to a modulation index, m, of 0.3, and whose phase angle, α, is 45°.

Solution: Figure 7.73 shows a fragment of interest of the diagram of per-unit voltage vectors of an inverter with the input voltage, V_i, taken as the base (see Figure 7.55). The reference voltage vector, $\vec{v}^*$, lies between vectors $\vec{V}_9$ (or $\vec{V}_{22}$) and $\vec{V}_{12}$ (or $\vec{V}_{25}$).

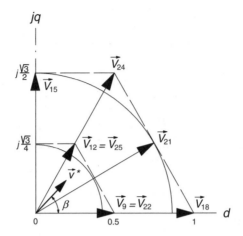

Figure 7.73   Per-unit voltage vectors of the three-level inverter in Example 7.3.

In state 9, $abc = 100$ (as $1 \times 3^2 + 0 \times 3^1 + 0 \times 3^0 = 9$), and in state 12, $abc = 110$, while $abc = 211$ in state 22 and $abc = 221$ in state 25. Both pairs of stationary voltage vectors can be employed, as the transition between states 9 and 12 and between states 22 and 25 changes only one switching variable, b. Let us assume that vectors $\vec{V}_9$ and $\vec{V}_{12}$ will be used along the zero vector $\vec{V}_{13}$ generated when $abc = 111$.

The modulation index $m = 0.3$ implies that the absolute magnitude of the reference voltage vector equals 0.3 of the maximum possible magnitude, which is $(\sqrt{3}/2)V_i$. Thus, in the per-unit format,

$$\vec{v}^* = 0.3 \times \frac{\sqrt{3}}{2} \cos(45°) + j0.3 \times \frac{\sqrt{3}}{2} \sin(45°) = 0.184 + j0.184$$

while

$$\vec{V}_9 = 0.5 + j0$$

and

$$\vec{V}_{12} = 0.25 + j\frac{\sqrt{3}}{4} = 0.25 + j0.433.$$

Denoting by d_9, d_{12}, and d_{13} duty ratios of states 9, 12, and 13, the following three equations must be satisfied:

$$d_9 \times 0 + d_{12} \times 0.5 = 0.184$$

$$j(d_9 \times 0 + d_{12} \times 0.433) = j0.184$$

$$d_9 + d_{12} + d_{13} = 1$$

which, when solved, yield $d_9 = 0.155$, $d_{12} = 0.424$, and $d_{13} = 0.421$.

The same results can be obtained using Eqs. (4.77) to (4.79) employed in the control of two-level PWM rectifiers and inverters. However, because low-voltage vectors are used, $2m$ should be substituted for m. Then

$$d_9 = d_X = 2 \times 0.3 \times \sin(60° - 45°) = 0.155$$
$$d_{12} = d_Y = 2 \times 0.3 \times \sin(45°) = 0.424$$
$$d_{13} = 1 - 0.155 - 0.424 = 0.421.$$

The example illustrates the feasibility of space vector PWM techniques for multilevel inverters. Only minor modifications must be made; for example, should the medium-voltage vectors be used for synthesis of the reference voltage vector, $2/\sqrt{3}m$ would have to be substituted for m in Eqs. (4.77) to (4.79).

Example 7.4   A resonant dc-link inverter is supplied from a 200-V dc source and operates with a switching frequency of 10 kHz. The inductance, L, and capacitance, C, of the resonant circuit are 0.24 mH and 1 μF, respectively, and the circuit resistance is negligible. The current, I_o, drawn by the inverter is constant at 200 A and the input current, I_2, at the beginning of the operation cycle of the dc link, is 209 A. Find:

(a) The average output voltage, V_o, of the resonant dc link when the pulses of the voltage are not clipped

(b) The same voltage when an active clamp shaves off the peaks of the voltage pulses to the level of 1.3 of the dc input voltage, V_i, to the resonant dc link

Solution:

(a) With the assumed absence of damping in the resonant circuit, the resonance frequency, ω_r, can be calculated as

$$\omega_r = \frac{1}{\sqrt{LC}} = \frac{1}{\sqrt{0.24 \times 10^{-3} \times 10^{-6}}} = 64{,}550 \text{ rad/s}$$

and the term $e^{-\alpha t}$ can be removed from expression (7.46) for the output voltage, v_o, of the resonant dc link. Parameters V and φ_2 in that expression, as given by Eq. (7.48), are

$$V = \sqrt{200^2 + \left(\frac{209 - 200}{10^{-6} \times 64{,}550}\right)^2} = 243.8 \text{ V}$$

and

$$\varphi_2 = \tan^{-1}\frac{209 - 200}{10^{-6} \times 64{,}550 \times 200} = 0.609 \text{ rad.}$$

Thus, the output voltage of the link applied to the inverter is

$$v_o = 200 - 243.8 \, \cos(64{,}550t + 0.609) \text{ V}.$$

Based on the equation above, the length, t_2 (see Figure 7.58), of the resonance subcycle of operation of the link can be determined as

$$t_2 = \frac{2\pi - \cos^{-1}(200/243.8) - 0.609}{64{,}550} = 7.85 \times 10^{-5}\text{s} = 78.5 \, \mu\text{s}$$

and the average value, V_o, of v_o can be calculated as

$$V_o = \frac{1}{T_{sw}} \int_0^{t_2} v_0 \, dt$$

where T_{sw} is the switching period, equal $1/10 \text{ kHz} = 10^{-4} \text{ s} = 100 \, \mu\text{s}$. Hence,

$$V_o = \frac{1}{10^{-4}} \left\{ 200 \times 7.85 \times 10^{-5} \right.$$
$$\left. - \frac{243.8}{64{,}550} [\sin(64{,}550 \times 7.85 \times 10^{-5} + 0.609) - \sin(0.609)] \right\}$$
$$= 200 \text{ V} = V_i.$$

The result is not surprising in view of the purported lossless operation of the link.

(b) When the peak of the voltage pulse is shaved off to the level of $1.3V_i$, that is, to 260 V, the resulting voltage reduction, ΔV_o, is given by

$$\Delta V_o = \frac{1}{T_{SW}} \int_{t_A}^{t_B} (v_o - 1.3V_i) \, dt$$
$$= \frac{1}{10^{-4}} \int_{t_A}^{t_B} [200 - 243.8 \, \cos(64{,}550t + 0.609) - 260] \, dt$$

where t_A and t_B are instants at which the voltage clipping begins and ends, respectively. They can be determined as

$$t_A = \frac{\cos^{-1}(-60/243.8) - 0.609}{64{,}550} = 1.88 \times 10^{-5}\text{s} = 18.8 \, \mu\text{s}$$

and

$$t_B = \frac{2\pi - \cos^{-1}(-60/243.8) - 0.609}{64{,}550} = 5.97 \times 10^{-5}\text{s} = 59.7 \, \mu\text{s}$$

yielding

$$\Delta V_o = \frac{1}{10^{-4}} \{-60 \times (5.97 - 1.88) \times 10^{-5}$$

$$-\frac{243.8}{64{,}550}[\sin(64{,}550 \times 5.97 \times 10^{-5} + 0.609)]$$

$$-\sin(64{,}550 \times 1.88 \times 10^{-5} + 0.690)]\} = 48.6 \text{ V}.$$

As a result of the clamping,

$$V_o - \Delta V_o = 200 - 48.6 = 151.4 \text{ V}$$

that is, the voltage gain of the entire inverter is reduced by about 24%.

Example 7.5   A 150-kVA PWM voltage-source inverter designed for unidirectional power flow is supplied from a six-pulse diode rectifier fed from a 460-V line. Assuming no voltage drops in the system and using safety margins of 0.25 for the rated current and 0.4 for the rated voltage, find the minimum required ratings of switches and diodes for the inverter.

Solution: According to Eq. (4.4), the average dc voltage provided by the rectifier as an input voltage, V_i, of the inverter is

$$V_i = \frac{3}{\pi} \times \sqrt{2} \times 460 = 641 \text{ V}.$$

Assuming a unity voltage gain of the inverter, the maximum available peak value, $V_{LL,1p}$, of the fundamental line-to-line output voltage is also 621 V, and the rated rms value, $V_{LL,1(rat)}$, of this voltage is $621/\sqrt{2} = 439$ V. Thus, the rated rms fundamental line output current, $I_{L,1(rat)}$, can be calculated from the rated power of the inverter as

$$I_{L,1(rat)} = \frac{150 \times 10^3}{\sqrt{3} \times 439} = 197 \text{ A}.$$

In agreement with Eq. (7.54), the rated voltage, V_{rat}, of the semiconductor power switches and diodes must satisfy the condition

$$V_{rat} \geq (1 + 0.4) \times \sqrt{2} \times 460 = 911 \text{ V}$$

while condition (7.55) for the rated current, $I_{S(rat)}$, of the switches yields

$$I_{S(rat)} \geq \frac{\sqrt{2}}{\pi}(1 + 0.25) \times 197 = 111 \text{ A}.$$

The rated current, $I_{D(rat)}$, of the freewheeling diodes can be taken as 50% of $I_{S(rat)}$, that is,

$$I_{D(rat)} \geq 0.5 \times 111 = 56 \text{ A.}$$

PROBLEMS

P7.1   A single-phase voltage-source inverter is supplied from a 310-V dc source. Find the fundamental output voltage when the inverter operates in the simple and optimal square-wave modes.

P7.2   Using the input dc voltage as a base, calculate the per-unit peak values of the first, fifth, and seventh harmonics of the output voltage in a single-phase voltage source inverter in the simple and optimal square-wave modes.

P7.3   A three-phase voltage-source inverter is supplied from a 620-V dc source and operates in the square-wave mode. Find the rms values of fundamental line-to-line and line-to-neutral output voltages of the inverter.

P7.4   Starting with state 5, determine the state sequence of a three-phase voltage source inverter that would result in a negative phase sequence of output voltages.

P7.5   A three-phase voltage-source inverter operates with a switching frequency of 4 kHz, using the voltage space vector PWM technique. The inverter is supplied from a 620-V dc source, and it is to produce a line-to-line output voltage of 400 V with a frequency of 95 Hz. Within a certain switching interval, the reference voltage vector has a phase angle of 200°. Determine the high-quality sequence and durations of inverter states in that interval.

P7.6   Repeat Problem 7.5 for the high-efficiency state sequence.

P7.7   Which switches are not switched in the interval considered in Problem 7.6?

P7.8   When the programmed, harmonic elimination PWM strategy is used with a magnitude control ratio of 1, the primary switching angles are 9.48°, 14.80°, 87.93°, and 89.07°. Determine and sketch the full-cycle switching pattern for phase A.

P7.9   A three-phase current-source inverter, supplied from a 200-A dc current source, operates in the square-wave mode. The wye-connected load of the inverter represents a 2-Ω/phase resistance in series with a 5-mH/phase inductance. Find rms values of the fundamental output current and fundamental line-to-line and line-to-neutral output voltages.

P7.10   A three-phase PWM current-source inverter is controlled using the harmonic-elimination PWM technique with three primary switching angles. For a single cycle of operation, find all switching angles for both switches in phase A.

P7.11 A three-level neutral-clamped inverter is supplied from a 1.2-kV ac line via a six-pulse diode rectifier. Assume that the input dc voltage of the inverter equals the average output voltage of the rectifier and find rms values of the line-to-line and line-to-neutral output voltages of the inverter in the square-wave operation mode.

P7.12 Which switches in a three-level neutral-clamped inverter are on and which are off when the inverter is in state 20? Also, taking the input voltage as unity, determine values of all the line-to-line and line-to-neutral output voltages of the inverter in that state.

P7.13 Repeat Example 7.4 for $m = 0.8$ and $\alpha = 110°$. High-voltage vectors are to be used for synthesis of the reference voltage vector.

P7.14 Calculate the total harmonic distortion of the line-to-line and line-to-neutral output voltages of a three-level neutral-clamped inverter in the square-wave mode.

P7.15 Consider the resonant dc link inverter in Example 7.4 and determine the maximum voltage stress on semiconductor devices of the inverter if:

(a) The inverter operates with the input dc voltage of 200 A and no clamp in the resonant dc link circuit

(b) The inverter operates with the clamp as specified in the example, but the input dc voltage has been increased to compensate for the clipping of the link voltage waveform

P7.16 A 60-kVA three-phase voltage-source inverter is supplied from a 460-V ac line through a six-pulse diode rectifier. The inverter is designed for a unidirectional power transfer, from dc to ac. Assume a voltage safety margin of 40%, a current safety margin of 25%, and determine the minimum required voltage and current ratings of switches and diodes in the inverter.

P7.17 A 200-kVA three-level neutral-clamped inverter is supplied from a 2.4-kV ac line through a six-pulse diode rectifier. Assuming the same operating conditions and safety margins as in Problem 7.16, determine the minimum required voltage and current ratings of switches and diodes in the inverter.

COMPUTER ASSIGNMENTS

*CA7.1 Run PSpice programs *Sqr_Wv_VSI_1ph.cir* and *Opt_Sqr_Wv_VSI_1ph.cir* for a single-phase voltage-source inverter in the simple and optimal square-wave modes. For each mode and for both the output voltage and current, find:

(a) The rms value

(b) The rms value of the fundamental

(c) The total harmonic distortion

Observe oscillograms of the input current.

*CA7.2   Run PSpice program *Sqr_Wv_VSI_3ph.cir* for a three-phase voltage-source inverter in the square-wave mode. For the output voltages (line-to-line and line-to-neutral) and current, find:

(a) The rms value

(b) The rms value of the fundamental

(c) The total harmonic distortion

Observe oscillograms of the input current.

CA7.3   Develop a computer program for calculation of switching angles in a PWM voltage-source inverter employing the voltage space vector PWM technique. For given values of the switching frequency (for convenience, it can be a multiple of the output frequency) and magnitude control ratio, the program should determine pulse trains of all three switching variables of the inverter. Provide an option for selection of either the high-quality state sequence or the high-efficiency sequence.

*CA7.4   Run PSpice programs *PWM_VSI_9.cir* and *PWM_VSI_18.cir* for a three-phase voltage-source with 9 and 18 switching intervals per cycle, respectively. In both cases, determine for the line output current:

(a) The most prominent higher harmonic

(b) The rms value

(c) The rms value of the fundamental

(d) The total harmonic distortion

Observe oscillograms of the input current.

*CA7.5   Run PSpice program *Progr_PWM.cir* for a three-phase voltage-source inverter with programmed PWM (harmonic elimination), with four primary switching angles. Measure the harmonic spectrum of the line-to-neutral output voltage and find which low-order harmonics have been eliminated. For the line output current, find:

(a) The rms value

(b) The rms value of the fundamental

(c) The total harmonic distortion

Note that the number of pulses of each switching variable is nine per cycle and, for comparison, repeat the measurements for the respective inverter in Computer Assignment 7.6.

*CA7.6   Run PSpice program *Hyster_Curr_Contr.cir* for a three-phase voltage-source inverter with hysteresis current control. Set the width of the tolerance band to 5% of the peak reference current and determine:

(a) The number of pulses of a switching variable per cycle

(b) The rms value of the line output current

(c) The rms value of the fundamental line output current

(d) The total harmonic distortion of the line output current

Repeat the measurements for the width of the tolerance band of 10% of the peak reference current.

*CA7.7   Run PSpice program *Sqr_Wv_CSI.cir* for a three-phase current-source inverter in the square-wave mode. For the line output current, find:

(a) The rms value

(b) The rms value of the fundamental

(c) The total harmonic distortion

Observe oscillograms of the output voltage and load EMF.

*CA7.8   Run PSpice program *PWM_CSI.cir* for a three-phase PWM current-source inverter. For the line output current, find:

(a) The rms value

(b) The rms value of the fundamental

(c) The total harmonic distortion

Observe oscillograms of the voltages and currents in the inverter.

*CA7.9   Run PSpice program *Half_Brdg_Inv.cir* for a half-bridge voltage-source inverter with hysteresis current control. For the output current, determine:

(a) The rms value

(b) The rms value of the fundamental

(c) The total harmonic distortion

Observe power spectra of the output voltage and current. Notice that although the current ripple is sharply defined, the spectrum is not, because of the randomness of individual switching instants.

*CA7.10   Run PSpice program *Three_Lev_Inv.cir* for a three-level neutral-clamped inverter. For the line-to-neutral output voltage and line output current, determine:

(a) The rms value

(b) The rms value of the fundamental

(c) The total harmonic distortion

Compare the results with those for the two-level inverter in Computer Assignment 7.2.

*CA7.11   Run PSpice program *Reson_DC_Lnk.cir* for a resonant dc-link network. Observe oscillograms of the voltages and currents. Check if the pulsed voltage waveform conforms to Eq. (7.46).

LITERATURE

[1] De Doncker, R. W., and Lyons, J. P., The auxiliary resonant commutated pole converter, *Conference Record of the 1990 IEEE Industry Applications Society Annual Meeting*, pp. 1228–1235, 1990.

[2] Divan, D. M., The resonant dc link converter: a new concept in static power conversion, *IEEE Transactions on Industry Applications*, vol. 25, no. 2, pp. 317–325, 1989.

[3] Enjeti, P. N., Ziogas, P. D., and Lindsay, J. F., Programmed PWM techniques to eliminate harmonics: a critical evaluation, *IEEE Transactions on Industry Applications*, vol. 26, no. 2, pp. 302–316, 1990.

[4] Holtz, J., Pulsewidth modulation: a survey, *IEEE Transactions on Industrial Electronics*, vol. 39, no. 5, pp. 410–420, 1992.

[5] Rashid, M. H., *Power Electronics Handbook*, 2nd ed., Academic Press, San Diego, CA, 2007, Chaps. 15, 16, and 17.

[6] Trzynadlowski, A. M., Borisov, K., Li, Y., and Qin, L., A novel random PWM technique with low computational overhead and constant sampling frequency for high-volume, low-cost applications, *IEEE Transactions on Power Electronics*, vol. 20, no. 1, pp. 116–122, 2005.

8 Switching Power Supplies

Switching dc-to-dc power supplies are presented in this chapter, and the distinction between switched-mode and resonant dc-to-dc converters is explained. Basic types of non-isolated switched-mode converters are analyzed in detail, and their isolated counterparts and extensions are shown. The concept of resonant switches is introduced and use of these switches in quasi-resonant dc-to-dc converters is illustrated. Series–loaded, parallel-loaded, and series-parallel resonant converters are described.

8.1 BASIC TYPES OF SWITCHING POWER SUPPLIES

Proliferation of personal computers, small communication devices, and automotive electronics has spurred intensive research and development activity in the area of switching power supplies (precisely, switching *dc* power supplies). The shrinking size of electronic equipment demands ever-increasing *power density* of the supply systems. Power density, meant as the ratio of available power to volume, or weight, of the power supply, can be high only in highly efficient systems. Otherwise, excessive power losses would cause unacceptable heat stresses on system components, whose compactness severely limits their thermal capacity.

The input voltage for switching power supplies is obtained from batteries, photovoltaic or fuel cells, or, via a rectifier, from utility lines, and as such it may be subject to fluctuations. The current drawn by the load can vary, too, affecting the output voltage. In certain applications, such as laboratory power supplies, wide-range adjustability of the output voltage or current is required. Therefore, a switching power supply is usually equipped with a closed-loop system for control of the output quantities. As explained in Section 1.4, switching converters, based on the principle of pulse width modulation, are better suited for efficient power control than the once popular linear voltage regulators operated, in essence, as controlled resistors.

Typical switching power supplies are low-power electronic systems and it could be argued whether they truly belong in the mainstream power electronics. However, as switching is again used for conversion of electric power, the switching power supplies are related to such medium- and high-power electronic converters as those covered in Chapters 4 through 7, particularly with choppers. Actually, as shown later, the distinction between choppers and certain types of low-power dc-to-dc converters used in switching power supplies is blurred. Technical journals and conferences

Introduction to Modern Power Electronics, Second Edition, by Andrzej M. Trzynadlowski
Copyright © 2010 John Wiley & Sons, Inc.

devoted to power electronics routinely include papers on switching power supplies. Therefore, a brief review of these systems deserves a place in this book.

Two basic structures of switching power supplies reflect the fact that for safety reasons a transformer can be required to isolate the input from the output. The transformer also provides the desired ratio between the output and input voltages. Therefore, switching power supplies can be divided into two classes: *nonisolated* and *isolated*, depending on whether a transformer is incorporated into the power conversion scheme. Bipolar junction transistors and power MOSFETs are most commonly used as semiconductor power switches.

The dc-to-dc converters used in switching power supplies can be classified as *switched-mode* or *resonant* converters. Switched-mode converters are characterized by hard switching (see Section 7.4), while the phenomenon of electric resonance is utilized in resonant converters to provide conditions for zero-voltage switching (ZVS) or zero-current switching (ZCS). Soft switching makes the resonant converters more efficient, but their control ranges are usually narrower than those of switched-mode converters.

8.2   NONISOLATED SWITCHED-MODE DC-TO-DC CONVERTERS

Basic nonisolated switched-mode dc-to-dc converters are single-switch networks that, besides the switch, contain a diode, one or two inductors, and one or two capacitors, one of them connected across the output terminals. Similar to choppers, switched-mode dc-to-dc converters operate in the steady state with constant duty ratios of their switches. The output capacitor smoothes the output voltage, so that in a properly designed converter the voltage ripple is negligible and the voltage can be considered to have the ideal dc quality.

As shown later, increasing the switching frequency allows reduction in the size of the capacitor. This observation also holds true for other elements, such as inductors and, in isolated converters, transformers. Again, as in the case of other PWM power converters, the switching frequency employed represents a trade-off between the quality and efficiency of converter performance. Ultrasonic (above 20 kHz) switching frequencies are typically used to eliminate acoustic noise emitted by the magnetic components. In contrast to most power electronic converters covered in Chapters 4 through 7, the switched-mode dc-to-dc converters cannot be thought of as simple networks of switches, because of the vital role played by energy storage elements.

A quality converter should be simple, reliable, efficient, compact, and characterized by low ripple of the output voltage and, possibly, of the input current. In certain applications, such as controlled dc current sources, a fast response to the magnitude control command is an important consideration as well. The most common topologies of nonisolated switched-mode dc-to-dc converters are *buck*, *boost*, *buck–boost*, *Ćuk*, *SEPIC*, and *Zeta*. The individual topologies are described in detail in the following five sections. Considerations are limited to the continuous operation mode, which is a more common condition than the discontinuous mode, which is employed at the low end of the output power range. Similarly to choppers, dc-to-dc converters are supplied from a diode rectifier or a battery via a capacitive dc link (not included in subsequent circuit diagrams).

8.2.1 Buck Converter

The circuit diagram of the buck (step-down) converter shown in Figure 8.1 is identical to that of the step-down, first-quadrant chopper (see Figure 6.7) with an addition of a low-pass LC filter at the load terminals. In all subsequent considerations, difference equations are employed to describe switched-mode converters (see Section 1.6.1). Additional assumptions include ideal circuit components and negligible ripple of the output voltage.

The difference equation for the inductor can be written as

$$\Delta i_{\mathrm{L}} = \frac{v_{\mathrm{D}} - V_o}{L} \Delta t \qquad (8.1)$$

where Δi_{L} denotes the change in inductor current, i_{L}, within the time interval Δt, v_{D} is the voltage across the diode, V_o is the constant output voltage, and L is the coefficient of inductance of the inductor. During the time when the switch is on, v_{D} equals the input voltage, V_i, and the inductor current increases by

$$\Delta I_{\mathrm{L(ON)}} = \frac{V_i - V_o}{L} d T_{\mathrm{sw}} \qquad (8.2)$$

where d denotes the duty ratio of the switch (not the differentiation operator!) and T_{sw} is the switching period, a reciprocal of the switching frequency, f_{sw}. When the switch is off during the $\Delta t = (1 - d)T_{\mathrm{sw}}$ interval, the diode is forced to carry the inductor current, so $v_{\mathrm{D}} = 0$. The inductor current decreases, and by the end of the switching cycle it has changed by

$$\Delta I_{\mathrm{L(OFF)}} = -\frac{V_o}{L} (1 - d) T_{\mathrm{sw}} \qquad (8.3)$$

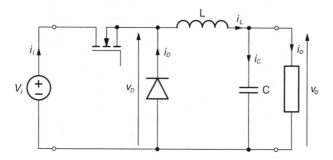

Figure 8.1   Buck converter.

In the steady state, $\Delta I_{L(ON)} = -\Delta I_{L(OFF)} = \Delta I_L$, where ΔI_L denotes the peak-to-peak amplitude of the inductor current ripple, and the resulting equation

$$\frac{V_i - V_o}{L} d T_{sw} = \frac{V_o}{L} (1 - d) T_{sw} \tag{8.4}$$

can be solved for V_o to give

$$V_o = d V_i. \tag{8.5}$$

The result is not surprising, as it has already been obtained for the first-quadrant chopper [see Eq. (6.6)].

For conciseness, in subsequent considerations the subscript "dc" is disposed of in symbols of dc components (average values). A *continuous conduction mode* of the converter is assumed, which means there is continuous current in the inductor. It must be pointed out that the average voltage across an inductor, and the average current through a capacitor are zero when averaged over an integer number of cycles. If so, then the average inductor current, I_L, in the buck converter equals the output current, I_o, which can be determined easily if the load is known. This allows finding all the other quantities in the converter from circuit equations. For example, Eqs. (8.2) and (8.3) and relation $I_L = I_o$ determine the $i_L(t)$ waveform. With $i_L(t)$ known, the capacitor current, $i_C(t)$, can be found as $i_L(t) - I_o$. Finally, the input current, $i_i(t)$, and diode current, $i_D(t)$, equal $x i_L(t)$ and $(1 - x) i_L(t)$, respectively, where x denotes the switching variable associated with the switch. Waveforms of individual currents when the switch is off can be found in a similar way. Selected waveforms, specifically those of the inductor current, i_L, capacitor current, i_C, diode current, i_D, input current, i_i, and voltage, v_D, across the diode are shown in Figure 8.2.

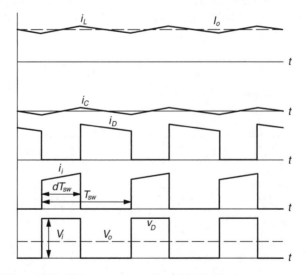

Figure 8.2 Voltage and current waveforms in a buck converter.

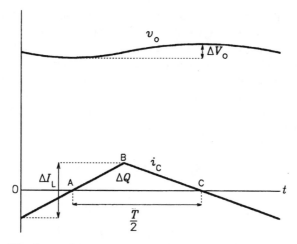

Figure 8.3   Waveforms of the capacitor current and output voltage in a buck converter.

An ideal dc output voltage assumed so far would require an infinitely high capacitance, C, of the capacitor or an infinitely high switching frequency, f_{sw}. Therefore, in practice, the output voltage, v_o, is somewhat rippled. Waveforms of the capacitor current and output voltage, the latter given by

$$v_o(t) = v_0(0) + \frac{1}{C} \int_0^t i_C(t)\,dt \tag{8.6}$$

are shown in Figure 8.3. The peak-to-peak amplitude, ΔV_o, of the voltage ripple can be calculated as

$$\Delta V_o = \frac{\Delta Q}{C} \tag{8.7}$$

where ΔQ is the charge increment in the capacitor, represented in Figure 8.3 by the triangle ABC. Thus,

$$\Delta Q = \frac{1}{2}\frac{T_{sw}}{2}\frac{\Delta I_L}{2} = \frac{T_{sw}}{8}\frac{V_i - V_o}{L}dT_{sw} = \frac{(1-d)V_i}{8L}dT_{sw}^2 = \frac{(1-d)V_0}{8Lf_{sw}^2} \tag{8.8}$$

and, from Eq. (8.7),

$$\frac{\Delta V_o}{V_o} = \frac{1-d}{8LCf_{sw}^2} \tag{8.9}$$

Equation (8.9) quantifies the impact of parameters L and C and the switching frequency on ripple of the output voltage. Clearly, increasing the switching frequency allows reducing the sizes of the inductor and capacitor. In isolated switched-mode converters, the same observation applies to the transformer.

Certain control requirements may result in such low values of the duty ratio, d, that the inductor current becomes discontinuous. As seen in Figure 8.2, this would happen if one-half of the peak-to-peak ripple, ΔI_L, of the inductor current exceeded the average value of this current, I_L. Since $\Delta I_L = |\Delta I_{L(OFF)}|$ and, as already mentioned, $I_L = I_o$, then, based on Eq. (8.3), the condition for the continuous conduction mode can be written as

$$\frac{V_o}{2L}(1-d)T_{sw} < I_o \tag{8.10}$$

or

$$Lf_{sw} > \frac{V_o}{2I_o}(1-d). \tag{8.11}$$

When designing a buck converter, Eqs. (8.9) and (8.11) facilitate proper selection of the inductance, capacitance, and switching frequency.

8.2.2   Boost Converter

The boost (step-up) converter shown in Figure 8.4 has the same topology as that of the step-up chopper (see Figure 6.22) with an output capacitor. The capacitor smoothes the output voltage, which, as shown in Section 6.3 [see Eq. (6.40)], is given by

$$V_o = \frac{V_i}{1-d} \tag{8.12}$$

that is, always higher than V_i. According to Eq. (6.35), the peak-to-peak amplitude, ΔI_i, of ripple of the input current, i_i, can be expressed as

$$\Delta I_i = \frac{V_i}{L}dT_{sw} \tag{8.13}$$

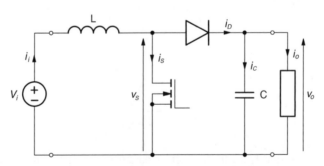

Figure 8.4   Boost converter.

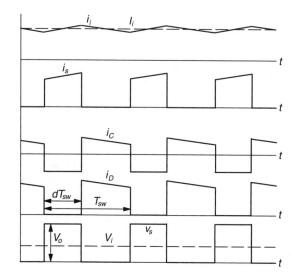

Figure 8.5   Voltage and current waveforms in a boost converter.

while the average input current, I_i, can be calculated from the power balance

$$V_i I_i = V_o I_o. \tag{8.14}$$

Taking Eq. (8.12) into account, I_i is found to be

$$I_i = \frac{I_o}{1 - d} \tag{8.15}$$

which, with Eq. (8.13), allows determination of the input current waveform, $i_i(t)$. Other currents and voltages can easily be found as being related to the input and output quantities. Waveforms of the switch current, i_S, capacitor current, i_C, diode current, i_D, input current, i_i, and voltage, v_S, across the switch are shown in Figure 8.5.

Figure 8.6 illustrates the impact of capacitor current on the ripple of the output voltage. Equation (8.7) can be again be used, in which ΔQ, represented by the rectangle ABC0, is given by

$$\Delta Q = d T_{sw} I_o. \tag{8.16}$$

Hence,

$$\Delta V_o = \frac{d I_o}{C f_{sw}} \tag{8.17}$$

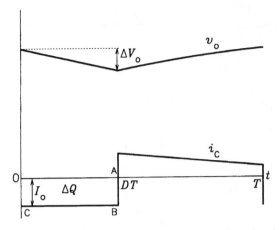

Figure 8.6 Waveform of the capacitor current and output voltage in a boost converter.

or, for a passive load (no load EMF),

$$\frac{\Delta V_o}{V_o} = \frac{d}{RCf_{\text{sw}}} \tag{8.18}$$

where R denotes the load resistance.

For the continuous conduction mode, the average inductor current, I_i, must be greater than $\Delta I_i/2$ (see Figure 8.5). Based on Eqs. (8.12), (8.13), and (8.15), the continuous conduction condition can thus be expressed as

$$\frac{I_o}{1-d} > \frac{(1-d)\,V_o}{2L}dT_{\text{sw}} \tag{8.19}$$

or, after rearrangement,

$$Lf_{\text{sw}} > \frac{d\,(1-d)^2\,V_o}{2I_o}. \tag{8.20}$$

As noted in Section 6.3, when d approaches unity, the magnitude of the output voltage saturates at a finite level, depending on the resistances of the converter elements, particularly that of the inductor. For example, for a boost converter with a passive load, the output voltage can be shown to be

$$V_o = \frac{V_i}{1-d+r_{\text{L}}/(1-d)} \tag{8.21}$$

where r_{L} denotes the ratio of resistance of the inductor to the load resistance. The voltage gain, $K_V \equiv V_o/V_i$, as a function of d for various values of r_{L} is shown in Figure 8.7.

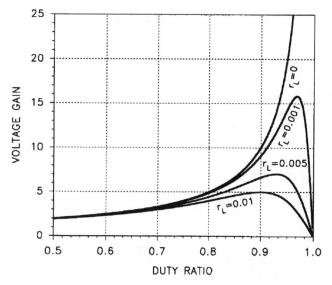

Figure 8.7 Impact of the inductor resistance on the voltage gain of a boost converter.

8.2.3 Buck–Boost Converter

The buck and boost converters described in the preceding sections have limited ranges of control of the output voltage. In a buck–boost converter (Figure 8.8) the output voltage can be made less than, equal to, or greater than the input voltage.

Similar to the boost converter, turning the switch on causes the inductor current, i_L, to increase by

$$\Delta I_{L(ON)} = \frac{V_i}{L} dT_{sw} \tag{8.22}$$

(see Eq. (8.13)). When the switch turns off, the diode is forced to conduct the inductor current, and the output voltage, V_o, appears across the inductor. Now, by the end of

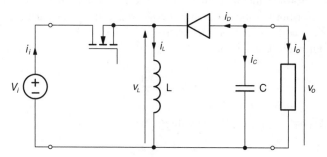

Figure 8.8 Buck–boost converter.

the switching cycle, the inductor current decreases by

$$\Delta I_{L(OFF)} = \frac{V_o}{L}(1 - d)T_{sw}. \tag{8.23}$$

In the steady state, $\Delta I_{L(ON)} = -\Delta I_{L(OFF)} = \Delta I_L$. Thus,

$$\frac{V_i}{L}dT_{sw} = -\frac{V_o}{L}(1 - d)T_{sw} \tag{8.24}$$

which yields

$$V_o = -\frac{d}{1 - d}V_i. \tag{8.25}$$

As seen from Eq. (8.25), the output voltage is negative, that is, inverted with respect to the input voltage. When the duty ratio, d, of the switch is adjusted from 0 to 0.5, the magnitude of the output voltage changes from zero to the input voltage. Increasing d from 0.5 up, causes the output voltage to grow further. As in the boost converter, the maximum voltage gain is limited by the resistances of elements (see Figure 8.7).

The average inductor current, I_L, determined from equations

$$I_i = dI_L \tag{8.26}$$

$$V_i I_i = V_o I_o \tag{8.27}$$

and Eq. (8.25), is given by

$$I_L = -\frac{I_o}{1 - d}. \tag{8.28}$$

Waveforms of the inductor current, i_L, capacitor current, i_C, diode current, i_D, input current, i_i, and voltage, v_L, across the inductor are shown in Figure 8.9. Notice the apparent similarity of Figures 8.5 and 8.9. In particular, the capacitor current has the same waveform as that in the boost converter (see Figure 8.5). Consequently, the output voltage ripple in the buck−boost converter can be described by Eqs. (8.17) and (8.18) derived for the boost converter.

In the continuous conduction mode,

$$I_L > \frac{\Delta I_L}{2} \tag{8.29}$$

which, as $\Delta I_L = \Delta I_{L(ON)}$, is tantamount to

$$\frac{|I_o|}{1 - d} > \frac{V_i}{2L}dT_{sw} \tag{8.30}$$

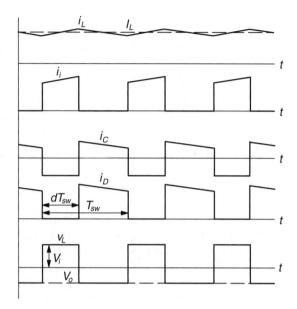

Figure 8.9   Voltage and current waveforms in a buck–boost converter.

because $I_o < 0$. Taking into account that

$$V_i = \frac{1-d}{d} |V_o| \tag{8.31}$$

condition (8.30) can be rearranged to

$$Lf_{sw} > \frac{(1-d)^2 V_o}{2I_o}. \tag{8.32}$$

8.2.4   Ĉuk Converter

A buck–boost converter allows wide-range control of the output voltage, but the input current is discontinuous and its ripple is high, which in many applications is undesirable. For example, when the converter is supplied from a rectifier fed from the utility grid, the high-frequency ac component of the drawn current can cause serious electromagnetic interference in the vicinity of transmission lines. To prevent it, a low-pass filter must be installed between the rectifier and the converter or between the grid and the rectifier.

In contrast to the buck–boost converter, instead of the inductor, the Ĉuk converter (so called after its inventor) uses an extra capacitor for storage and transfer of energy. As seen in Figure 8.10, two inductors are employed to smooth the input current, i_i, and the current, i_{L2}, supplying the output stage of the converter. This allows

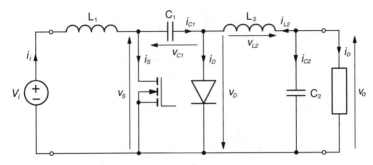

Figure 8.10 Ĉuk converter.

significant reduction of the output capacitance, C_2, and of the possible line filter. The two inductors are wound on the same core, and the converter is remarkably compact.

The Ĉuk converter is sometimes said to have an optimum topology as voltages and currents at the both ends of the converter are true dc quantities and, as shown later, the voltage gain can theoretically be controlled from zero to infinity. The converter also has other advantages, such as the highest efficiency of all the switched-mode dc-to-dc converters, or the grounded cathode of the switch.

As the initial conditions, constant currents in the inductors and constant voltages across the capacitors are assumed. Specifically, $i_{L1}(t) = I_i$, $i_{L2}(t) = I_{L2}$, $v_{C1}(t) = V_{C1}$, and $v_{C2}(t) = V_o$. With the switch on, the diode is off and

$$i_{C1(ON)} = -I_{L2}. \tag{8.33}$$

When the switch turns off, the diode is forced to conduct the inductor currents, i_{L1} and i_{L2}, and

$$i_{C1(OFF)} = I_i. \tag{8.34}$$

In the steady state, the average charge received by capacitor C_1 over a switching cycle is zero, that is,

$$dT_{sw}i_{C1(ON)} + (1-d)T_{sw}i_{C1(OFF)} = 0 \tag{8.35}$$

and, after rearrangement,

$$\frac{i_{C1(ON)}}{i_{C1(OFF)}} = \frac{d}{1-d}. \tag{8.36}$$

Equations (8.33), (8.34), and (8.36) yield

$$-\frac{I_i}{I_{L2}} = \frac{d}{1-d}. \tag{8.37}$$

The average input power,

$$P_i = V_i I_i \tag{8.38}$$

equals the average power, P_o, delivered to the output stage of the converter and given by

$$P_o = V_o I_{L2}. \tag{8.39}$$

Comparing Eqs. (8.38) and (8.39) yields

$$\frac{V_o}{V_i} = \frac{I_i}{I_{L2}} \tag{8.40}$$

and, using Eq. (8.37), we have

$$V_o = -\frac{d}{1-d} V_i. \tag{8.41}$$

Relation (8.41) is identical with that for the buck–boost converter [see Eq. 8.25].

When the switch is on, the input voltage, V_i, is impressed across inductor L_1 for the dT_{sw} period of time, causing current i_i to increase by

$$\Delta I_i = \frac{V_i}{L_1} dT_{sw}. \tag{8.42}$$

At the same time, the voltage across inductor L_2 is

$$v_{L2} = V_o + v_{C1} \tag{8.43}$$

where the voltage across capacitor C_1 equals

$$v_{C1} = V_i + V_o \tag{8.44}$$

as average voltages across the inductors in the steady state of the converter are zero. Consequently, $v_{L2} = V_i$ and current i_{L2} increases by

$$\Delta I_{L2} = \frac{V_i}{L_2} dT_{sw} \tag{8.45}$$

Note that when the switch is off, the voltage, v_S, across the switch equals v_{C1}, given by Eq. (8.45). Thus,

$$v_S = V_i - V_o = \frac{V_i}{1-d}. \tag{8.46}$$

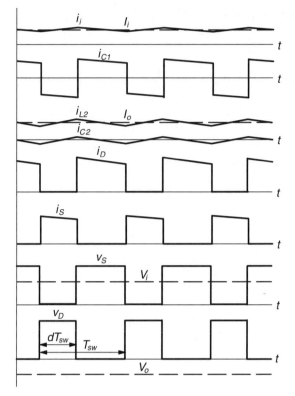

Figure 8.11   Voltage and current waveforms in a Ĉuk converter.

The same voltage appears across the diode when the switch is on. Current and voltage waveforms in the Ĉuk converter are shown in Figure 8.11.

The output part of the Ĉuk converter, composed of inductor L_2, capacitor C_2, and the load, is identical with that of the buck converter. Also, current i_{L2} in that inductor has a continuous waveform with a sawtooth ripple, which is similar to that of inductor current, i_L, in the buck converter. Therefore, Eq. (8.9) for the output voltage ripple in the buck converter is also valid for a Ĉuk converter. In a sense, the Ĉuk converter can be considered a single-switch cascade of the boost and buck converters.

As in the other converters, for continuous conduction, the average current in the inductors must be greater than half of the respective ripple amplitude, ΔI_i and ΔI_{L2}. Based on Eqs. (8.37), (8.41), (8.42), and (8.45), conditions

$$I_i > \frac{\Delta I_i}{2} \tag{8.47}$$

and

$$I_{L2} = I_o > \frac{\Delta I_{L2}}{2} \tag{8.48}$$

can be rearranged to

$$L_1 f_{sw} > \frac{(1-d)^2 V_o}{2d I_o} \tag{8.49}$$

and

$$L_2 f_{sw} > \frac{(1-d)^2 V_o}{2 I_o}. \tag{8.50}$$

8.2.5 SEPIC and Zeta Converters

The SEPIC (single-ended primary inductor converter) and Zeta converters, shown in Figure 8.12, represent attempts at improvement of the Ĉuk converter. It can be seen that all the three converters have two inductors, L_1 and L_2, and a capacitor, C_1, separating the output from the input and providing protection from a shorted load. The SEPIC converter employs an N-channel MOSFET and the Zeta converter a P-channel MOSFET.

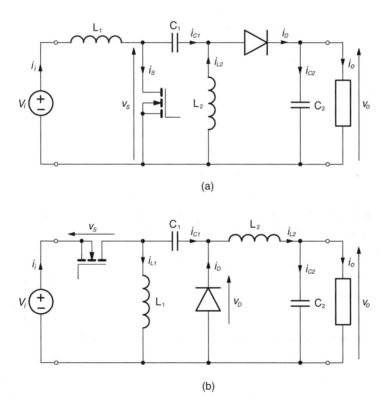

(a)

(b)

Figure 8.12 SEPIC (a) and Zeta (b) converters.

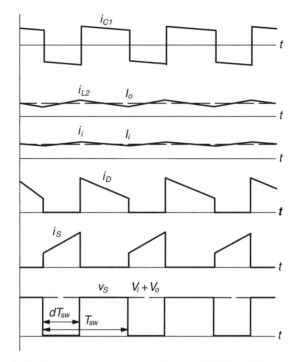

Figure 8.13   Voltage and current waveforms in SEPIC and Zeta converters.

Both the SEPIC and Zeta converters perform a noninverting buck–boost operation, that is,

$$V_o = \frac{d}{1-d} V_i. \tag{8.51}$$

However, Eq. (8.51) is only valid if a zero voltage drop across the diode is assumed (this comment applies to all dc-to-dc converters with a diode in series with the load). In reality, the output voltage is reduced by the amount of that voltage drop. This is an important consideration if a dc-to-dc converter is designed for application in a low-voltage system such as a cell phone. Selected voltage and current waveforms in the SEPIC and Zeta converters, which despite different topologies of these converters are similar, are shown in Figure 8.13.

8.2.6   Comparison of Nonisolated Switched-Mode DC-to-DC Converters

The dc-to-dc converters can be thought of as dc transformers with an adjustable voltage gain. Indeed, in the presence of only minor losses in a converter, its output power approximately equals the input power, while the output voltage can be adjusted, often in a wide range. The concept of the dc transformer applies especially well to

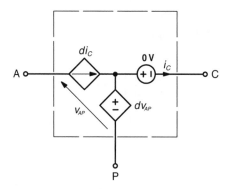

Figure 8.14 Vorperian's switch model.

the Ĉuk converter, in which the voltages and currents at the input and output are continuous, with only a minor ripple.

In many instances, a dc-to-dc converter forms a part of a larger system, usually with some type of closed-loop voltage or current control. To analyze or simulate such systems, it is convenient to use *averaged models* of converters, based on *Vorperian's switch model*. The averaged model of a converter produces ripple-free output current and voltage waveforms, representing values of these quantities averaged over consecutive switching cycles.

The Vorperian's switch model, depicted in Figure 8.14, describes the dc-transformer properties of the converters described in this chapter. The letters a, p, and c denote the "active", "passive", and "common" terminals. As shown in Figure 8.15, Vorperian's model is incorporated into a converter structure as an analog functional equivalent of the actual switching devices. To illustrate the concept of averaged converter model, waveforms of the output current and voltage in a freshly started buck converter are compared in Figure 8.16 with those in the equivalent averaged converter.

All the converters presented were analyzed assuming continuous currents in their inductors, which is a preferred mode of operation and which results in the voltage gain independent of the load. Clearly, under extreme operating conditions, such as I_o approaching zero, a continuous conduction mode cannot be maintained. Analysis of the discontinuous conduction mode is somewhat more involved, and interested readers are referred to the specialized literature.

When choosing a converter for a specific application, consideration must be given to various features, such as the voltage gain required, switch utilization, weight and volume of the converter, component cost, or transient response. Generally, if the rated output voltage is lower than the input voltage, a buck converter should be selected and, viceversa, in the case of the output voltage always higher than the input voltage, the boost converter constitutes the best choice. For a wide range of control of the output voltage, the Ĉuk converter is preferred over the buck−boost converter. Yet it is not free from certain disadvantages, such as the high component count, high switch current, and the required capability of capacitor C_1 to handle high-ripple current.

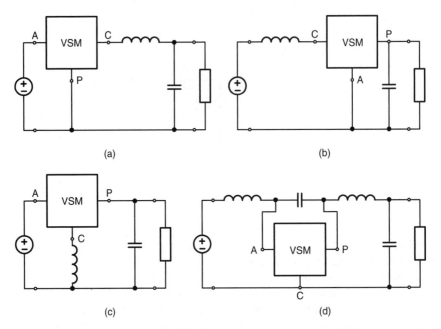

Figure 8.15   Averaged models of switched-mode dc-to-dc converters with Vorperian's switch model: (a) buck; (b) boost; (c) buck−boost; (d) Ĉuk.

Switch utilization, defined as the ratio of minimum required switch power rating (product of the rated voltage and current) to the rated power of the converter, depends on the duty ratio, but its average value for buck and boost converters is much higher than that for buck−boost and Ĉuk converters. Therefore, if sufficient for an application, buck and boost converters offer the highest power density. On the other

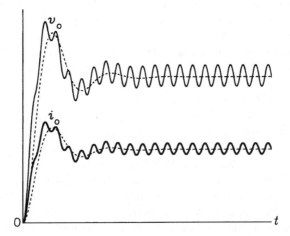

Figure 8.16   Waveforms of the output voltage and current in a buck converter: solid lines, actual converter; dashed lines, averaged model.

hand, the dynamic response to voltage control commands of boost and buck–boost converters is slower than that of comparable buck and Ĉuk converters.

8.3   ISOLATED SWITCHED-MODE DC-TO-DC CONVERTERS

Most practical switching power supplies are based on switched-mode dc-to-dc converters employing a transformer that provides isolation between the input and output of the converter. For safety and reliability reasons, it is necessary in converters fed from a power grid. As already mentioned, the transformer may also be used for matching the input and output voltages. For example, if the output voltage is to be much higher than the available input voltage, the voltage gain of the nonisolated boost, buck–boost, or Ĉuk converters may turn out to be insufficient (see Figure 8.7).

Isolated dc-to-dc converters can be classified as single-switch and multiple-switch converters. Single-switch converters use their magnetic components in a unipolar, single-quadrant mode and are generally restricted to low-power applications. In contrast, transformers in multiple-switch converters are fully utilized magnetically, which at higher power levels (above 100 W) results in a converter being reduced in size to almost half that of an equivalent single-switch converter.

Two equivalent circuits of a transformer are shown in Figure 8.17. The ideal transformer in Figure 8.17a, also used as a general transformer symbol in circuit diagrams, is represented as a pair of coupled coils and is described by the relation

$$\frac{v_1}{v_2} = \frac{i_2}{i_1} = \frac{N_1}{N_2} \tag{8.52}$$

where N_1 and N_2 denote numbers of turns in the primary and secondary winding, respectively. With the *polarity marks* placed as in Figure 8.17, the input and output voltages, v_1 and v_2, and currents, i_1 and i_2, are in phase. Generally, if the voltage at the dotted end of one winding is positive, the voltage at the dotted end of another winding is also positive.

For simulating and analyzing isolated switched-mode dc-to-dc converters, the equivalent circuit in Figure 8.17b is more practical, as it includes the magnetizing inductance, L_m. In this model of a transformer,

$$\frac{v_1}{v_2} = \frac{i_2}{i_1'} = \frac{N_1}{N_2} \tag{8.53}$$

where $i_2 = i_1 - i_m$.

8.3.1   Single-Switch Isolated DC-to-DC Converters

Single-switch converters are derived from buck, buck–boost, and Ĉuk converters. They are the *forward converter*, *flyback converter*, and *isolated Ĉuk converter*,

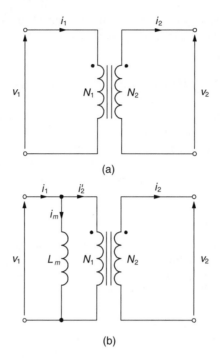

Figure 8.17 Equivalent circuits of a transformer: (a) ideal transformer; (b) transformer with magnetizing inductance included.

respectively. A two-switch *step-up flyback converter*, related to the boost converter, is feasible but little used and is not described.

Forward Converter The forward converter, which is an isolated version of the buck converter in Figure 8.1, is shown in Figure 8.18. When the switch is on, diode D1 is conducting and electrical energy is stored in the inductor. Diode D2 provides a current path for the release of that energy when the switch turns off. To remove the energy stored in the magnetizing inductance of the transformer, diode D3 connected to an extra winding is used.

The average output voltage of the forward converter in the continuous conduction mode is given by

$$V_o = k_N \, d \, V_i \tag{8.54}$$

where

$$k_N \equiv \frac{N_1}{N_2} \tag{8.55}$$

denotes the turn ratio of the transformer. Since the output stage of the forward converter is the same as that of the buck converter, the ripple of the output voltage in

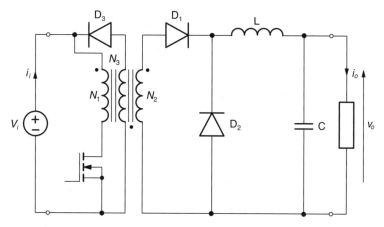

Figure 8.18 Forward converter.

both converters is expressed by Eq. (8.9). The same applies to condition (8.11) for continuous conduction. However, the maximum allowable value, d_{max}, of the duty ratio, d, must be less than $N_1/(N_1 + N_3)$ to ensure complete demagnetizing of the transformer core within each switching cycle. Otherwise, the core would saturate after several cycles, and the transformer would cease to work. In practice, usually $N_1 = N_3$ and $d_{max} = 0.5$.

Flyback Converter A derivative of the buck–boost converter, the flyback converter, is shown in Figure 8.19. The magnetizing inductance, L_m, of the transformer plays the role of inductor L in the circuit diagram of the buck–boost converter in Figure 8.8. Keep in mind that the magnetizing inductance is part of the equivalent circuit of a transformer, not a discrete component. Notice that placement of the polarity marks indicates that the transformer performs voltage inversion. Therefore, the diode is also inverted with respect to that in Figure 8.8. When the switch is turned on, energy is stored in the core of the transformer and, when the switch turns off, is released into the capacitor. An airgap is cut in the core to prevent magnetic saturation.

The output voltage of the flyback converter is given by

$$V_o - k_N \frac{d}{1-d} V_i \tag{8.56}$$

and the ripple of this voltage is given by Eqs. (8.17) and (8.18) as for the boost and buck–boost converters. However, condition (8.32) for the continuous conduction mode must be modified to

$$L_m f_{sw} > \frac{(1-d)^2 V_o}{2k_N^2 I_o} \tag{8.57}$$

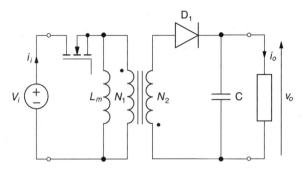

Figure 8.19   Flyback converter.

to account for referring the magnetizing inductance to the primary side of the transformer. The minus sign in Eq. (8.25) for the boost–buck converter is absent in Eq. (8.56), thanks to the voltage inversion in the transformer.

Isolated Ĉuk Converter   The isolated version of the Ĉuk converter is shown in Figure 8.20. Capacitor C_1 in Figure 8.10, depicting the nonisolated converter, is split here into two capacitors, C_{11} and C_{12}, large enough to stabilize the voltages across them. As in the flyback converter, the transformer inverts the secondary voltage, which results in the inverted diode and output voltage of the converter. Equations. (8.56) and (8.9) (the latter with $L = L_2$ and $C = C_2$) describe the output voltage and its ripple, respectively. The condition for continuous conduction of the current in inductor L_1 is

$$L_1 f_{sw} > \frac{(1-d)^2 V_o}{2k_N^2 I_o} \tag{8.58}$$

while Eq. (8.50) expresses the respective condition for inductor L_2. In practice, usually $N_1 = N_2$; that is, $k_N = 1$.

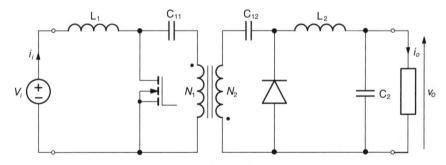

Figure 8.20   Isolated Ĉuk converter.

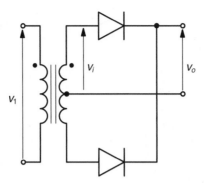

Figure 8.21   Midpoint rectifier.

8.3.2   Multiple-Switch Isolated DC-to-DC Converters

The three most common multiple-switch isolated dc-to-dc converters, the *push—pull converter*, *half-bridge converter*, and *full-bridge converter*, described in this section are related to the buck converter, and their output voltage is always less than the secondary voltage of the transformer. Each of these converters can be considered a cascade of an inverter, transformer, rectifier, and low-pass filter. A single-phase inverter converts the input dc voltage into a high-frequency square-wave ac voltage, which is next transformed, rectified in a diode rectifier, and smoothed in the filter.

In contrast to the two-pulse rectifier described in Section 1.6.3, rectifiers employed in multiple-switch dc-to-dc converters are based on two, not four, diodes. This *midpoint rectifier* is depicted in Figure 8.21. It needs a three-wire supply, which is provided by a transformer with a center-tapped secondary winding. If the voltage across a half of the secondary winding is considered as the input voltage, v_i, the output voltage of the rectifier is given by Eq. (4.5), that is, $V_{o,dc} = 2V_{i,p}/\pi \approx 0.637 V_{i,p}$, where $V_{i,p}$ denotes the peak value of v_i. It is the same as in the four-diode bridge rectifier in Figure 1.33, but the minimum required rated voltage of the diodes must be twice as high as that of diodes in the bridge arrangement. In typically low-voltage switching power supplies, it is not a serious problem, though.

Apart from rectifying diodes, circuits of multiple-switch dc-to-dc converters include freewheeling diodes connected antiparallel with the semiconductor switches. As explained in Section 7.1.1, these diodes are necessary for conducting reactive load currents in inverters. Here, they would not be required if the inverter fed an ideal transformer, since the current in the output-filter inductor is freewheeled by the rectifying diodes. However, the phenomenon of flux leakage in a practical transformer, irrelevant for the dc-to-dc power conversion, makes the transformer appear as a slightly inductive load for the inverter. By providing a path for the resulting reactive currents, the freewheeling diodes protect the switches from overvoltages resulting from interrupting such currents. In practice, the freewheeling diodes do not have to be separate devices, but switches with internal diodes, such as power MOSFETs or switch modules, can be used.

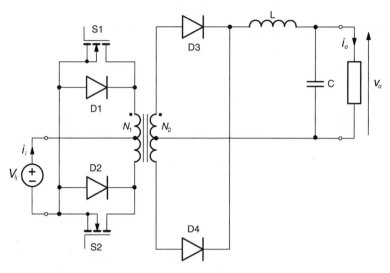

Figure 8.22   Push–pull converter.

Push–Pull Converter   Figure 8.22 is a circuit diagram of a push–pull converter. Switches S1 and S2, both protected by freewheeling diodes D1 and D2, operate with the duty ratio, d, and their switching patterns are shifted by one-half of the switching cycle. The duty ratio is adjustable from zero to 0.5, so that the switches are never on simultaneously. Diodes D3 and D4 rectify the secondary current of the transformer, supplying dc power to the output stage of the converter. When both switches are off and the secondary voltage of the transformer is zero, the rectifying diodes freewheel the inductor current. It can be shown that with a continuous current in inductor L, the average output voltage is given by

$$V_o = 2k_N \, d \, V_i \tag{8.59}$$

and the output voltage ripple by

$$\frac{\Delta V_o}{V_o} = \frac{1 - 2d}{32LCf_{\mathrm{sw}}^2}. \tag{8.60}$$

Half-Bridge Converter   *Half-bridge* in the converter name comes from the half-bridge inverter on the primary side of the transformer. The half-bridge inverter was mentioned in Section 7.3 and depicted in Figure 7.52. Figure 8.23 is a circuit diagram of a half-bridge dc-to-dc converter. Capacitors C_1 and C_2 form the input voltage divider. Switches S1 and S2 operate in the same way as those in the push–pull converter. The maximum allowable duty ratio of 0.5 prevents shorting of the source by both switches being on simultaneously. Diodes D1 and D2 freewheel reactive

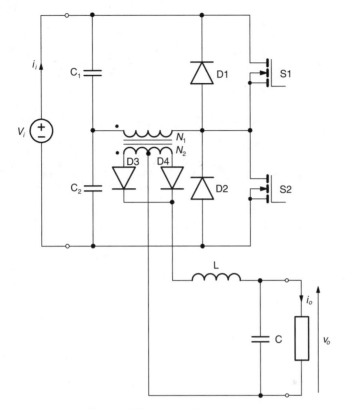

Figure 8.23   Half-bridge converter.

currents due to the flux leakage in the transformer, and diodes D3 and D4 make up
the rectifier on the secondary side. The output voltage is given by

$$V_o = k_N \, d \, V_i \tag{8.61}$$

and since the circuit on the secondary side of the transformer is the same as in a
push–pull converter, the ripple voltage complies with Eq. (8.60).

Full-Bridge Converter   Analogously to the half-bridge converter, the full-bridge
converter in Figure 8.24 is named after its constituent single-phase bridge inverter,
which was described in Section 7.1.1 and shown in Figure 7.2. The switch pairs
S1–S4 and S2–S3 are switched alternately with the duty ratio not exceeding 0.5.
The twice as many semiconductor devices as in the half-bridge inverter result in a
doubled output voltage, which is given by Eq. (8.59). The ripple of this voltage is
again expressed by Eq. (8.60).

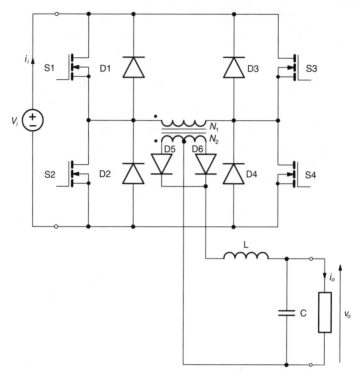

Figure 8.24 Full-bridge converter.

8.3.3 Comparison of Isolated Switched-Mode DC-to-DC Converters

None of the isolated switched-mode dc-to-dc converters presented can be considered definitely superior or inferior to the others, and it is the specific application that dictates selection of the most suitable type. Generally, single-switch converters are employed at lower power levels than are the multiple-switch converters. The flyback converter, with the lowest component count, is simple and very popular, but its transformer is relatively large and voltage stress on the switch is high, at twice the input voltage, V_i. Typical power levels are up to 100 W. The forward converter is used for somewhat higher power requirements, since the extra inductor allows for a smaller transformer. The Ĉuk, SEPIC, and Zeta converters are highly efficient, and their continuous currents result in reduced input and output filter requirements, but the component count is high.

The push−pull converter, suitable for medium-power applications, in a typical range of up to 500 W, is simple and inexpensive, but the switches and freewheeling diodes are subjected to high voltage stresses of $2V_i$. In addition, a dc imbalance caused by even slightly differing switching patterns of switches is likely to lead to saturation of the transformer core. The half-bridge converter, with the voltage stresses of V_i, is the most common switched-mode dc-to-dc converter in the medium power range, while the use of full-bridge converters is usually reserved for powers above 500 W.

Regarding other design considerations, the feasibility of grounded, as opposed to floating, drivers for switches is an important advantage, since it simplifies the overall layout of the converter. Selection of the switching frequency depends mostly on the semiconductor switches used, and the frequencies in switched-mode dc-to-dc converter range from tens to hundreds of kilohertz. As already stressed, high frequencies allow reduction in size of the electromagnetic components, but they also produce high switching losses. The presence of a transformer in the isolated converters allows for *multiple-output converters*, as several output stages can be supplied from multiple secondary windings.

8.4   RESONANT DC-TO-DC CONVERTERS

Analysis of resonant dc-to-dc converters, many types of which have been developed, takes considerable time and effort, as their power circuits change their topology several times during a single switching cycle. Therefore, only the most common types and general operating principles of those converters are subsequently described. Specialized literature is recommended for detailed analytical considerations, while the enclosed PSpice circuit files allow close inspection of voltage and current waveforms in selected converters.

The interest in resonant converters arose from the quest for very high switching frequencies. In switched-mode dc-to-dc converters, these have been mostly limited to tens of kilohertz as a trade-off between the optimal size, weight, efficiency, reliability, and cost of the converter. In certain applications, such as miniature electronic equipment or space technology, much higher switching frequencies are desired. Appropriate semiconductor switches are available, as, for example, MOSFETs can be switched with frequencies of tens of megahertz. The major obstacles are switching losses and switching stresses.

Parasitic inductances in the converters, such as the leakage inductances in transformers and the junction capacitance in the switches, cause the switches to operate with *inductive turn-off* and *capacitive turn-on*. When a switch turns off with an inductive load, high-voltage spikes are generated by the high *di/dt* values. Conversely, when turning on, the energy stored in the junction capacitance is trapped and dissipated in the device. Also, a switching noise is induced that may interfere with proper operation of the converter itself or neighboring sensitive systems. At high switching frequency levels, resonant soft-switching converters offer an attractive alternative to hard-switching switched-mode converters.

Although many solutions have been proposed, two classes stand out in the field of resonant power conversion: *quasi-resonant* or *resonant-switch* converters and *load-resonant* converters. These are presented in subsequent sections.

8.4.1   Quasi-Resonant Converters

A resonant switch is a semiconductor switch, typically a power MOSFET or a BJT, with an LC resonant circuit (*LC tank*) and, sometimes, a diode interconnected with

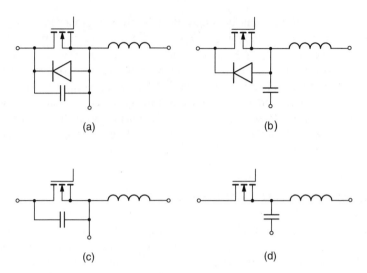

Figure 8.25 Voltage-mode resonant switches: (a) L-type, half-wave; (b) M-type, half-wave; (c) L-type, full-wave; (d) M-type, full-wave.

the switch. The resulting network makes either the voltage across the switch, or the current through it, acquire a sinusoidal waveform instead of the square-wave one typical for switched-mode converters. There are voltage- and current-mode resonant switches, each class subdivided into L- or M-type switches. In turn, each L- or M-type switch can be of the half- or full-wave variety. The eight resulting topologies are shown in Figures 8.25 and 8.26.

As known from circuit theory, application of a dc voltage, V, to an undamped (lossless) series LC circuit at $t = t_0$ produces a sinusoidal current in the inductor,

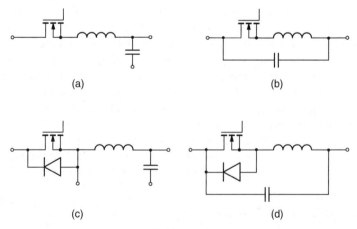

Figure 8.26 Current-mode resonant switches: (a) L-type, half-wave; (b) M-type, half-wave; (c) L-type, full-wave; (d) M-type, full-wave.

given by

$$i_L = \frac{V - v_C(t_0)}{Z_0} \sin[\omega_0(t - t_0)] + i_L(t_0) \cos[\omega_0(t - t_0)] \tag{8.62}$$

where $v_C(t_0)$ and $i_L(t_0)$ denote the initial voltage across the capacitor and initial inductor current, respectively. The characteristic impedance, Z_0, and ω_0 are given by

$$Z_0 = \sqrt{\frac{L_r}{C_r}} \tag{8.63}$$

and

$$\omega_0 = \frac{1}{\sqrt{L_r C_r}} \tag{8.64}$$

where L_r and C_r denote the respective values of inductance and capacitance of the resonant circuit. The capacitor voltage is also sinusoidal and it leads the current waveform by 90°, as

$$v_C(t) = V - [V - v_C(t_0)] \cos[\omega_0(t - t_0)] + Z_0 i_L(t_0) \sin[\omega_0(t - t_0)] \tag{8.65}$$

Selected $v_C(t)$ and $i_L(t)$ waveforms, with $i_L(t_0) > 0$ and $v_C(t_0) < 0$, are shown in Figure 8.27.

The voltage-mode resonant switches in Figure 8.25 have the resonant capacitor connected in parallel with the switching device. In half-wave switches, the clamping diode prevents the voltage, v_C, across the resonant capacitor from becoming negative, and the resonance ends in the midcycle. However, in the full-wave switches, the voltage can change the polarity.

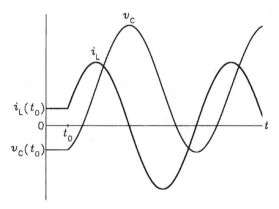

Figure 8.27 Waveforms of the inductor current and capacitor voltage in an undamped resonant circuit.

Turning a switch off initiates the resonance process, and the voltage across the switch acquires a sinusoidal waveform. When, after half a cycle, the voltage reaches zero, the switch can be turned on again under ZVS conditions. It is particularly advantageous in power MOSFET-based converters operating with very high, megahertz-range, switching frequencies, as the capacitive turn-on stresses on the devices are greatly alleviated.

In the current-mode resonant switches in Figure 8.26, the inductor is connected in series with the switch in order to shape the switched current. The inductor and capacitor connected in parallel with the switch form a series resonant circuit, whose resonance is triggered by the switch turning on.

In half-wave switches, the unidirectional conduction ability of the semiconductor power switch prevents the current from changing its polarity. Therefore, only a half cycle of the resonance can be executed, with the current, i_L, flowing in the direction of the load. In full-wave switches, an antiparallel diode is connected across the power switch so that the current can flow to the source.

When the semiconductor power switch turns on at $t = t_0$, the resonant current conducted increases gradually from zero, following the sinusoidal wave shape. After a half-cycle, the current reaches zero and the switch is naturally turned off (commutated). Consequently, both turn-on and turn-off occur under ZCS conditions, and the switching losses and switching stresses are reduced significantly.

An additional advantage of the resonant switches in high-frequency converters consists in the possibility of utilization of parasitic inductances and capacitances. Indeed, since the resonance frequency must be higher than the switching frequency, the required values of the resonant inductance and capacitance become very low. For example, a resonance frequency of 1 MHz can be obtained using a 1-mH inductance and a 1nF capacitance.

All the basic switched-mode dc-to-dc converters described in Sections 8.2 and 8.3 can be converted into quasi-resonant converters by replacing the regular switches with resonant switches. An example ZVS buck converter with an L-type half-wave switch is shown in Figure 8.28, and a ZCS quasi-resonant boost converter with an M-type full-wave resonant switch is illustrated in Figure 8.29.

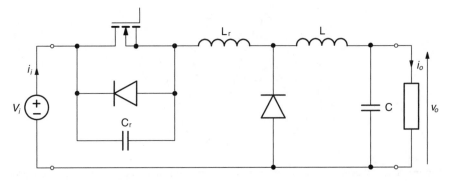

Figure 8.28   Quasi-resonant ZVS buck converter with L-type half-wave switch.

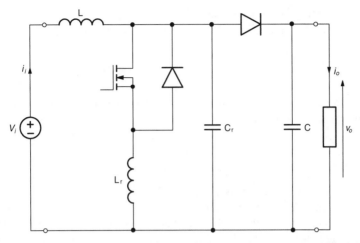

Figure 8.29   Quasi-resonant ZCS boost converter with M-type full-wave switch.

Because of the fixed resonance frequency, control of the duty ratio of a resonant switch is performed by changing the switching frequency. The control range is narrower than that in switched-mode converters, and the EMI, although reduced considerably, is difficult to contain. In general, at high switching frequencies, zero-voltage switching is preferable to zero-current switching. It must also be pointed out that the resonant switches have to carry higher peak currents than their counterparts in switched-mode dc-to-dc converters. This may not only require higher current ratings in semiconductor devices, but also cause increased conduction losses. As with any engineering system, selection of a given converter type must be based on the best trade-off between technical and economical advantages and disadvantages.

Distinguishing properties of ZVS and ZCS quasi-resonant converters are listed in Table 8.1. Generally, the full-wave switches make the converter operation load-insensitive. However, the voltage-mode full-wave switches have the disadvantage of trapping energy in the junction capacitance of the semiconductor device. The resulting capacitive turn-on makes those switches inappropriate for converters with

TABLE 8.1   Comparison of the ZVS and ZCS Resonant-Switch Converters

Property	ZCS Converters	ZVS Converters
Voltage gain		
Buck converter	k_f	$1 - k_f$
Boost converter	$1/(1 - k_f)$	$1/k_f$
Buck–boost converter	$k_f/(1 - k_f)$	$1/k_f - 1$
Control	constant t_{ON}	constant t_{OFF}
	variable t_{OFF}	variable t_{ON}
Waveform of voltage across a switch	square	sinusoidal
Waveform of current through a switch	sinusoidal	square
Load range	$1 < r < \infty$	$0 < r < 1$

very high switching frequencies. The voltage gain, $K_V \equiv V_o/V_i$, of a converter with a full-wave switch is a function of the *frequency ratio*, k_f, defined as

$$k_f \equiv \frac{f_{sw}}{f_0} \tag{8.66}$$

where f_0 denotes the frequency (in hertz). In converters with half-wave switches, K_V also depends (nonlinearly) on the *normalized load resistance*, r, defined as

$$r \equiv \frac{R}{Z_0}. \tag{8.67}$$

In circuit theory, r represents a reciprocal of the quality factor, Q, of a series resonant circuit. In all converters with half-wave switches, the voltage gain increases with r and equals that of the full-wave switch converter only when $r = 1$. When inspecting Table 8.1, note the allowable ranges of r for ZVS and ZCS converters.

8.4.2 Load-Resonant Converters

In load-resonant converters, an LC resonant circuit causes the load voltage and current to oscillate, creating opportunities for zero-voltage and/or zero-current switching. The three most common topologies are the *series-loaded*, *parallel-loaded*, and *series–parallel* converters. All are based on an inverter–rectifier cascade appended with a resonant circuit between these two subconverters.

Series-Loaded Converter   The circuit diagram of a half-bridge series-loaded converter is shown in Figure 8.30. A half-bridge inverter, based on two semiconductor switches, S1 and S2, two freewheeling diodes, D1 and D2, and a capacitive voltage divider, C1 and C2, converts the input dc voltage, V_i, into a square-wave voltage. The

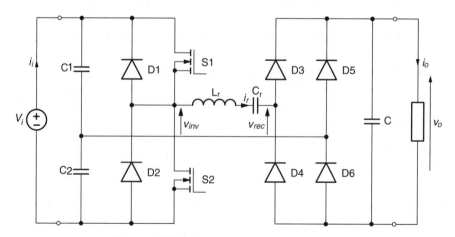

Figure 8.30   Series-loaded resonant converter.

inverter is connected via a series resonant circuit, L_r and C_r, to a two-pulse rectifier, composed of four diodes, D3 through D6, and an output capacitor, C. The output capacitor is sufficiently large to allow treating the rectifier–load circuit as a source of fixed voltage, V_o.

Three basic modes of the multiple-stage operation of the converter can be distinguished. If the switching frequency, f_{sw}, is less than half of the resonance frequency, f_0, the converter operates in a *discontinuous mode*, with a discontinuous current in the resonant inductor. The inverter switches turn on with zero current and turn off with both zero current and voltage, making it possible to use SCRs in certain high-power, low-frequency applications. If $f_0/2 < f_{sw} < f_0$, the converter operates in the *below-resonant continuous mode*, with the switches turning on at nonzero voltage and current, thus causing turn-on switching losses. However, the turn-off occurs under the zero-voltage and zero-current conditions, so that SCRs can again be used. Finally, in the *above-resonant continuous mode* of operation, with $f_{sw} > f_0$, the switches turn on with zero voltage and current, but they are turned off in the presence of the voltage and current, producing turn-off switching losses. Clearly, this operating mode requires fully controlled switches.

To derive an approximate relation for the voltage gain, K_V, a simple ac model of the converter can be used. The actual output voltage, $v_{inv}(t)$, of the inverter is an ac square wave with the peak value of $V_i/2$. Analogously, the input voltage, $v_{rec}(t)$, of the rectifier is an ac quasi-square wave with the peak value of V_o. However, current in the resonant circuit, $i_r(t)$, is practically sinusoidal. All three waveforms have a fundamental frequency of f_{sw}. Replacing $v_{inv}(t)$ and $v_{rec}(t)$ with their fundamentals, $v_{inv,1}(t)$ and $v_{rec,1}(t)$, an ac equivalent circuit of the converter, shown in Figure 8.31, is obtained. The peak value $V_{inv,1,p}$ of $v_{inv,1}(t)$ is given by

$$V_{inv,1,p} = \frac{4}{\pi} \times \frac{V_i}{2} = \frac{2}{\pi} V_i \tag{8.68}$$

and the peak value $V_{rec,1,p}$ of $v_{rec,1}(t)$, by

$$V_{rec,1,p} = \frac{4}{\pi} V_o \tag{8.69}$$

(see Eq. (1.28)).

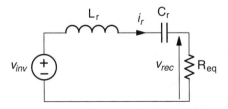

Figure 8.31 Ac equivalent circuit of the series-loaded resonant converter.

Thanks to the practically ideal dc output voltage, the output current can also be assumed to be of pure dc quality; that is, $i_o(t) = I_o$. It is obtained by rectification of $i_r(t)$, so that

$$I_o = \frac{2}{\pi} I_{r,p} \tag{8.70}$$

where $I_{r,p}$ denotes peak value of $i_r(t)$ [see Eq. (1.15)]. Dividing $V_{\text{rec},1,p}$ by $I_{r,p}$ yields the equivalent resistance, R_{eq}, of the load,

$$R_{\text{eq}} = \frac{V_{\text{rec},1,p}}{I_{r,p}} = \frac{4/\pi V_o}{\pi/2V_o} = \frac{8}{\pi^2} R = \frac{8}{\pi^2} r Z_0 \tag{8.71}$$

as seen from the input terminals of the rectifier.

From the equivalent circuit in Figure 8.31,

$$\frac{V_{\text{rec},1,p}}{V_{\text{inv},1,p}} = \frac{R_{\text{eq}}}{\left| R_{\text{eq}} + j\left(X_{\text{L}} - X_{\text{C}}\right) \right|} \tag{8.72}$$

where

$$X_{\text{L}} = 2\pi f_{\text{sw}} L_r = k_f Z_0 \tag{8.73}$$

and

$$X_{\text{C}} = \frac{1}{2\pi f_{\text{sw}} C_r} = \frac{Z_0}{k_f}. \tag{8.74}$$

Based on Eqs. (8.68) and (8.69), Eq. (8.72) can be rearranged to

$$K_V = \frac{V_o}{V_i} = \frac{1}{2\sqrt{1 + (X_{\text{L}} - X_{\text{C}})/R_{\text{eq}}}} \tag{8.75}$$

and substituting Eqs. (8.71), (8.73), and (8.74), to

$$K_V = \frac{1}{2\sqrt{1 + \left[\pi^2/8r\left(k_f - 1/k_f\right)\right]^2}} \tag{8.76}$$

A three-dimensional graph of the control characteristic (8.76) is shown in Figure 8.32. The highest voltage gain of 0.5 is obtained when $k_f = 1$; that is, $f_{\text{sw}} = f_0$. The discontinuous mode is most advantageous because no switching losses occur in the converter.

Alternative topologies of the series-loaded resonant dc-to-dc converter include converters with a full-bridge inverter and converters with a transformer between the resonant circuit and the rectifier. A *load-commutated inverter* is obtained by

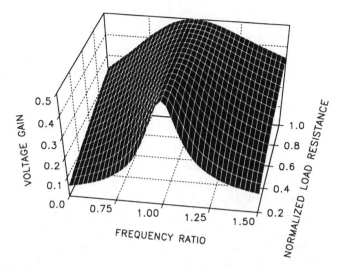

Figure 8.32   Control characteristic of the series-loaded resonant converter.

eliminating the rectifier and connecting the inverter to the load directly, via a resonant circuit. Such an inverter can be based on SCRs, which are being naturally commutated, while the resonant circuit results in quasi-sinusoidal waveforms of the output voltage and current.

Parallel-Loaded Converter   The parallel-loaded resonant dc-to-dc converter, whose circuit diagram is shown in Figure 8.33, differs from the series-loaded converter by the connection of the resonant capacitor across the input terminals of the rectifier. Another difference consists in an output inductor, L, that smoothes and stabilizes the

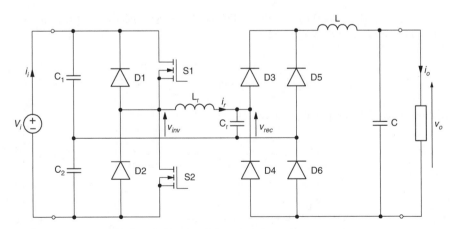

Figure 8.33   Parallel-loaded resonant converter.

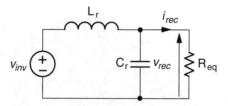

Figure 8.34   Ac equivalent circuit of the parallel-loaded resonant converter.

current supplied to the final stage of the converter. Consequently, the rectifier can be considered a source of a constant current, I_o.

Similarly to a series-loaded resonant converter, a parallel-loaded converter operates in the discontinuous mode when $f_{sw} < f_0/2$, in the below-resonant continuous mode when $f_0/2 < f_{sw} < f_0$, and in the above-resonant continuous mode when $f_{sw} > f_0$. Again, the discontinuous mode is characterized by no switching losses, the below-resonant continuous mode by no turn-off losses, and the above-resonant mode by no turn-on losses. The ac equivalent circuit of the converter is shown in Figure 8.34. In the actual converter, $v_{rec}(t)$ and $i_{rec}(t)$ have ac square waveforms, while the waveform of $v_{inv}(t)$ is sinusoidal. Consequently, Eq. (8.68) is valid again, and, since V_o is the average value of rectified $v_{rec}(t)$, then

$$V_o = \frac{2}{\pi} V_{rec,1,p}. \tag{8.77}$$

The peak value of $i_{rec}(t)$ equals the output current, I_o; thus the peak value, $I_{rec,1,p}$, of the fundamental of $i_{rec}(t)$ is given by

$$I_{rec,1,p} = \frac{4}{\pi} I_o. \tag{8.78}$$

Based on Eqs. (8.77) and (8.78), the equivalent load resistance, R_{eq}, of the parallel-loaded converter can be expressed as

$$R_{eq} = \frac{V_{rec,1,p}}{I_{rec,1,p}} = \frac{(\pi/2) V_o}{(4/\pi) I_o} = \frac{\pi^2}{8} R = \frac{\pi^2}{8} r Z_0. \tag{8.79}$$

Finally, from the ac equivalent circuit of Figure 8.34,

$$\frac{V_{rec,1,p}}{V_{inv,1,p}} = \frac{1}{\left| 1 - X_L/X_C + j \left(X_L/R_{eq} \right) \right|} \tag{8.80}$$

or

$$K_V = \frac{V_o}{V_i} = \frac{4}{\pi^2 \sqrt{(1 - X_L/X_C)^2 + \left(X_L/R_{eq} \right)^2}} \tag{8.81}$$

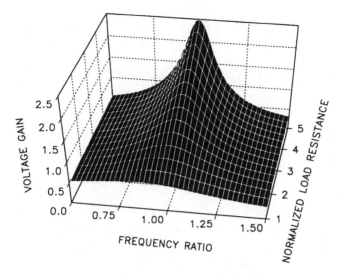

Figure 8.35 Control characteristic of the parallel-loaded resonant converter.

which, after substituting Eqs. (8.73), (8.74), and (8.79), yields

$$K_V = \frac{1}{(\pi^2/4)\sqrt{(1 - k_f)^2 + 8k_f/\pi^2 r}}. \tag{8.82}$$

Control characteristic (8.82) of the parallel-loaded resonant converter is illustrated in Figure 8.35. Similarly to a series-loaded converter, a half-bridge inverter can be replaced with a full-bridge inverter and a transformer can be added to provide electrical isolation between the input and output.

Series–Parallel Converter   A clone of the parallel-loaded resonant converter, called a *series–parallel resonant converter*, is shown in Figure 8.36. Addition of the series resonant capacitor increases the range of control of the output voltage and improves the efficiency of the converter with light loads. If $C_{r1} = C_{r2} = 1/\omega L_r$, the approximate control characteristic of the series–parallel converter is given by

$$K_V = \frac{1}{(\pi^2/4)\sqrt{\left(2 - k_f^2\right)^2 + \left[(8/\pi^2)\left(k_f - 1/k_f\right)\right]^{-2}}} \tag{8.83}$$

and illustrated in Figure 8.37.

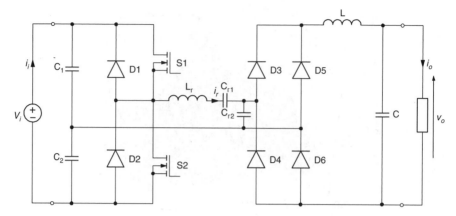

Figure 8.36 Series—parallel resonant converter.

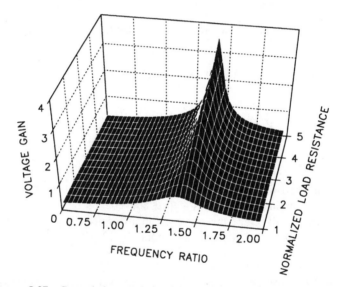

Figure 8.37 Control characteristic of the series—parallel resonant converter.

8.4.3 Comparison of Resonant DC-to-DC Converters

The analogy between the single-switch switched-mode dc-to-dc converters and quasi-resonant converters is obvious, as the latter are obtained by replacing regular switches by ZVS or ZCS resonant switches. There is also an analogy between multiple-switch switched-mode converters and load-resonant converters, as both classes are based on a single-phase inverter feeding, indirectly, a two-pulse rectifier. Consequently, the analogous converters tend to share similar applications. Generally, single-switch dc-to-dc converters are used at low power levels, while the multiple-switch converter are used primarily for medium-power loads, typically on the order of 1 kW and more.

Comparing the individual types of load-resonant converters, it must be stressed that Eqs. (8.76), (8.82), and (8.83), based on representation of square-wave quantities by their fundamentals, are not exact. Nevertheless, important functional differences between series- and parallel-loaded dc-to-dc converters can be discerned from the control characteristics in Figures 8.30 and 8.33. First, the voltage gain of a series-loaded converter is limited to the range zero to 0.5, while that of a parallel-loaded converter can exceed unity, depending on the load. Second, to allow control of the output voltage, the load resistance, R, should be less than the characteristic impedance, Z_0, in a series-loaded converter, and higher than the characteristic impedance in a parallel-loaded converter. Finally, the load resistance affects the voltage gain of the converters in different ways: although the maximum voltage gain of 0.5 in a series-loaded converter can be realized with any load, the control range of the output voltage increases when the load decreases. In contrast, in a parallel-loaded converter, the control range and maximum voltage gain increase with the load. In the continuous operating modes, the resonant converters perform "firm" switching, that is, the switching losses are reduced compared with the hard-switching converters, but not completely eliminated.

By exact analysis of operation, it can be shown that the series-loaded converter in the preferred discontinuous mode is impervious to short circuits in the load. However, the converter loses its control capability with light loads, including the no-load situation. On the other hand, a parallel-loaded converter in the transformer version is better suited for multiple outputs, but with light loads its efficiency is poor. A series−parallel converter offers certain improvements over the functionally similar parallel-loaded converter.

8.5 SUMMARY

Switching power supplies are employed as sources of stabilized or adjustable dc voltage in numerous applications, mainly those involving electronic circuits. The input dc voltage is obtained from batteries, photovoltaic or fuel cells, or rectifiers, and controlled in switching dc-to-dc converters. These can be divided into hard-switching switched-mode converters and soft-switching (or "firm-switching") resonant converters.

Switched-mode dc-to-dc converters are built in nonisolated and isolated versions. A transformer is used to provide electrical isolation and, if required, voltage matching between the input and output terminals. Also, multiple output stages can be supplied from multiple secondary windings of the transformer. The basic topologies of nonisolated converters such as the buck, boost, buck−boost, and Ĉuk converters employ a single semiconductor switch, whereas certain isolated converters, such as the push-pull, half-bridge, and full-bridge converters are based on two or more switches. Multiple-switch converters are composed of a single-phase inverter coupled with a single-phase rectifier through a transformer.

High power density is often a very important consideration for switching power supplies, such as portable communication equipment or spacecraft power systems. High switching frequencies allow reduction of the magnetic and electrostatic components, but they cause high switching losses. As miniature converters cannot dissipate much heat, the size reduction must be accompanied by reduction of losses. This is the main rationale for resonant converters, in which the phenomenon of electric resonance is harnessed to provide zero-current switching (ZCS) and zero-voltage switching (ZVS) conditions. The resonant converters can be classified as quasi-resonant (resonant-switch) converters and load-resonant converters. Resonant switches, constructed by augmentation of regular semiconductor switches with an LC tank and a diode, come in various configurations. Employing a resonant switch in a switched-mode topology results in a ZVS or ZCS resonant converter.

Similarly to multiple-switch switched-mode converters, load-resonant converters are based on the inverter-rectifier cascade, with a resonant LC circuit inserted between these two subconverters. An isolation transformer can be used at the input to the rectifier.

Because of the variety of types of switching power supplies, it is difficult to formulate general rules for selection of semiconductor devices. As usual, the rated voltage should exceed the highest voltage expected anywhere in the converter. Certain types of converters employ the chopper and inverter topologies covered in previous chapters, so that current ratings of the switches can be selected similarly. In resonant converters, the semiconductor devices are often forced to carry low-average but high-peak currents. Specialized software and PSpice simulations are recommended for the design of converters.

It must be stressed that to cover the rich field of switching power supplies in some depth, an entire book, not just a book chapter, is required. Indeed, for many practicing engineers, power electronics means just that field and not much else. Numerous topics, such as power-factor correction preregulators and Luo converters, have been omitted here for lack of space and because the general focus of the book is on medium- and high-power switching converters. For interested readers, several good sources are listed in the Literature section.

EXAMPLES

Example 8.1 A buck converter supplied from a 12-V dc source operates with a switching frequency of 10 kHz. The maximum load resistance is 6 Ω, and the output capacitor has a capacitance of 20 μF. Find:

(a) The minimum inductance of the output inductor required for continuous conduction

(b) The duty ratio of the switch required to produce an output voltage of 9 V

(c) The peak-to-peak amplitude of ripple of the output voltage

Solution:

(a) If condition (8.11) is satisfied for $d = 0$, it is satisfied for any other values of the duty ratio. Substituting the maximum load resistance, $R_{max} = 6$ Ω, for the

V_o/I_o ratio yields

$$L > \frac{R_{max}}{2 f_{sw}} = \frac{6}{2 \times 10 \times 10^3} = 3 \times 10^{-3} \text{H} = 3 \text{ mH}.$$

(b) From Eq. (8.5),

$$d \frac{V_o}{V_i} = \frac{9}{12} = 0.75.$$

(c) From Eq. (8.9),

$$\Delta V_o = \frac{1-d}{8LCf_{sw}^2} V_o = \frac{1 - 0.75}{8 \times 3 \times 10^{-3} \times 20 \times 10 \times 10^{-6} \times \left(10 \times 10^3\right)^2} \times 9$$

$$= 0.047 \text{ V}$$

$$= 47 \text{ mV}.$$

At about 0.5% of the output voltage, the ripple is practically negligible.

Example 8.2   A boost converter supplied from a 6-V dc source is to produce an output voltage of 15 V. Assuming an ideal, lossless inductor, find the required duty ratio of the switch. What is the output voltage if the resistance of the inductor is 2% of that of the load?

Solution: From relation (8.12) for an ideal converter,

$$d = 1 - \frac{V_i}{V_o} = 1 - \frac{6}{15} = 0.6.$$

If the inductor resistance is taken into account, Eq. (8.21) should be used instead, giving

$$V_o = \frac{6}{(1 - 0.6)\left[1 + 0.02/(1 - 0.6)^2\right]} = 13.3 \text{ V}.$$

Thus, the inductor resistance causes reduction of the output voltage by about 11% from the ideal value.

Example 8.3   A half-bridge converter is fed from a 120-V ac line via a two-pulse rectifier with a capacitive output filter such that the average input voltage to the converter is 165 V. The output voltage of the converter is 24 V, and the turns ratio, N_2/N_1, of the isolation transformer is 1 : 3. Values of the inductance and capacitance of the output filter are 1 mH and 25 μF, respectively, and the converter operates with a switching frequency of 20 kHz. Find the required duty ratio of the converter switches and the peak-to-peak amplitude of ripple of the output voltage.

Solution: From Eq. (8.61),

$$d = \frac{V_o}{k_N V_i} = \frac{24}{\frac{1}{3} \times 165} = 0.436$$

which is a feasible value, as the duty ratio must be less than 0.5. The amplitude of the voltage ripple can be found from Eq. (8.59) as

$$\Delta V_o = \frac{1 - 2 \times 0.436}{32 \times 1 \times 10^{-3} \times 25 \times 10^{-6} \times \left(20 \times 10^3\right)^2} \times 24 = 0.0096\,\text{V} = 9.6\,\text{mV}$$

amounting to only 0.04% of the magnitude of the output voltage.

Example 8.4   Draw a circuit diagram of a ZCS quasi-resonant buck converter with an L-type, full-wave switch.

Solution: To obtain the converter in question, the regular semiconductor switch in the switched-mode buck converter in Figure 8.1 is replaced by the current-mode resonant switch shown in Figure 8.26c. The resulting circuit is shown in Figure 8.38.

Example 8.5   A parallel-loaded resonant converter supplied from a 100-V dc source operates with a switching frequency of 5 kHz. The resonant circuit of the converter is composed of a 0.1-mH inductor and a 4-μF capacitor, and the load resistance is 10 Ω. Determine the operating mode of the converter and estimate the average output voltage.

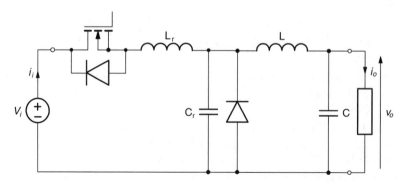

Figure 8.38   Quasi-resonant ZCS buck converter with an L-type full-wave switch.

Solution: The characteristic impedance and resonance frequency of the resonant circuit of the converter, calculated from Eqs. (8.63) and (8.64), are

$$Z_0 = \sqrt{\frac{0.1 \times 10^{-3}}{4 \times 10^{-6}}} = 5\,\Omega$$

and

$$f_0 = \frac{1}{2\pi \sqrt{0.1 \times 10^{-3} \times 4 \times 10^{-6}}} = 7958\,\text{Hz} \approx 8\,\text{kHz}$$

respectively. The switching frequency of 5 kHz is less than the resonance frequency but higher than half of the resonance frequency; thus, the converter operates in the continuous below-resonant mode. According to definitions (8.66) and (8.67), the frequency ratio, k_f, and normalized load resistance, r, are

$$k_f = \frac{5}{8} = 0.625$$

and

$$r = \frac{10}{5} = 2.$$

Now, the average output voltage can be estimated from Eq. (8.82) as

$$V_o = \frac{100}{(\pi^2/4)\sqrt{(1 - 0.625^2)^2 + [(8 \times 0.625)/(\pi^2 \times 2)]^2}} = 61.4\,\text{V}.$$

PROBLEMS

P8.1 A buck converter produces an average output voltage of 12 V. The on-time of the switch is 32 μs and the off-time is 8 μs. Find the input voltage and switching frequency of the converter.

P8.2 A buck converter supplied from a 50-V dc source operates with the switching frequency of 15 kHz, a duty ratio of 0.6, and a 10-Ω load. A 50-μF output capacitor is employed, and the inductor has the inductance twice as high as the minimum required for the continuous conduction mode. Find the average output voltage and the amplitude of its ripple.

P8.3 Repeat Problem 8.1 for a boost converter.

P8.4 Repeat Problem 8.1 for a boost−buck converter.

P8.5 Design a buck converter (select the inductance, capacitance, and switching frequency values) that will convert a 12-V input voltage into a 6-V output voltage with the ripple amplitude less than 1% of the output voltage (many solutions are possible).

P8.6 Repeat Problem 8.5 for a boost converter.

P8.7 Repeat Problem 8.5 for a buck–boost converter.

P8.8 A forward converter supplied from a 15-V dc source operates with the switching frequency of 25 kHz and duty ratio of 0.4. The turns ratio of the transformer is $1:1$, the output capacitance is 160 μF, and the load draws a current of 2 A. Select the minimum inductance for the continuous conduction mode and determine the average value and ripple amplitude of the output voltage.

P8.9 A flyback converter is supplied from a 60-V dc source and produces an output voltage of 100 V across a 25-Ω load. The transformer turns ratio is $1:2$, and the magnetizing inductance and output capacitance are 0.13 mH and 0.2 mF, respectively. What is the minimum switching frequency that will ensure the continuous conduction mode, and what is the corresponding amplitude of ripple of the output voltage?

P8.10 An isolated Ĉuk converter supplied from a 100-V dc source operates with a switching frequency of 30 kHz and a duty ratio of 0.35. The transformer turns ratio is $1:1$ and the load resistance is 20 Ω. Assuming inductances of the converter at twice the minimum values for continuous conduction, find the average value and ripple of the output voltage if the output capacitance is 25 μF.

P8.11 A push-pull converter supplied from a 50-V dc source operates with a switching frequency of 20 kHz. The transformer turns ratio is $2:1$, and the inductance and capacitance of the output filter are 0.1 mH and 20 μF, respectively. Determine the duty ratio of the converter switches required to obtain an average output voltage of 80 V. What is the ripple amplitude of that voltage?

P8.12 A half-bridge converter is to produce a 100-V average output voltage. The turns ratio of the isolation transformer is $2:1$. What is the minimum required input dc voltage?

P8.13 Repeat Problem 8.12 for a full-bridge converter.

P8.14 Draw a circuit diagram of a quasi-resonant ZCS buck–boost converter with an L-type half-wave resonant switch.

P8.15 Draw a circuit diagram of a quasi-resonant ZCS flyback converter with an M-type full-wave resonant switch.

P8.16 Draw a circuit diagram of a quasi-resonant ZVS Ĉuk converter with an L-type full-wave resonant switch.

P8.17 The resonant inductance and capacitance in certain resonant dc-to-dc converter are 0.1 mH and 0.25 μF, respectively, the switching frequency is 35 kHz, and the load resistance is 15 Ω. Determine the resonance frequency, characteristic impedance, frequency ratio, and normalized load resistance of the converter.

P8.18   A series-loaded resonant converter supplied from a 120-V dc source operates with a switching frequency of 24 kHz and a resistive load of 20 Ω. The resonant inductance and capacitance are 60 μH and 120 nF, respectively. Estimate the output voltage and current of the converter.

P8.19   A 50-V 500-VA parallel-loaded resonant converter is designed to operate with a switching frequency of 30 kHz. The resonant inductance and capacitance are 12 μH and 3.3 μF, respectively. Estimate the required input voltage (round it up to the nearest 10 V).

P8.20   A series–parallel resonant converter with two identical resonant capacitors of 1 μF each is to operate with a switching frequency of 18 kHz. Assuming a normalized load resistance of 4, select the resonant inductance that should result in the highest voltage gain of the converter (this problem may require use of a computer).

COMPUTER ASSIGNMENTS

*CA8.1   Run PSpice program *Buck_Conv.cir* for a buck converter. Observe the oscillograms of the voltages and currents, and find the peak-to-peak amplitude of ripple of the output voltage for two different values of the switching frequency.

*CA8.2   Run PSpice program *Boost_Conv.cir* for a boost converter. Observe the oscillograms of the voltages and currents, and determine the average output voltage for two different values of duty ratio of the switch.

*CA8.3   Run PSpice program *Buck-Boost_Conv.cir* for a buck–boost converter with two values of the switch duty ratio such that the average output voltage is less than and greater than the input voltage. In both cases, find the average output voltage and peak-to-peak amplitude of its ripple.

*CA8.4   Run PSpice program *Cuk_Conv.cir* for a Ĉuk converter. Determine the peak-to-peak amplitudes of ripple of the input current and output voltage. Observe other voltage and current waveforms.

*CA8.5   Run PSpice program *Buck_Conv.cir* for a buck converter and program *Buck_Aver_Model.cir* for the average model of the same converter using Vorperian's switch model. Compare the output voltage waveforms.

CA8.6   Use file *Boost_Conv.cir* to develop a PSpice circuit file for the average model of the boost converter employing Vorperian's switch model.

*CA8.7   Run PSpice program *Forward_Conv.cir* for a forward converter. Find the average output voltage and peak-to-peak amplitude of its ripple. Observe other voltage and current waveforms.

*CA8.8   Run PSpice program *Flyback_Conv.cir* for a flyback converter. Find the average value and the peak-to-peak amplitude of ripple of output voltage. Observe other voltage and current waveforms.

*CA8.9   Run PSpice programs *Push_Pull_Conv.cir*, *Half_Brdg_Conv.cir*, and *Full_Brdg_Conv.cir* for push−pull, half-bridge, and full-bridge converters. Note that all these converters are supplied with the same voltage, have the same output filter parameters, and operate with the same switching frequency and duty ratio of switches. Compare the output voltage waveforms by determining the average values and ripple amplitudes. Observe other voltage and current waveforms.

*CA8.10   Run PSpice program *Reson_Buck_Conv.cir* for a quasi-resonant ZVS buck converter. Find the average value and the peak-to-peak amplitude of ripple of the output voltage. Observe, simultaneously, oscillograms of the voltage, $V(A,C)$, across the switch and the gate signal, $V(G,C)$, that well illustrate the ZVS conditions.

*CA8.11   Run PSpice program *Reson_Boost_Conv.cir* for a quasi-resonant ZCS boost converter. Find the average value and the peak-to-peak amplitude of ripple of the output voltage. Observe, simultaneously, oscillograms of the current, $I(Vsense)$, in the switch and the gate signal, $V(G,C)$, that well illustrate the ZCS conditions.

*CA8.12   Run PSpice program *Series_Load_Conv.cir* for a series-loaded resonant converter for all three conduction modes (comment out the unneeded .PARAM statements). Observe oscillograms of the switch current, $I(Vsense)$, and gate signal, $V(G1,C)$ that well illustrate the ZCS conditions.

*CA8.13   Repeat Computer Assignment 8.13 for a parallel-loaded resonant converter in PSpice file *Parallel_Load_Conv.cir.*

*CA8.14   Based on the PSpice *Parallel_Load_Conv.cir* file, develop a similar file for the series−parallel resonant converter.

LITERATURE

[1] Kazimierczuk, M. K., and Czarkowski, D., *Resonant Power Converters*, Wiley, New York, 1995.

[2] Luo, F. L., and Ye, H., *Advanced DC/DC Converters*, CRC Press, Boca Raton, FL, 2004.

[3] Maniktala, S., *Switching Power Supplies A to Z*, Newnes, Oxford, UK, 2006.

[4] Pressman, A. I., Billings, K., and Taylor, M., *Switching Power Supply Design*, 3rd ed., McGraw-Hill, New York, 2009.

[5] Rashid, M. H., *Power Electronics Handbook*, 2nd ed., Academic Press, San Diego, CA, 2007, Chaps. 13 and 14.

9 Power Electronics and Clean Energy

An overview of "green" applications of power electronics is presented in this chapter. Use of power converters in renewable energy sources and distributed generation systems is outlined. Electric and hybrid cars are described. The role of power electronics in energy conservation by power conversion and control is discussed.

9.1 WHY IS POWER ELECTRONICS INDISPENSABLE IN CLEAN ENERGY SYSTEMS?

The rapidly growing interest in various aspects of clean energy is a reaction to the major energy-related challenges facing humanity. The grim reality of climate changes, environmental pollution, and limited resources of fossil fuels have spawned massive research and development efforts devoted to "clean," or "green," energy. These efforts involve renewable energy sources, distributed generation systems, and conservation of energy in a variety of electromechanical systems such as industrial drives and electric and hybrid cars.

Renewable energy sources can be classified as sustained and intermittent. Typical examples of sustained sources include geothermal and hydroelectric plants or biofuel distilleries. The energy is generated, or concentrated, in a fashion similar to that in traditional power plants or oil refineries, and the direct role of power electronics is limited. Intermittent sources include the increasingly common wind and photovoltaic systems, whose randomly varying output makes power electronic converters a vital part. To interface a renewable energy source efficiently with a load or power grid, the converters perform power conditioning, voltage boosting, and control of the flow of power.

As an aside, it is worth mentioning that the intermittency of sunlight-dependent sources of renewable energy is alleviated significantly in *concentrated solar power plants*, in which parabolic mirrors focus the sunlight to heat a liquid medium, such as fluoride salts, to high temperatures (above 700°C). This makes it possible to store thermal energy, which can be used around the clock to produce steam for electric turbogenerators. Such plants are somewhat similar to binary-cycle geothermal

Introduction to Modern Power Electronics, Second Edition, by Andrzej M. Trzynadlowski
Copyright © 2010 John Wiley & Sons, Inc.

plants, in which heat from underground hot water is transferred to fluid hydrocarbon (isobutane or isopentane), whose vapors drive the turbine of a generator. Power electronics does not play a significant role in those sustained sources of renewable energy.

Maximum power point (MPP) tracking constitutes an important part of power conversion and control. At a given illumination, the MPP on the voltage–current characteristics of a solar cell marks the maximum product of the voltage and current, that is, the maximum power drawn from the cell. Similarly, at a given wind speed, the power coefficient of a wind turbine reaches the maximum at a specific value of the tip speed ratio and, if applicable, the blade pitch angle. The control system of a photovoltaic or wind-power source tracks the MPP continuously, to draw the maximum power from the sun or wind.

Traditional power systems are based on large electric power plants located as close as possible to fossil-fuel sources such as coal mines or natural gas fields. The energy is then delivered to consumption centers over high-voltage transmission lines. Recently, the power infrastructure has begun to change. In addition to gas- and coal-fired plants, distributed generation systems employ a variety of renewable energy sources, reducing the environmental damage and the demand for fossil fuels. In areas without access to the power grid, such as remote locations of North America or numerous underdeveloped countries, power microgrids are formed from renewable and nonrenewable sources and energy storage devices. Diesel generators and electric batteries are commonly used to smooth temporal variations in the power supply. For a large number of sources of varying type, ratings, and intermittency to operate efficiently and harmoniously, the electric power generated by each must undergo various types of conversion and control.

Since the beginnings of modern power electronics, energy conservation has been an important factor supporting growth of that discipline. Most of the electric power consumed in industry is spent in drives of such fluid-handling machinery as pumps, fans, blowers, and compressors. If, for example, a fixed-speed motor powers a pump in a hydraulic system, the flow intensity must be controlled by valves. If weak flow is required, the valve is almost closed and the pump basically mixes the stagnant fluid, wasting most of the energy drawn by the motor. Controlling the flow by adjusting the pump speed in a variable-speed drive is a much more efficient solution. Huge energy savings in industrial plants provided the principal motivation for the rapid growth of power electronics and electric drives in the 1970s and 1980s. The trend toward replacing fixed-speed drives with variable-speed drives continues unabated. Recently, the variable-speed drives have entered the automobile world, appearing in electric and hybrid cars. Apart from some niche applications, the variable-speed drives are of the ac type because ac motors are distinctly less expensive and more robust then their dc counterparts. Therefore, as noted in Chapter 7, inverters constitute one of the most popular types of power electronic converters.

As is clear from this overview, power electronics constitute a vital component of most clean energy systems. More details of applications of power converters in those systems are described in subsequent sections.

9.2   SOLAR AND WIND RENEWABLE ENERGY SYSTEMS

The role of power electronic converters in solar and wind renewable energy systems depends on the type and ratings of the source as well as the type of load. Most often, the load is the electric power grid, which collects and distributes electrical energy supplied from a number of sources. Sometimes, though, a renewable energy source directly feeds a specific load, such as a battery pack supplying an off-grid household. In each case, the source voltage varies randomly, while the output of the system must provide the fixed voltage and frequency required by the load. Maximum power tracking must constitute a part of the control strategy.

In a distributed generation system, where a variety of "small" sources (e.g., photovoltaic, wind, small hydro, cogeneration) provide electrical power along large traditional generators, care must be taken to detect and prevent *islanding*. If a fault occurs on the grid, all the small sources must disconnect from the grid, leaving only large centrally controlled power plants. Otherwise, the small sources would back-feed into the fault and worsen the situation. Also, people's safety, especially that of the maintenance crews, could be threatened. It is also necessary that the small sources cease to operate in an unintentional "island" separated from the grid. However, to supply certain loads when the grid has failed, intentional islands are allowed after taking steps to prevent the grid from back-energizing. Islanding can be detected by sensing sudden changes in system frequency, voltage, or real and reactive power output.

9.2.1   Solar Energy Systems

The quantum-mechanical photovoltaic effect is utilized in solar cells, most commonly based on silicon semiconductor material, to convert sunlight to electricity. The open-circuit voltage, V_{oc}, of a solar cell is typically 0.6 to 0.7 V and the short-circuit current, I_{sc}, is 20 to 40 mA/cm^2. To increase the output voltage, a number of cells, usually 36 or 72, are connected in series to form a solar, or photovoltaic (PV), module. A collection of modules forms a PV panel, which constitutes a mechanical and electrical entity. A PV array is a collection of several panels. For maximum clarity, in the remainder of the chapter the term *array* is used rather loosely, to mean any set of PV modules connected in series and parallel to form a source of dc voltage.

Voltage and power as functions of the current drawn from a solar array are illustrated in Figure 9.1. It can be seen that the power curve has a sharp peak, so that even a small deviation from the MPP reduces the power yield significantly. Note that a specific voltage−current characteristic depends on the irradiation and temperature of the cells. Therefore, MPP tracking usually involves some type of perturb-and-observe technique. Other approaches have also been proposed. Once the MPP is located, the power electronic interface must maintain the current close to the I_{MMP} level.

PV arrays come in various sizes, the small ones often mounted on buildings, including houses. Typically, the owner of the array sells the energy to the local utility by interfacing the array to the grid, usually through a single-phase power line. Large PV arrays can form PV power plants, which are connected to the grid via three-phase power lines. Without a loss of generality a three-phase connection to the grid is shown

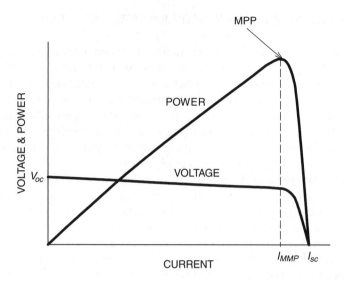

Figure 9.1   Voltage and power as functions of the current drawn from a solar array.

in subsequent diagrams of the interfaces, as the grid-side ac-output converter can be of the three- or single-phase variety.

The simplest solution for a PV array interface with the grid is shown in Figure 9.2. The array is connected to the grid through a dc link, voltage-source inverter, and a filter that smoothes the current ripple. Clearly, the array's voltage must be sufficiently high to match the grid voltage. Otherwise, a transformer is needed, as in the system shown in Figure 7.65. A high-frequency transformer can also be embedded in an isolated dc-to-dc converter (see Section 8.3) following the array. If galvanic isolation is not required, the voltage of the PV array can be increased in a transformer-less dc-to-dc boost converter, as illustrated in Figure 9.3. Another solution, shown in Figure 9.4, involves a current-source inverter. Thanks to the inductive dc link, the inverter provides a boost to the output voltage. The filter includes capacitors,

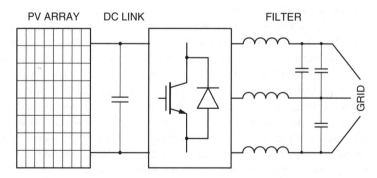

Figure 9.2   PV array-to-grid interface with a voltage-source inverter.

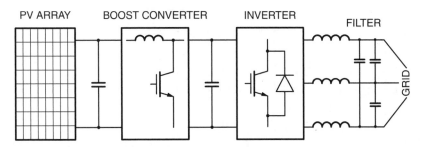

Figure 9.3 PV array-to-grid interface with a dc-to-dc boost converter and voltage-source inverter.

to avoid the dangerous connection that occurs when two inductors carry different currents.

Connecting a PV array to a single converter creates an inflexible structure, making expansion (or reduction) of an array difficult without retrofitting the entire system. Therefore, the recent tendency in PV systems is to use strings of series-connected PV modules, with each string feeding its own inverter. The number of modules in the string is large enough to avoid the necessity of voltage boosting. More radical solutions are:

- A multistring interface, in which each string of modules is equipped with its own dc-to-dc converter interfaced to a single inverter
- An ac-module system, in which a power electronic interface is integrated into each module
- A single-cell system with an inverter interface employing a large PV cell (e.g., of the photo electrochemical type), which can be made arbitrarily large.

Multilevel inverters are well suited as PV array-to-grid interfaces. Various levels of dc input voltage can easily be set up using an appropriate structure of PV modules. The more levels that are need, the higher the quality of output voltage obtained in the

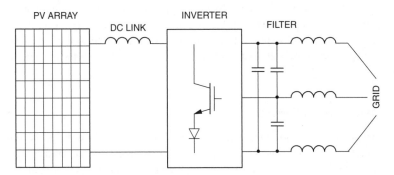

Figure 9.4 PV array-to-grid interface with a boost current-source inverter.

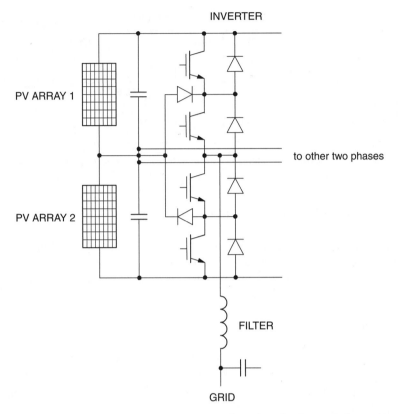

Figure 9.5 PV array-to-grid interface with a three-phase three-level neutral-clamped inverter (one leg shown only).

low-loss square-wave mode or a PWM mode with low switching frequency. One leg of a three-phase three-level neutral-clamped inverter supplied from two PV arrays is shown in Figure 9.5, while Figure 9.6 depicts a single-phase l-level cascaded H-bridge inverter with a number of arrays equal $(l-1)/2$ (see Section 7.3).

Various types of inverters, especially dc-to-dc converters, some of them developed specifically for this application, can be used in the interfaces. Therefore, the review presented is far from complete. Interested readers are referred to the many existing specialized publications—alternative energy is one of the hottest topics in engineering today.

9.2.2 Wind Energy Systems

Wind energy systems enjoy rapid growth and, as of now, produce much more power than do photovoltaic systems. Wind as an energy source is less intermittent than sunlight, and a wind turbine occupies less space than a solar array of comparable power. Large three-blade horizontal-axis turbines, with a rated power of several megawatts

INVERTER

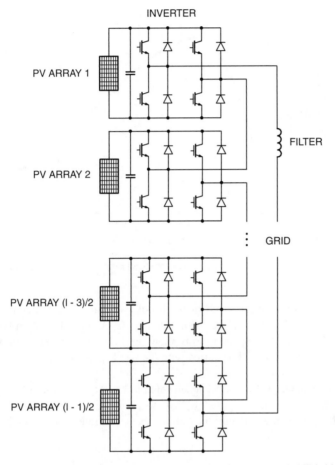

Figure 9.6 PV array-to-grid interface with a single-phase *l*-level cascaded H-bridge inverter.

each, are usually grouped into commercial *wind farms*. Low-power turbines of various designs are slowly making their way into residential market of renewable energy, but they are still relatively rare in practice.

The aerodynamic power, P_a, of a wind turbine, (i.e., the amount of power extracted from wind) is given by

$$P_a = \frac{1}{2}\rho C_p A v_w^3 \tag{9.1}$$

where ρ denotes air density, C_p is the power coefficient, A is the area swept by the blades, and v_w is the wind speed. The power coefficient depends strongly on the tip speed ratio, λ, defined as

$$\lambda = \frac{v_t}{v_w} \tag{9.2}$$

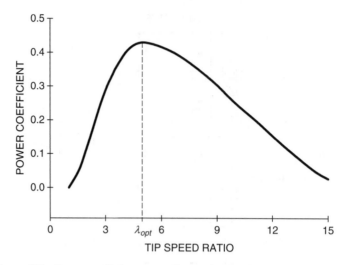

Figure 9.7   Power coefficient versus tip speed ratio of a typical wind turbine.

where v_t denotes the linear speed of the tip of the blade. In pitch-controlled wind turbines, C_p also depends on the blade pitch angle. A typical relation between the power coefficient and tip speed ratio is shown in Figure 9.7. Clearly, to best utilize the turbine, its speed should be varied with the wind speed to maintain the optimum value λ_{opt}, of that ratio.

The output power versus wind speed relation for a typical variable-speed wind-turbine system is shown in Figure 9.8. With the increase in wind speed, the optimum tip speed ratio is maintained until the turbine reaches its rated speed and the generator

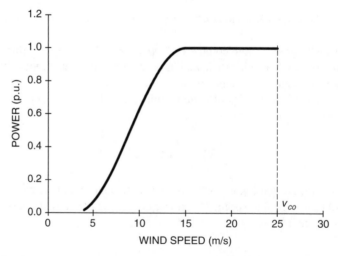

Figure 9.8   Output power of a typical wind-turbine system as a function of the wind speed.

driven by the turbine produces the rated amount of electrical power. Further increase in turbine speed must be prevented to protect the generator from overloading. At this point, active control of the pitch of blades is employed to reduce the energy capture capability of the turbine. In turbines without pitch control, aerodynamic stall produces a similar result. The blades are so shaped that the smooth flow of air around the blades becomes turbulent when the wind speed exceeds a specified threshold and the forces driving the blades decrease. If the wind speed exceeds the cutoff value, v_{co}, the turbine is stopped, to avoid structural damage.

Fixed-Speed Systems   If an ac squirrel-cage induction machine is employed as a generator, the speed of the turbine is practically constant, varying only within the limits of slip of the machine: that is, a few percentage points. The generator is driven by the turbine via a gearbox and, typically, is connected to the grid through a transformer. The frequency of the grid voltage dictates the speeds of the generator and turbine. To improve the power factor at the point of connection with the grid, a capacitor bank is employed as a reactive power compensator. The pitch control or aerodynamic stall limits the power output at high speeds. To limit the current at system startup, a soft-starter is employed. Figure 9.9 is a block diagram of the system described. The speed range can be somewhat increased by replacing the squirrel-cage generator with a wound-rotor generator and adding controlled resistances to rotor windings.

Variable-Speed Systems   Conditioning the power produced by a generator allows significant improvements in the operation of wind energy systems in terms of MPP tracking and control of the real and reactive powers. MPP tracking requires a variable speed of the turbine, which means that the frequency of the generator voltage is floating. Therefore, a frequency changer is needed between the generator and the transformer. As shown in Figure 9.10, the frequency changer consists of a rectifier, a

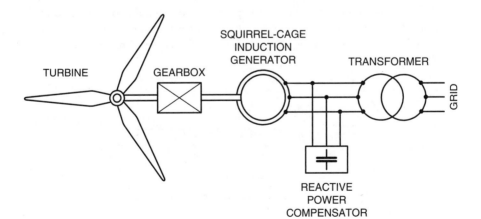

Figure 9.9   Wind-turbine system with induction generator (soft-starter not shown).

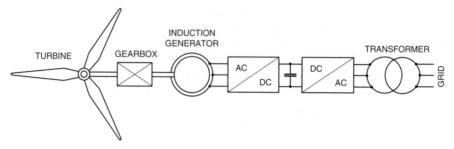

Figure 9.10 Wind-turbine system with induction generator and frequency changer.

dc link, and an inverter (also see Figure 7.70b). Control of real power is performed in the rectifier and that of reactive power in the inverter. A matrix converter can be used in place of the rectifier—inverter cascade.

In high-power systems, an electrically excited ac synchronous generator can be employed. Both the real and reactive powers are controlled in the inverter. As shown in Figure 9.11, dc voltage must be provided to field winding to produce a magnetic field in the generator. The rated power of the rectifier employed for that purpose is much lower than that of the frequency changer.

Using a low-speed generator, that is, one with a large number of magnetic poles makes it possible to dispose of a gearbox and improve on the cost and reliability of the system. In low- and medium-power systems, the permanent-magnet synchronous generators employed do not need an electrical excitation circuit.

In the scheme depicted in Figure 9.12, use of a doubly-fed wound-rotor induction generator allows a reduction in the rated power of the power electronic converters employed. Recall that the ac-to-dc-to-ac converter cascade can pass power in both directions. Indeed, if the generator runs with a super-synchronous speed, that is, the speed of the rotor is higher than that of the stator field, the electrical power is delivered, via the transformer, to the grid from both the stator and rotor. However, a subsynchronous speed makes the electric power to flow into the rotor. As a result, a wide range of turbine speed is achieved, while the converter can be rated at a fraction (25 to 30%) of the rated power of the generator. The real and reactive powers are

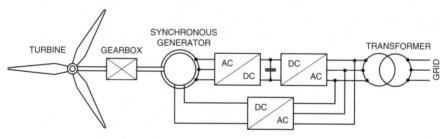

Figure 9.11 Wind-turbine system with electrically excited synchronous generator and frequency changer.

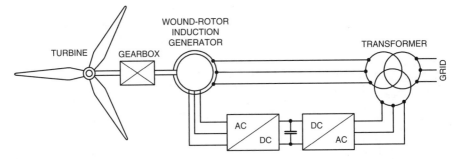

Figure 9.12 Wind-turbine system with doubly-fed induction generator.

controlled in the converter connected to the rotor. This scheme is now used most often in high-power wind-turbine systems.

Wind-turbine systems produce an increasing share of the total electric power in many countries, with Denmark leading the way. It is expected that in the near future as much as half of the total power in that country will be generated by wind farms, most of them offshore. Wind farms, which are self-contained power plants, must satisfy strict tolerances as to frequency and voltage levels, which requires precise real and reactive power control. Also, quick response to transients is necessary to ensure the stability of the grid. Different configurations of wind farms are used, such as a local ac network or a local dc network. Various arrangements of control and delivery systems, such as high-voltage dc transmission lines for distant offshore farms, are employed, depending on the specifics of the farm.

9.3   FUEL CELL ENERGY SYSTEMS

A fuel cell (FC) converts the chemical energy of a fuel directly to electrical energy, and it is considered a much cleaner source of power than the commonly used electric generator driven by an internal combustion engine, usually of the diesel type. The principle of energy conversion is similar to that of the electric battery, with the fuel oxidized at the anode and an oxidant reduced at the cathode. The ions produced are exchanged through an electrolyte, and electrons through the load circuit.

At the rated current, a single cell generates about 0.6 to 0.7 V. Therefore, a practical FC constitutes a stack of cells. Various types of FCs have been developed, the important distinction being the type of fuel used: liquid or gaseous. Liquid fuels include various hydrocarbons (e.g., methanol) or alcohols, with chlorine and chlorine dioxide as oxidants. If gaseous hydrogen is fed to an FC, oxygen serves as the oxidant.

Increasingly, FCs find applications in transportation and as portable and distributed power, the type of FC depending on the application requirements, such as specific weight, power density, operating temperature, and startup speed. For obvious reasons, the output power must be conditioned. The conditioning requirements include the power and current ranges allowed, as well as the change rate, polarity (negative current is forbidden), and the current ripple.

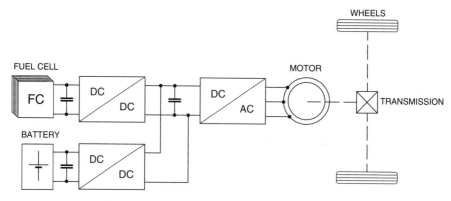

Figure 9.13 FC-powered drive system of a vehicle.

A power system for an FC-powered vehicle is shown in Figure 9.13. Front-wheel drive is assumed. The motor driving wheels of the vehicle are of the ac type (permanent-magnet synchronous motor or induction motor), and the normally charged battery allows bursts of power at acceleration. The converters allow for regenerative braking of the vehicle, with the power recovered charging the battery. Consequently, the dc-to-dc converter connected to the battery must be of the two-quadrant type. A supercapacitor pack can be employed in place of the battery. In practice, two or more wheels of a vehicle (automobile, bus, locomotive) can be driven separately using the system described for each of them. FC drives are also employed in aircraft, mostly of the unmanned type.

FCs are very well suited for use in distributed generation systems because of their nonpolluting operation. In one version of the clean energy policy of the future, hydrogen is employed as a medium for energy storage and transport. It can be produced from water using electrolysis fed by excess energy from PV or wind sources. When power from renewable energy sources decreases (e.g., after sunset), the same hydrogen in an FC is oxidized back into water and the energy produced is transferred to the grid. If FC voltage is not much lower than that of the grid, the system shown in Figure 9.14 and consisting of a boost dc-to-dc converter, dc link, and

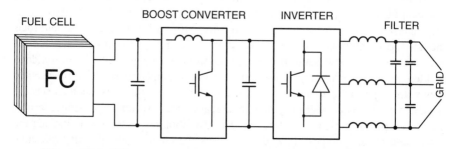

Figure 9.14 FC power system with boost converter for distributed generation.

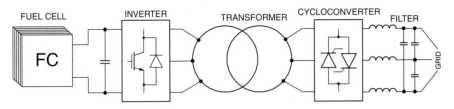

Figure 9.15   FC power system with transformer isolation for distributed generation.

voltage-source inverter can be used. The boost converter provides stable and suitably increased dc voltage to the inverter, which converts it to ac voltage of the magnitude and frequency required by the grid. Notice the similarity of the systems shown in Figures 9.3 and 9.14, which is obvious, as both the PV arrays and FCs are sources of low dc voltage.

If the grid voltage is much higher than the FC voltage, and/or transformer isolation is desired, the system shown in Figure 9.15 can be used. The inverter produces high-frequency voltage to minimize the transformer size. The secondary voltage of the transformer is then converted to a grid-frequency voltage using a cycloconverter. If the FC is to supply power to a dc network, the system in Figure 9.14 can be modified by elimination of the load-side inverter (and dc link). Similarly, the cycloconverter in the system in Figure 9.15 can be replaced with a rectifier.

9.4   ELECTRIC AND HYBRID CARS

Because of the limited amount of energy that today's batteries are able to store, purely electric cars have not yet been able to gain a solid foothold on the vast automobile market. In 1996, General Motors (GM) Corporation introduced EV1, which was to be the first modern electric car (the EV1 program was limited to leasing). The two-seater EV1, heavily loaded with lead−acid or nickel−metal hybrid batteries, was capable of rapid acceleration from standstill thanks to the high starting torque of its electric motor. The car was quiet and nonpolluting. However, the short driving range, inconvenient and lengthy recharging of the battery, high production costs, and the low public demand caused GM to cease production in 2003. Subsequently, to avoid potential liability problems, all EV1s were removed from streets, dismantled, and crushed (some nondrivable pieces were donated to universities and museums).

It must be stressed that apart from the battery issue, an electric vehicle (EV) represents an attractive proposition, due mostly to the fact that the average efficiency of electric machines is three times higher than that of internal combustion engines. A simplified block diagram of an advanced version of an EV powertrain is shown in Figure 9.16. Each of the four wheels of the vehicle is driven separately by a low-speed high-torque ac motor fed from an inverter. The inverters are supplied from a battery through a capacitive dc link. The battery is protected from overcharging, e.g., when the vehicle coasts down a long slope by a switched braking resistance (see Figure 6.25).

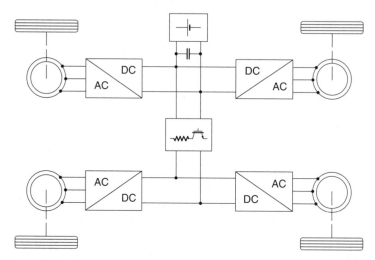

Figure 9.16 Simplified block diagram of the powertrain of an electric vehicle.

The decentralized structure of the power scheme allows independent control of the torque and speed of each wheel, eliminating the need for mechanical transmission, differential, and gears. All sophisticated drive functions, such as stability control, nonskid regenerative braking, and optimal distribution of the torque between wheels on slippery roads, can be implemented in a microcomputer-based digital control system.

It is hoped that progress in technologies of batteries (e.g., the lithium-ion batteries) and fuel cells will in the foreseeable future allow production of marketable electric automobiles. Indeed, Tesla Motors in California recently started selling the fully electric Tesla Roadster sports car. Its performance is impressive, but the price of about $130,000 makes it a niche, not a mainstream, automobile. A prototype of an inexpensive electric car capable of a 100-km (60-mile) driving range was recently unveiled in Japan.

The current opinion of most automobile experts is that hybrid cars make much better economic sense at present and that electric cars will have to wait until radically more advanced and inexpensive batteries or fuel cells are developed. Several models of hybrid cars have already reached the stage of mass production. Even the acclaimed Chevy Volt, touted as an electric automobile, is in fact a hybrid car, although the electric charge of the battery is said to be sufficient for commutes of up to 40 miles. According to the International Electrotechnical Commission, a hybrid electric vehicle (HEV) is "one in which propulsion energy . . . is available from two or more kinds or types of energy stores, sources, or converters. . . . " In practice, today's hybrid cars combine a battery, an electric motor, and an internal combustion engine to provide driving power. The battery is often supplemented with a supercapacitor bank, which allows short bursts of extra energy for acceleration.

In a hybrid electric drive system, advantage is taken of the difference between the mechanical characteristics of electric motors and internal combustion engines. For

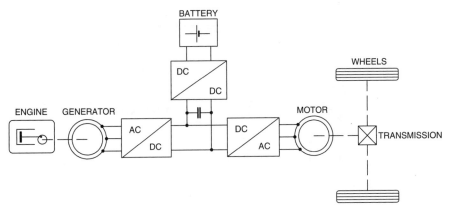

Figure 9.17   Series electric hybrid powertrain.

example, an electric motor can generate full torque at any speed, even at standstill, while the engines require a considerable speed to achieve maximum torque ability. The efficiency of an electric machine is high under all normal operating conditions, while internal combustion engines operate most efficiently at midrange speeds and high torque levels. The driving process involves a number of modes, such as acceleration, constant-speed, coasting, braking, downhill driving, or uphill driving. Varying conditions include the amount of load, state of the road surface, wind speed and direction, or ambient temperature. A micropocessor-based control system, equipped with sophisticated sensors and intelligent operating algorithms, can take advantage of the flexibility of the drive to ensure best dynamic performance while minimizing fuel consumption.

Three basic architectures of the HEV powertrain are *series*, *parallel*, and *series–parallel*. A series powertrain, illustrated in Figure 9.17, uses an electric motor to drive the wheels (as in Figure 9.13, a front-wheel drive is assumed). A battery and an electric generator driven by an internal combustion engine are two independent sources of electrical energy. Depending on operating conditions, the motor is fed from the battery (with the engine stopped), from the generator, or from the both sources. Also, if needed, the battery can be charged from the generator, or vice versa, the battery can supply power to the generator when it operates as a starter motor for the engine. This flexibility helps to save significant amounts of fuel, especially during city driving. There, frequent starts and stops allow the best use of the torque–speed characteristics of the motor, and the recovery of the vehicles kinetic energy during regenerative braking. Also, the engine can operate most of the time under most fuel-efficient conditions.

Compared with internal combustion engines, electric machines have lower power density, that is, they are heavier than their internal combustion counterparts of comparable power. Therefore, series hybrid powertrains employing two electric machines are now used primarily in heavy commercial and military vehicles, buses, and small locomotives. However, as explained later, the series powertrain is employed in the Chevy Volt, a four-passenger sedan that will appear on the market as a 2011 model.

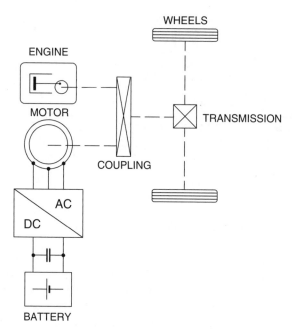

Figure 9.18   Parallel electric hybrid powertrain.

The parallel hybrid drive system shown in Figure 9.18 fits better in existing cars, as it does not require a generator. The engine and electric motor combine their torques through a mechanical coupling (gearbox, pulley or chain unit, or a common axle). Torque is then supplied to the wheels via a mechanical transmission. Most hybrid passenger cars, such as Honda Insight and Ford Escape, use this powertrain configuration.

The series–parallel hybrid powertrain depicted in Figure 9.19 and employed in the popular Toyota Prius has higher operational flexibility than that of the series and parallel drive systems described above. Its planetary *power-split device* divides the engine power between the wheels and generator, whose main function is to charge the battery. The same generator serves as a starter motor. The driving motor, fed from the dc bus, provides additional torque to the wheels. With the battery fully charged and the generator disconnected mechanically from the engine, this is the regular operating mode. The generator is activated if the battery needs adding some charge. Other modes are also feasible, with the microprocessor-based control system selecting the one that is most appropriate for given driving conditions and requirements. For example, while braking, the motor acts as a generator, opposing the motion of wheels and charging the battery. It is even possible to have the generator, driven by the engine, to feed the motor, which then provides the power needed, the wheels with all the no direct help from the engine.

Another classification of HEV systems for passenger cars distinguishes between *micro*, *mild*, and *full hybrid*, depending on what percentage of total power is delivered

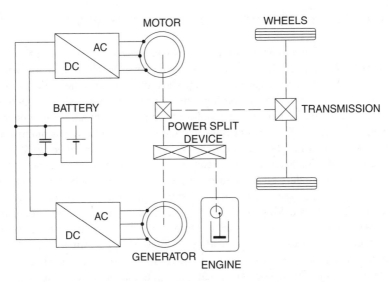

Figure 9.19 Series–parallel electric hybrid powertrain.

by the electrical portion of the drive. Levels of electric power in micro hybrids are low, typically not more than 5 kW. Fuel consumption is 5 to 10% lower than that in regular cars of comparable power. Mild hybrids have 7 to 12 kW of electric power, and their fuel economy gains are on the order of 10 to 15%. Full hybrids, utilizing series–parallel architecture with a power split device and two electric machines, can save as much as 40% of fuel without compromising the driving performance. The machines are rated at several tens of kilowatts. For example, the 2009 Toyota Prius has a 60-kW permanent-magnet synchronous motor.

In the Chevy Volt, which is expected to help revive the declining U.S. automobile industry, a 1.4-liter engine is used to recharge the 16-kWh battery when it runs low. The Volt can best be utilized as an electric commuter car, recharged overnight from the grid, and using gasoline only when exceeding a 40-mile range. Its permanent-magnet synchronous motor produces up to 111 kW (149 hp) of power and 371 Nm (273 lb-ft) of torque, which allows for a driving performance comparable to that of regular cars of similar-size. Despite its series–hybrid powertrain, the Volt is often called an *extended-range electric car*, without the adjective *hybrid*. Nevertheless, complete elimination of the engine as an electricity-generating unit seems unlikely in the short term. Extended-range electric cars are at the forefront of advanced automobile technologies of today, including an interesting vehicle such as the Karma by Fisker Automotive, an elegant, powerful (402 hp), and sophisticated passenger car for less than $90,000.

Hybrid electric cars seem to have entered the automobile market for good, the more so that gasoline prices are not forecast to drop significantly in the foreseeable future. Of about 60 million cars produced in the global economy, HEVs are expected to represent a 10 to 15% share. Thus, power electronics have found a large application area there.

9.5 POWER ELECTRONICS AND ENERGY CONSERVATION

Compared with other possible methods of power conversion and control, the use of power electronic converters invariably saves energy. The electromachine rotating power converters of the past were significantly less efficient than the static converters of today, and many processes involving control of power flow, whether hydraulic, pneumatic, mechanical, or electrical, employed some sort of "choking." For example, as explained in Section 9.1, flow of a fluid was traditionally controlled by manipulating a valve, such as that in a faucet, and flow of a current was controlled by a rheostat (see Section 1.4). Such an approach can be compared to driving a car with the transmission and accelerator stuck in fixed positions, with only brakes available to slow or accelerate the vehicle. In such controlled systems, the efficiency decreases with the load, as the "choking" is especially severe at low, or zero, flow intensities.

The use of power electronic converters allows wide-range torque and speed control in adjustable-speed drive systems, leading to huge energy savings in a variety of commercial enterprises, such as the manufacturing and food-processing industries, or electric transport. The majority of electric drives use ac motors, primarily squirrel-cage induction and permanent-magnet synchronous motors, supplied from inverters. However, rectifier-fed dc drives, which offer more advanced performance than ac drives, still find applications in high-precision positioning systems. Many countries impose high efficiency standards on appliances, forcing manufacturers to install adjustable-speed drives in "white goods," such as refrigerators, washers, and dryers.

Power electronic converters in the national grid improve the power factor and facilitate control of the real and reactive power, raising the efficiency of transmission and distribution of electric energy. FACTS (flexible ac transmission systems) devices, such as STATCOMs (static synchronous compensators), SVRs (static VAr compensators), TCPARs (thyristor-controlled phase angle regulators), TCRs (thyristor-controlled reactors), or TSCs (thyristor-switched capacitors), are increasingly common in the grid. Converters find numerous applications in homes and other buildings as part of HVAC (heat, ventilation, and air conditioning) and lighting systems. Switching power supplies in computer equipment, radios and TV sets, cell phones and portable media players, are so designed as to provide high-quality power. The high input power factor, obtained by employing preregulators and sophisticated control schemes, minimizes the current drawn from the supply source and flow and diminishes the associated ohmic losses.

The increasing importance of power electronics in modern, clean, energy-conscious societies cannot be understated, and most engineers encounter this fact in their everyday practice. Developing countries and poor societies also take advantage of technological progress in the fields of power electronics and renewable energy sources, although many communities in the world lack access to an electric grid. To make things worse, the fuel used in diesel generators, which are in common use, tends to be more expensive in the developing world than in the more affluent nations. Therefore, solar panels and wind turbines are springing up in many places, primarily to provide light and power for basic appliances. The hope for a better future brought about by the simple means of electricity production is arguably as important as the tangible results of the availability of cheap energy.

9.6 SUMMARY

Power electronics and clean energy are closely tied. Alternative energy sources such as solar arrays, wind turbines and fuel cells require interfacing with the electric grid or another type of load. Power electronic converters provide the necessary power conditioning to match the source with the load. Depending on the types of source and load, various configurations of power electronic interfaces have been developed for solar, wind, and fuel-cell systems. Bidirectional ac-to-dc and dc-to-dc converters, with possible transformer isolation, are the most common components of these interfaces.

Purely electric cars are still impractical because of their limited energy storage capability and the inconvenience in charging today's batteries. However, hybrid electric automobiles enjoy a healthy public acceptance and robust growth. In a hybrid vehicle, an internal combustion engine is supplemented with an electric supply of driving power to ensure optimal use of fuel under varying driving conditions. Batteries, often complemented by supercapacitors, constitute the source of electric energy, and a generator driven by the engine provides charging power, especially during regenerative braking. In advanced hybrids, the power of the engine and the power of the motor are combined in a sophisticated power-split device which, in addition to two electric machines, allows highly flexible operation of the powertrain. Extended-range electric cars represent the most recent trend. They are basically series-hybrid vehicles, but the internal combustion engine is relatively small and its major role is to supply power to the battery when it runs low.

Power electronic converters appear in modern electric grids, electric drives, buildings, appliances, computer and communication equipment, and other applications where electrical power is transmitted or converted. Enormous amounts of energy are saved thanks to the efficiency-optimized operation of those systems. Both highly developed countries and poor societies enjoy the benefits of clean electric energy assisted by power electronics.

LITERATURE

[1] Blaabjerg, F., Chen, Z., and Kjaer, S. B., Power electronics as efficient interface in dispersed power generation systems, *IEEE Transactions on Power Electronics*, vol. 19, no. 5, pp. 1184–1194, 2004.

[2] Blaabjerg, F., Consoli, A., Ferreira, J. A., and van Wyk, J. D., The future of electronic power processing and conversion, *IEEE Transactions on Power Electronics*, vol. 20, no. 3, pp. 715–720, 2005.

[3] Ehsani, M., Gao, Y., and Miller, J. M., Hybrid electric vehicles: architecture and motor drives," *Proceedings of the IEEE*, vol. 95, no. 4, pp. 719–728, 2007.

[4] Quiroga, T., " 2011 Chevrolet Volt," *Car and Driver*, Aug. 2009, pp. 49–52.

[5] Rashid, M. H., *Power Electronics Handbook*, 2nd ed., Academic Press, San Diego, CA, 2007, Chaps. 25 to 29 and 31 to 33.

[6] Strzelecki, R., and Benysek, G. (Eds.), *Power Electronics in Smart Electrical Energy Networks*, Springer, London, 2008.

APPENDIX A
PSpice Simulations

For a better understanding of the operating principles of power electronic converters, 48 PSpice circuit text files have been developed to be used with this book. They are available on the Internet at ftp://ftp.wiley.com/public/sci_tech_med/power_electronics. For all files, the name identifies clearly the topic, e.g., the *Boost_Conv.cir* file models the dc-to-dc boost converter. An asterisk after the sequential number of a computer assignment indicates the existence of an accompanying circuit file. For example, assignment CA8.2*, which involves simulation of the boost converter, requires the use of the already mentioned *Boost_Conv.cir* file.

Simple switches based on the generic voltage-controlled switches available in PSpice are employed in all converter models. Circuit diagrams of four such switches appearing in the circuit files, specifically (a) single-gate switch, SGSW, (b) two-gate switch, TGSW, (c) unidirectional switch, UDSW, and (d) generic SCR, are shown in Figure A.1. In some circuit files, additional components, such as a series RC snubber circuit, are connected to the switches to improve operating conditions and secure the convergence of computations. The hysteresis current control in the three-phase voltage-source inverter in the *Hyster_Curr_Contr.cir* file is realized by subcircuit HCC as shown in Figure A.2. The inductor-coupling PSpice device, K, is used to model transformers in isolated dc-to-dc converters.

The PSpice was originally developed for simulation of analog electronic circuits, and switched networks are sometimes handled poorly. Therefore, certain files contain the message "RUN AS IS!" to indicate that any parameter change will endanger the convergence. The reader can also notice that the RON and ROFF resistances of the voltage-controlled switches, Sgen, vary from file to file. The same applies to the .OPTIONS statement. Those variations have resulted from trials to get the simulations to run smoothly. Generally, the convergence can be improved by changing (usually relaxing) the tolerances in the .OPTIONS statement and reducing the ROFF/RON ratio in the Sgen model. Regrettably, some less important waveforms are still marred by "spikes" resulting mainly from the idealized nature of the assumed switches. To view such waveforms, the amplitude (Y axis) scale should be reduced manually.

No effort was spared to make the simulations user friendly. In particular, text strings, not numbers, are used to designate nodes, e.g., "OUT" for an output terminal.

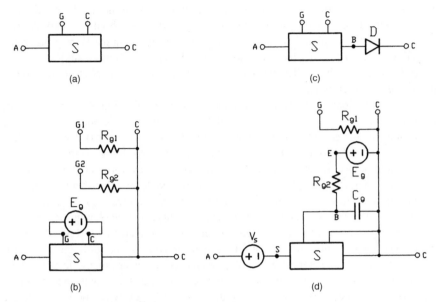

Figure A.1 Generic switches used in the PSpice files: (a) single-gate switch, SGSW, (b) two-gate switch, TGSW, (c) unidirectional switch, UDSW, (d) generic SCR.

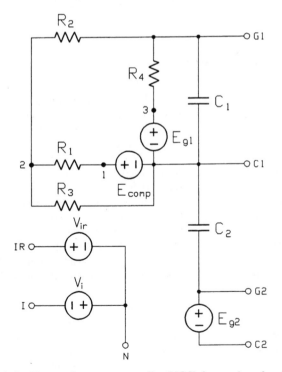

Figure A.2 Hysteresis current controller (HCC) for one leg of an inverter.

This facilitates selection of waveforms to be displayed using the mock-oscilloscope Probe program. Also, each circuit file contains information about the textbook figure depicting the simulated circuit. Each assignment involving the PSpice programs should begin with copying the corresponding figure and marking the nodes in accordance with the circuit file. The work involved will pay off by faster memorization of power electronic circuits.

For measurement of waveform components and determination of figures of merit, the $avg(\ldots)$ and $rms(\ldots)$ operators should be used. The average and rms values are computed with the progress of the simulation time (running averages). Therefore, they should be read at the right end of the screen. Use the cursor for accurate coordinates of measured points. To determine the fundamental of a waveform, change the X axis from time to frequency employing the "Fourier" option. Remember that the harmonic spectra show peak, not rms, values of the waveform harmonics.

The provided PSpice circuit files represent, in a sense, a virtual laboratory of power electronics. Performing the simulations, observing the waveforms and power spectra of voltages and currents, and determining waveform components and figures of merit should give the reader better feel and appreciation of power electronic converters. Wherever possible, compare the simulation results with theoretical ones given by equations in the book. By no means should the learning exercises be limited to the suggested assignments. Students (and instructors) are encouraged to "tinker" with the modeled converters and expand the set of "virtual experiments" by developing new files.

LIST OF THE PSPICE CIRCUIT FILES

1. *AC_Chopp.cir*   Single-phase ac chopper with an input filter
2. *AC_Volt_Contr_1ph.cir*   Single-phase ac voltage controller
3. *AC_Volt_Contr_3ph.cir*   Three-phase ac voltage controller
4. *Boost_Conv.cir*   Boost converter
5. *Buck_Aver_Model.cir*   Averaged model of buck converter
6. *Buck_Conv.cir*   Buck converter
7. *Buck-Boost_Conv.cir*   Buck-boost converter
8. *Chopp_12Q.cir*   First-and-second quadrant chopper
9. *Chopp_1Q.cir*   First-quadrant chopper
10. *Chopp_2Q.cir*   Second-quadrant chopper
11. *Contr_Rect_1P.cir*   Phase-controlled single-pulse rectifier
12. *Contr_Rect_6P.cir*   Phase-controlled six-pulse rectifier
13. *Contr_Rect_6P_F.cir*   Phase-controlled six-pulse rectifier with an input filter
14. *Cuk_Conv.cir*   Ĉuk converter
15. *Cycloconv.cir*   Single-phase six-pulse cycloconverter
16. *DC_Switch.cir*   SCR-based static dc switch
17. *Diode_Rect_1P.cir*   Single-pulse diode rectifier
18. *Diode_Rect_3P.cir*   Three-pulse diode rectifier
19. *DiodeRect_6P_F.cir*   Six-pulse diode rectifier with an input filter
20. *Dual_Conv.cir*   Six-pulse circulating current-conducting dual converter

21. *Flyback_Conv.cir*   Flyback converter

22. *Forward_Conv.cir*   Forward converter

23. *Full_Brdg_Conv.cir*   Full-bridge converter

24. *Gen_Ph-Contr_Rect.cir*   Generic phase-controlled rectifier

25. *Gen_PWM_Rect.cir*   Generic PWM rectifier

26. *Half_Brdg_Conv.cir*   Half-bridge converter

27. *Half_Brdg_Inv.cir*   Half-bridge voltage-source inverter with hysteresis current control

28. *Hyster_Curr_Contr.cir*   Three-phase voltage-source inverter with hysteresis current control

29. *Opt_Sqr_Wv_VSI_1ph*   Single-phase voltage-source inverter in the optimal square-wave mode

30. *Parallel_Load_Conv.cir*   Parallel-loaded resonant converter

31. *Progr_PWM.cir*   Three-phase voltage-source inverter with programmed PWM

32. *Push_Pull_Conv.cir*   Push-pull converter

33. *PWM_CSI.cir*   Three-phase PWM current-source inverter

34. *PWM_Rect_CT.cir*   Current-type PWM rectifier with an input filter

35. *PWM_Rect_CT_NF.cir*   Current-type PWM rectifier without input filter

36. *PWM_Rect_VT.cir*   Voltage-type PWM rectifier

37. *PWM_VSI_18.cir*   Three-phase voltage-source inverter in the PWM mode ($f_{sw}/f_o = 18$)

38. *PWM_VSI_9.cir*   Three-phase voltage-source inverter in the PWM mode ($f_{sw}/f_o = 9$)

39. *Rect_Source_Induct.cir*   Six-pulse controlled rectifier supplied from a source with inductance

40. *Reson_Boost_Conv.cir*   Quasi-resonant ZCS boost converter

41. *Reson_Buck_Conv.cir*   Quasi-resonant ZVS buck converter

42. *Reson_DC_Link.cir*   Resonant dc link network

43. *Series_Load_Conv.cir*   Series-loaded resonant converter

44. *Sqr_Wv_CSI.cir*   Three-phase current-source inverter in the square-wave mode

45. *Sqr_Wv_VSI_1ph.cir*   Single-phase voltage-source inverter in the simple square-wave mode

46. *Sqr_Wv_VSI_3ph.cir*   Three-phase voltage-source inverter in the simple square-wave mode

47. *Step_Up_Chopp.cir*   Step-up chopper

48. *Three_Lev_Inv.cir*   Three-level neutral-clamped inverter

APPENDIX B
Fourier Series

A function $\psi(t)$ is called *periodic* if

$$\psi(t + T) = \psi(t) \tag{B.1}$$

where T is the *period* of $\psi(t)$. The *fundamental frequency*, f_1, is defined as

$$f_1 \equiv \frac{1}{T} \ Hz \tag{B.2}$$

and the fundamental radian frequency, ω_1, as

$$\omega_1 \equiv \frac{2\pi}{T} = 2\pi f_1 \ rad/s. \tag{B.3}$$

A periodic function can be represented by an infinite *Fourier series* as

$$\psi(t) = a_0 + \sum_{k=1}^{\infty} [a_k \cos(k\omega_1 t) - b_k \sin(k\omega_1 t)] \tag{B.4}$$

where

$$a_0 = \frac{1}{T} \int_0^T \psi(t) dt \tag{B.5}$$

$$a_k = \frac{2}{T} \int_0^T \psi(t) \cos(k\omega_1 t) dt \tag{B.6}$$

$$b_k = \frac{2}{T} \int_0^T \psi(t) \sin(k\omega_1 t) dt. \tag{B.7}$$

Introduction to Modern Power Electronics, Second Edition, by Andrzej M. Trzynadlowski
Copyright © 2010 John Wiley & Sons, Inc.

Alternately, $\psi(t)$ can be expressed as

$$\psi(t) = c_0 + \sum_{K=1}^{\infty} c_k \cos(k\omega_1 t + \theta_k)dt \qquad \text{(B.8)}$$

where

$$c_0 = a_0 \qquad \text{(B.9)}$$

$$c_k = \sqrt{a_k^2 - b_k^2} \qquad \text{(B.10)}$$

$$\theta_k = \begin{cases} -\tan^{-1}\left(\dfrac{b_k}{a_k}\right) & \text{if } a_k \geq 0 \\[2mm] -\tan^{-1}\left(\dfrac{b_k}{a_k}\right) \pm \pi & \text{if } a_k \geq 0. \end{cases} \qquad \text{(B.11)}$$

Coefficient c_0 constitutes the *average value* of $\psi(t)$, and c_1 is the *peak value of the fundamental* of $\psi(t)$. If $\psi(t)$ represents a voltage or current, terms "dc component" for c_0 and "peak value of fundamental ac component" for c_1 are also in use. Coefficients c2, c3, ..., are peak values of *higher harmonics* of $\psi(t)$, the subscript denoting the *harmonic number*. Thus, the kth harmonic is a sinusoidal component of $\psi(t)$ with the peak value of c_k and radian frequency of $k\omega_1$. The harmonic angle, θ_k, is of minor importance in power electronics, except for the fundamental angle, θ_1.

Many practical waveforms are characterized by certain symmetries, which allow reducing the computational effort required for determination of Fourier series. To take advantage of those symmetries, the average value, if any, should first to be subtracted from the analyzed function $\psi(t)$, leaving a periodic function $\vartheta(t)$ given by

$$\vartheta(t) = \psi(t) - a_0 \qquad \text{(B.12)}$$

and whose average value is zero. Clearly, coefficients a_k and b_k are the same for both $\psi(t)$ and $\vartheta(t)$. Thus, the latter function can be used in place of $\psi(t)$ in Eqs. (B.6) and (B.7) to determine those coefficients.

The most common symmetries are:

(1) *Even symmetry*, when

$$\vartheta(-t) = \vartheta(t). \qquad \text{(B.13)}$$

Then,

$$b_k = 0 \qquad \text{(B.14)}$$

$$c_k = a_k \qquad \text{(B.15)}$$

for all values of k from 1 to ∞. In particular, the cosine function has the even symmetry, and that is why the sine coefficients, b_k, of the Fourier series are nulled.

(2) *Odd symmetry*, when

$$\vartheta(-t) = -\vartheta(t). \tag{B.16}$$

Then,

$$a_k = 0 \tag{B.17}$$

$$c_k = b_k \tag{B.18}$$

for all values of k from 1 to ∞. In particular, the sine function has the odd symmetry, and that is why the cosine coefficients, a_k, of the Fourier series are nulled.

(3) *Half-wave symmetry*, when

$$\vartheta\left(t + \frac{T}{2}\right) = -\vartheta(t). \tag{B.19}$$

Then $a_k = b_k = c_k = 0$ for $k = 2, 4, \ldots$, and the remaining coefficients of Fourier series can be calculated as

$$a_k = \frac{4}{T} \int_0^{T/2} \vartheta(t) \cos(k\omega_1 t) dt \tag{B.20}$$

$$b_k = \frac{4}{T} \int_0^{T/2} \vartheta(t) \sin(k\omega_1 t) dt \tag{B.21}$$

for odd values of k. Both the sine and cosine functions have the half-wave symmetry.

If $\vartheta(t)$ has both the even symmetry and half-wave symmetry, coefficients c_k can directly be found as

$$c_k = \begin{cases} \dfrac{8}{T} \displaystyle\int_0^{T/2} \vartheta(t) \cos(k\omega_1 t) dt & \text{for } k = 1, 3, \ldots \\ 0 & \text{for } k = 2, 4, \ldots \end{cases} \tag{B.22}$$

Analogously, if $\vartheta(t)$ has both the odd symmetry and half-wave symmetry, then

$$c_k = \begin{cases} \dfrac{8}{T} \displaystyle\int_0^{T/2} \vartheta(t) \sin(k\omega_1 t) dt & \text{for } k = 1, 3, \ldots \\ 0 \text{ for } k = 2, 4, \ldots \end{cases} \tag{B.23}$$

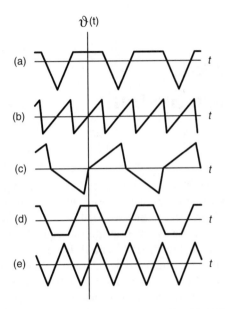

Figure B.1 Examples of various symmetries: (a) even, (b) odd, (c) half-wave, (d) even and half-wave, and (e) odd and half-wave.

Various symmetries are illustrated in Figure B.1. The equations presented can easily be adapted to functions expressed in the angle domain, that is, $\psi(\omega t)$ and $\vartheta(\omega t)$ instead of $\psi(t)$ and $\vartheta(t)$. The period T is then replaced with 2π and ω_1 with ω that constitutes the fundamental radian frequency of $\psi(\omega t)$ and $\vartheta(\omega t)$.

Certain waveforms, such as the optimal switching pattern in Figure 7.26, possess the so-called *quarter-wave symmetry*. It means that

$$\vartheta\left(t + \frac{T}{4}\right) = \vartheta\left(\frac{T}{4} - t\right)$$

(B.24)

and

$$\vartheta\left(t + \frac{3T}{4}\right) = \vartheta\left(\frac{3T}{4} - t\right).$$

(B.25)

APPENDIX C
Three-Phase Systems

Three-phase systems consist of three-phase sources and three-phase loads connected by three-wire or four-wire lines. A three-phase source is composed of three interconnected ac sources, and three interconnected loads comprise a three-phase load. Ideal sinusoidal voltage sources are assumed in the subsequent considerations. It is also assumed that the sources and loads are balanced. It means that the individual source voltages have the same value, differ from each other by the phase shift of 120°, and the load consists of three identical impedances with (if any) load EMFs satisfying the same balance conditions are the source voltages.

A three-phase source can be connected either in wye (Y) or in delta (Δ), as shown in Figure C.1. These two connections also apply to three-phase loads, depicted in Figure C.2. Clearly, there are four possible arrangements of a three-phase source supplying a three-phase load: Y-Y, Y-Δ, Δ-Y, and Δ-Δ. As illustrated in Figure C.3(a), three- or four-wire lines can be used between the source and load in the Y-Y system. The four-wire connection is not feasible in the other three systems, such as the Δ-Y system in Figure C.4(b).

Two types of voltage and two types of current, all indicated in Figure C.3, appear in three-phase systems. These are:

(1) *Line-to-neutral voltages*, v_{AN}, v_{BN}, and v_{CN}, given by

$$v_{AN} = \sqrt{2}V_{LN}\cos(\omega t)$$

$$v_{AN} = \sqrt{2}V_{LN}\cos\left(\omega t - \frac{2}{3}\right) \qquad \text{(C.1)}$$

$$v_{AN} = \sqrt{2}V_{LN}\cos\left(\omega t - \frac{4}{3}\right)$$

where V_{LN} denotes the rms value of these voltages. Somewhat imprecisely, the line-to-neutral voltages are sometimes called *phase voltages*.

Introduction to Modern Power Electronics, Second Edition, by Andrzej M. Trzynadlowski
Copyright © 2010 John Wiley & Sons, Inc.

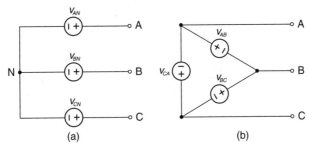

Figure C.1 Three-phase sources: (a) wye-connected and (b) delta-connected.

(2) *Line-to-line voltages,* v_{AB}, v_{BC}, and v_{CA}, given by

$$v_{AB} = \sqrt{2}V_{LL} \cos\left(\omega t - \frac{1}{6}\pi\right)$$

$$v_{AN} = \sqrt{2}V_{LN} \cos\left(\omega t - \frac{1}{2}\pi\right) \qquad \text{(C.2)}$$

$$v_{AN} = \sqrt{2}V_{LN} \cos\left(\omega t - \frac{7}{6}\pi\right)$$

where $V_{LL} = \sqrt{3}V_{LN}$ denotes the rms value of these voltages. It is worth stressing that the V_{LL} value is universally used as the rated voltage of three-phase lines and apparatus. Alternately, the line-to-line voltages are often called *line voltages.*

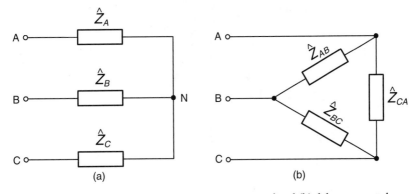

Figure C.2 Three-phase loads: (a) wye-connected and (b) delta-connected.

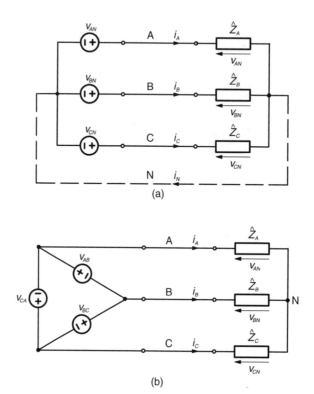

(a)

(b)

Figure C.3 Three-phase source-load systems: (a) Y-Y, (b) Δ-Y.

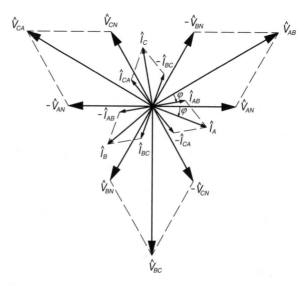

Figure C.4 Phasor diagram of voltages and currents in a three-phase system.

(3) Line currents, $i_A, i_B,$ and i_C, given by

$$i_A = \sqrt{2}I_L \cos(\omega t - \varphi)$$

$$i_B = \sqrt{2}I_L \cos\left(\omega t - \varphi - \frac{2}{3}\pi\right) \tag{C.3}$$

$$i_C = \sqrt{2}I_L \cos\left(\omega t - \varphi - \frac{4}{3}\pi\right)$$

where I_L denotes the rms value of these currents and φ is the *load angle*.

(4) *Line-to-line currents,* $i_{AB}, i_{BC},$ and i_{CA}, also known as *phase currents* and given by

$$i_{AB} = \sqrt{2}I_{LL} \cos\left(\omega t - \varphi - \frac{1}{6}\pi\right)$$

$$i_{BC} = \sqrt{2}I_{LL} \cos\left(\omega t - \varphi - \frac{1}{2}\pi\right) \tag{C.4}$$

$$i_{CA} = \sqrt{2}I_{LL} \cos\left(\omega t - \varphi - \frac{7}{6}\pi\right)$$

where the rms value, I_{LL}, of these currents equals $I_L/\sqrt{3}$.

Phasor diagram of the voltages and currents is shown in Figure C.4. The *apparent power*, S, expressed in volt-amperes (VA) and defined as

$$S \equiv V_{AN}I_A + V_{BN}I_B + V_{CN}I_C = V_{AB}I_{AB} + V_{BC}I_{BC} + V_{CA}I_{CA} \tag{C.5}$$

where the right-hand side quantities denote rms values of the respective voltages and currents, equals

$$S = 3V_{LN}I_L = 3V_{LL}I_{LL} = \sqrt{3}V_{LL}I_L \tag{C.6}$$

in a balanced system. The right-hand side expression is most commonly used because the line-to-line voltage and line current are easily available for measurement in both the three- and four-wire systems.

The *real power*, P, expressed in watts (W), transmitted from the source to the load, is given by

$$P = S \cos(\varphi) = \sqrt{3}V_{LL}I_L \cos(\varphi) \tag{C.7}$$

and the *reactive power*, Q, expressed in volt-amperes reactive (VAr), by

$$Q = S \sin(\varphi) = \sqrt{3}V_{LL}I_L \sin(\varphi). \tag{C.8}$$

Equations (C.6) through (C.8) are valid for all four source-load systems as shown in Figure C.3. The ratio of the real power to the apparent power is defined as the *power factor*, *PF*:

$$PF \equiv \frac{P}{S} = \cos(\varphi).\qquad\qquad(C.9)$$

It must be stressed that the power factor equals the cosine of load angle only in the case of sinusoidal voltages and currents, and identical load angles of all the three phase loads.

If P and V_{LL} are constant in Eq. C.7, then the low value of $\cos(\varphi)$ must be compensated by a high value of I_L. Thus, low power factor causes extra losses in the resistances of the power system. To recoup these loses, utility companies charge users for the amount of reactive energy (a time integral of reactive power). Indeed, according to Eqs. (C.8) and (C.9), a low power factor corresponds to a large power angle and high reactive power.

INDEX

Introduction to Modern Power Electronics, Second Edition, by Andrzej M. Trzynadlowski
Copyright © 2010 John Wiley & Sons, Inc.

431